Numerical Treatment of Coupled Systems

Edited by
Wolfgang Hackbusch and
Gabriel Wittum

Notes on Numerical Fluid Mechanics (NNFM) Volume 51

Volume 51 Numerical Treatment of Coupled Systems. Proceedings of the Eleventh GAMM-Seminar, Kiel, January 20–22, 1995 (W. Hackbusch / G. Wittum, Eds.)
Volume 50 Computational Fluid Dynamics on Parallel Systems. Proceedings of a CNRS-DFG Symposium in Stuttgart, December 9 and 10, 1993 (S. Wagner, Ed.)
Volume 49 Fast Solvers for Flow Problems. Proceedings of the Tenth GAMM-Seminar, Kiel, January 14–16, 1994 (W. Hackbusch / G. Wittum, Eds.)
Volume 48 Numerical Simulation in Science and Engineering. Proceedings of the FORTWIHR Symposium on High Performance Scientific Computing, München, June 17–18, 1993 (M. Griebel / Ch. Zenger, Eds.)
Volume 47 Numerical Methods for the Navier-Stokes Equations (F.-K. Hebeker, R. Rannacher, G. Wittum, Eds.)
Volume 46 Adaptive Methods – Algorithms, Theory and Applications. Proceedings of the Ninth GAMM-Seminar, Kiel, January 22–24, 1993 (W. Hackbusch / G. Wittum, Eds.)
Volume 45 Numerical Methods for Advection – Diffusion Problems (C. B. Vreugdenhil / B. Koren, Eds.)
Volume 44 Multiblock Grid Generation – Results of the EC/BRITE-EURAM Project EUROMESH, 1990–1992 (N. P. Weatherill / M. J. Marchant / D. A. King, Eds.)
Volume 43 Nonlinear Hyperbolic Problems: Theoretical, Applied, and Computational Aspects Proceedings of the Fourth International Conference on Hyperbolic Problems, Taormina, Italy, April 3 to 8, 1992 (A. Donato / F. Oliveri, Eds.)
Volume 42 EUROVAL – A European Initiative on Validation of CFD Codes (W. Haase / F. Brandsma / E. Elsholz / M. Leschziner / D. Schwamborn, Eds.)
Volume 41 Incomplete Decompositions (ILU) – Algorithms, Theory and Applications (W. Hackbusch / G. Wittum, Eds.)
Volume 40 Physics of Separated Flow – Numerical, Experimental, and Theoretical Aspects (K. Gersten, Ed.)
Volume 39 3-D Computation of Incompressible Internal Flows (G. Sottas / I. L. Ryhming, Eds.)
Volume 38 Flow Simulation on High-Performance Computers I (E. H. Hirschel, Ed.)
Volume 37 Supercomputers and Their Performance in Computational Fluid Mechanics (K. Fujii, Ed.)
Volume 36 Numerical Simulation of 3-D Incompressible Unsteady Viscous Laminar Flows (M. Deville / T.-H. Lê / Y. Morchoisne, Eds.)
Volume 35 Proceedings of the Ninth GAMM-Conference on Numerical Methods in Fluid Mechanics (J. B. Vos / A. Rizzi / I. L. Ryhming, Eds.)
Volume 34 Numerical Solutions of the Euler Equations for Steady Flow Problems (A. Eberle / A. Rizzi / E. H. Hirschel)
Volume 33 Numerical Techniques for Boundary Element Methods (W. Hackbusch, Ed.)
Volume 32 Adaptive Finite Element Solution Algorithm for the Euler Equations (R. A. Shapiro)
Volume 31 Parallel Algorithms for Partial Differential Equations (W. Hackbusch, Ed.)

Volumes 1 to 25 are out of print.
The addresses of the Editors and further titles of the series are listed at the end of the book.

Numerical Treatment of Coupled Systems

Proceedings of the Eleventh GAMM-Seminar
Kiel, January 20–22, 1995

Edited by
Wolfgang Hackbusch and
Gabriel Wittum

vieweg

Die Deutsche Bibliothek – CIP-Einheitsaufnahme

Gesellschaft für Angewandte Mathematik und Mechanik: Proceedings of the ... GAMM-Seminar ... –
Braunschweig; Wiesbaden: Vieweg
Früher Schriftenreihe

NE: HST

11. Numerical treatment of coupled systems. – 1995

Numerical treatment of coupled systems: Kiel,
January 20–22, 1995 / ed. by Wolfgang Hackbusch
and Gabriel Wittum. – Braunschweig; Wiesbaden:
Vieweg, 1995
(Proceedings of the ... GAMM-Seminar ...; 11)
(Notes on numerical fluid mechanics; Vol. 51)
ISBN-13: 978-3-322-86861-9 e-ISBN-13: 978-3-322-86859-6
DOI: 10.1007/978-3-322-86859-6
NE: Hackbusch, Wolfgang [Hrsg.]; 2. GT

Softcover reprint of the hardcover 1st edition 1995
Vieweg ist a subsidiary company of Bertelsmann Professional Information.

Produced by Langelüddecke, Braunschweig
Printed on acid-free paper

ISSN 0179-9614
ISBN-13: 978-3-322-86861-9

FOREWORD

The GAMM Committee for "Efficient Numerical Methods for Partial Differential Equations" organizes seminars and workshops on subjects concerning the algorithmic treatment of partial differential equations. The topics are discretisation methods like the finite element and the boundary element method for various type of applications in structural and fluid mechanics. Particular attention is devoted to the advanced solution methods.

The series of such seminars was continued in 1995, January 20–22, with the 11th Kiel-Seminar on the special topic

Numerical Treatment of Coupled Systems

at the Christian-Albrechts-University of Kiel. The seminar was attended by 100 scientist from 9 countries. 23 lectures were given, including two survey lectures.

Different kinds of couplings are considered in this volume. The coupling of different components may occur in the physical model. On the other hand, a coupling of subsystems can be generated by the numerical solution technique. General examples of the latter kind are the domain decomposition (see p. 128) or subspace decomposition (p. 117). The local defect correction method couples different discretizations of the same problem in order to improve the results, although the basic linear system to be solved remains unchanged (p. 47). In general, the aim of the numerical coupling is to make use of (efficient) subsystem solvers (p. 1). The combination of different discretization techniques is mentioned on page 59.

Various examples of coupling different physical components are discussed: the interaction of solid and fluid (p. 96, p. 175, and p. 198), the microelectronic systems (p. 11 and p. 86), and the coupled modelling in groundwater flow (p. 106 and p. 139).

A topic where the coupling may be of physical and numerical nature is the finite element boundary element coupling (FEM-BEM). This very interesting problem is discussed in four contributions (p. 25, p. 37, p. 73, p. 151).

Kiel, May 1995

W. Hackbusch
G. Wittum

Advanced information on the GAMM-seminars Kiel is electronically available via the WWW-site

`http://www.informatik.uni-kiel.de/~jb/gamm.html`

CONTENTS

NEWTON–COUPLING OF FIXED POINT ITERATIONS

Stefan Artlich and Wolfgang Mackens
Institut für Angewandte Mathematik, Universität Hamburg
Bundesstraße 55, D–20146 Hamburg, Germany

Summary

To solve a coupled system of two equations it may be intended not to use the Newton–Raphson method, for example due to the non–sparsity of the Jacobian of the entire system or because there exist solvers for the subsystems. For this type of problems we present an iterative Newton type method which requires only iterative solution steps for the single equations. The algorithm is based on a formal Block–Gauss elimination of the full Newton system and the solution of the resulting Schur complement equation by a Bi–CGSTAB iteration. No computation of the Jacobian of the whole system is necessary. Efficient alternative approaches demand that one of the incorporated systems has relatively small dimension. On the contrary our approach allows similar sizes of the subsystems. This is needed for the intended application which is a system of PDEs describing the behaviour of a chemical reactor for the combustion of coal. Our numerical examples deal with this combustion model.

1 Introduction

The combustion of coal in fluidized bed boilers represents the state of modern powerplant engineering. This type of boilers is superior by its low combustion temperature of about 850°C (compared to approximately 1100°C for pulverized coal combustion), which results in a low NO_x–emission rate. In addition, the SO_2–problem can be got under control by directly supplying limestone to the combustion chamber. Thus, a post combustion flue gas cleaning can be avoided. In order to increase the thermal efficiency recently the operation of fluidized bed combustors (FBC) under pressure has been intensified. In fig. 1 the sketch of a typical PFBC plant is shown.

The whole boiler is contained in a pressure vessel. The boiler is filled with the bed inventory consisting of coal and ash. The coal is supplied through the side–wall and the oxygen which is necessary for the combustion is injected through a distributor plate at the bottom of the reactor. The fluidization gas passes through the bed mostly as rising gas bubbles which causes an intensive motion of the bed material. The heat of reaction is well carried off through the reactor walls and particularly the immersed heat exchanger tubes.

For the simulation of the physical situation inside a PFBC plant a model has been developed which essentially takes into account the reaction $C + O_2 \to CO_2$ ([3]). Considering a two dimensional vertical section Ω of the reactor, this model is composed of two mass balances for the carbon concentration C_C and the oxygen concentration C_{O_2} and an enthalpy balance for the temperature T:

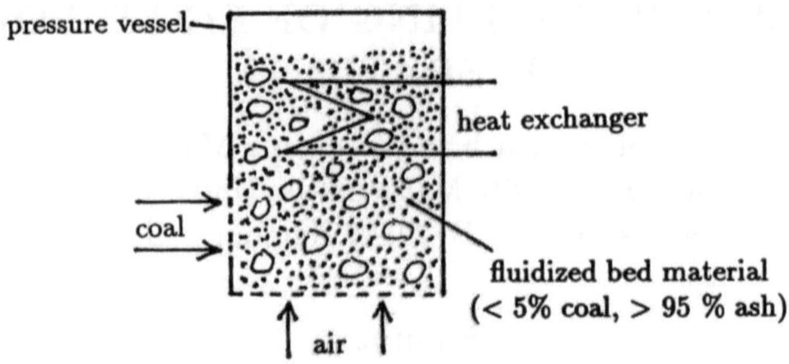

Figure 1: Sketch of a Pressurized Fluidized Bed Combustor (PFBC)

$$-\operatorname{div}(D\,\nabla C_C) + k_C(T)\,C_C\,C_{O_2} = 0 \qquad \text{in } \Omega \tag{1.1}$$

$$\frac{\partial}{\partial x_2} C_{O_2} + b_1\,k_C(T)\,C_C\,C_{O_2} = 0 \qquad \text{in } \Omega \tag{1.2}$$

$$\begin{aligned} &- c_1 \operatorname{div}(D\,\nabla T) \;-\; c_2 \operatorname{div}(T\,D\,\nabla C_C) \\ &+ c_3 \frac{\partial}{\partial x_2} T \;+\; c_4\,(T - T_{KM})\,\xi(x_2) \;=\; c_5\,k_C(T)\,C_C\,C_{O_2} \qquad \text{in } \Omega\;. \end{aligned} \tag{1.3}$$

The carbon mass balance (1.1) is of reaction–diffusion type with Neumann boundary conditions, whereas the enthalpy balance (1.3) is of convection–diffusion type with third type Dirichlet–Neumann boundary conditions. The O.D.E. (1.2) for the oxygen concentration is explicitly solvable in dependence of the carbon concentration and the temperature:

$$C_{O_2}(C_C, T)(x_1, x_2) = C_{O_2,0}\,\exp\left(-b_1 \int_0^{x_2} k_C(T(x_1, s))\,C_C(x_1, s)\,ds\right)\,. \tag{1.4}$$

The equations (1.1), (1.3), (1.4) are coupled through the temperature dependent reaction velocity $k_C(T)$ and the terms $\operatorname{div}(T\,D\nabla C_C)$, $\int k_C(T)C_C$. However, in the simplified case where k_C is independent of the temperature, problem (1.1)–(1.3) becomes reducible. This means that it can be solved in two successive steps where first the carbon and oxygen concentration and subsequently the temperature are computed. This fact will be used later in the numerical simulation of the model.

For the solution of the entire system (1.1), (1.3), (1.4) it suggests itself to use the Newton iteration. But in our case the resulting Jacobian matrix would be non–sparse due to the structure of (1.4). Therefore we discretized (1.1), (1.3) by means of the finite element method and afterwards solved these equations each individually. The latter was realized through two fixed point iterations for (1.1), (1.4) and (1.3) with fixed temperature $\overline{T}$ and concentrations $\overline{C}_C$, $\overline{C}_{O_2}$, respectively. For this we write (1.1) (including boundary conditions) as

$$\tilde{f}(C_C, C_{O_2}, T) = 0 \quad . \tag{1.5}$$

Furthermore we let

$$\begin{aligned}\tilde{g}_1(C_C, C_{O_2}, T) &:= \text{left hand side of (1.3)},\\ \tilde{g}_2(C_C, C_{O_2}, T) &:= \text{right hand side of (1.3)},\end{aligned}$$

such that (1.3) (including boundary conditions) can be written as

$$\tilde{g}_1(C_C, C_{O_2}, T) = \tilde{g}_2(C_C, C_{O_2}, T) \quad . \tag{1.6}$$

$\tilde{f}$ and $\tilde{g}_1$ are linear in C_C and T, respectively. Hence, it causes no problem to solve them for those quantities. Thus for given C_{O_2} and T equation(1.5) can be equivalently written as

$$C_C = \tilde{F}(C_{O_2}, T) \quad \text{with } \tilde{f}\big(\tilde{F}(C_{O_2}, T), C_{O_2}, T\big) = 0 \quad . \tag{1.7}$$

Analogously for given C_C, C_{O_2} and $\tilde{T}$ equation (1.6) can be written as

$$T = \tilde{G}(C_C, C_{O_2}, \tilde{T}) \quad \text{with } \tilde{g}_1\big(C_C, C_{O_2}, \tilde{G}(C_C, C_{O_2}, \tilde{T})\big) = \tilde{g}_2\big(C_C, C_{O_2}, \tilde{T}\big) \quad .$$

Inserting (1.4) into (1.7) finally produces the fixed point iterations (in C_C and T)

$$C_C^{k+1} = \tilde{F}\big(C_{O_2}(C_C^k, \overline{T}), \overline{T}\big) \quad (\overline{T} \text{ fixed}), \tag{1.8}$$

$$T^{n+1} = \tilde{G}\big(\overline{C}_C, \overline{C}_{O_2}, T^n\big) \quad (\overline{C}_C, \overline{C}_{O_2} \text{ fixed}), \tag{1.9}$$

respectively. Instead of using the finite element functions (C_C, C_{O_2}) and T we turn over to the corresponding coefficient vectors $x \in \mathbb{R}^n$ (for (C_C, C_{O_2})) and $y \in \mathbb{R}^m$ (for T) which represent the function values at the grid points of the finite element triangular mesh. With these notations the iterations (1.8), (1.9) now change to the (contractive) iterations

$$\begin{aligned} x^{k+1} &= F(x^k, y) \quad \text{to solve } 0 = f(x, y) \text{ for } x \text{ with fixed } y,\\ y^{k+1} &= G(x, y^k) \quad \text{to solve } 0 = g(x, y) \text{ for } y \text{ with fixed } x,\end{aligned}$$

where $f : \mathbb{R}^n \times \mathbb{R}^m \to \mathbb{R}^n$, $g : \mathbb{R}^n \times \mathbb{R}^m \to \mathbb{R}^m$.

2 Solution of Coupled Systems

As a result of the above discussion we assume the following starting situation for our development: Given is a system of nonlinear equations

$$\begin{aligned} f(x, y) &= 0\,, \\ g(x, y) &= 0\,, \end{aligned} \quad x \in \mathbb{R}^n,\ y \in \mathbb{R}^m, \quad \begin{aligned} f &: \mathbb{R}^{n+m} \to \mathbb{R}^n\,,\\ g &: \mathbb{R}^{n+m} \to \mathbb{R}^m\,, \end{aligned} \tag{2.1}$$

together with (contractive) fixed point iterations

$$x^{k+1} = F(x^k, y) \quad \text{to solve (2.1a) for } x \text{ with } y \text{ fixed}, \tag{2.2}$$

$$y^{k+1} = G(x, y^k) \quad \text{to solve (2.1b) for } y \text{ with } x \text{ fixed}, \tag{2.3}$$

respectively.

Notice that any exact solver of one of the subsystems fits into this framework since an exact solver is an iterative solver with contraction rate zero, which provides the solution after just one step. Thus, our aim is indeed to couple arbitrary solvers. As will become clear in section 2.3 we need a further prerequisite on the structure of f and its solver F for the derivation of our algorithm, namely that the identity

$$f(x,y) = x - F(x,y) \tag{2.4}$$

holds. But this is not really a restriction because having an arbitary exact solver $x = F(y)$ to solve $f(x,y) = 0$, this is

$$f(F(y),y) = 0 \quad , \tag{2.5}$$

we can define $\tilde{f}$ by

$$\tilde{f}(x,y) := x - F(y) \quad , \tag{2.6}$$

and the zeros of (2.6) are just those of (2.5). Thus for any solver the prerequisite (2.4) can be fulfilled by turning over from f to $\tilde{f}$. It should be noted that for one function evaluation of $\tilde{f}$ just one evaluation of F is necessary.

2.1 Jacobi– and Gauss–Seidel iteration

The easiest coupling algorithms for the solution of (2.1) are the

Jacobi iteration

$$\left.\begin{array}{rcl} x^{k+1} & = & F(x^k, y^k), \\ y^{k+1} & = & G(x^k, y^k), \end{array}\right\} \quad k = 0,1,2,\ldots \tag{2.7}$$

and the

Gauss–Seidel iteration

$$\left.\begin{array}{rcl} x^{k+1} & = & F(x^k \quad, y^k), \\ y^{k+1} & = & G(x^{k+1}, y^k), \end{array}\right\} \quad k = 0,1,2,\ldots \tag{2.8}$$

However, as is well known and is confirmed by our numerical examples convergence of these methods is very slow or even cannot be obtained.

2.2 Approximate Newton Method (Chan [2])

Our coupling method will be a modification of an algorithm by Chan [2], which in turn is Newton's method with an approximate version of Keller's algorithm ([4]) to solve the Newton step equation .

Writing Newton's method for the solution of (2.1) at the current position (x^c, y^c) the system of equations

$$\begin{pmatrix} f_x(x^c,y^c) & f_y(x^c,y^c) \\ g_x(x^c,y^c) & g_y(x^c,y^c) \end{pmatrix} \begin{pmatrix} \Delta x \\ \Delta y \end{pmatrix} = - \begin{pmatrix} f(x^c,y^c) \\ g(x^c,y^c) \end{pmatrix} \tag{2.9}$$

would have to be solved. Applying Block–Gauss elimination to this system leads to

$$\begin{pmatrix} f_x(x^c,y^c) & f_y(x^c,y^c) \\ 0 & S(x^c,y^c) \end{pmatrix} \begin{pmatrix} \Delta x \\ \Delta y \end{pmatrix} = - \begin{pmatrix} f(x^c,y^c) \\ R(x^c,y^c) \end{pmatrix}, \tag{2.10}$$

with the Schur complement and the modified right hand side

$$S := g_y - g_x \underbrace{f_x^{-1} f_y}_{=:\,C} \quad \text{and} \quad R := g - g_x \underbrace{f_x^{-1} f}_{=:\,p} \quad ,$$

respectively. From these equations and a backward solution of (2.10) Keller's Block Elimination algorithm to solve (2.9) arises:

Block Elimination Algorithm

1. Solve $f_x\, p = f$ for $p \in \mathrm{IR}^n$.
2. Solve $f_x\, C = f_y$ for $C \in \mathrm{IR}^{n\times m}$.
3. Compute the Schur complement $S = g_y - g_x\, C \in \mathrm{IR}^{m\times m}$.
4. Compute the modified right hand side $R = g - g_x\, p \in \mathrm{IR}^m$.
5. Solve $S\,\Delta y = -R$ for $\Delta y \in \mathrm{IR}^m$.
6. Compute $\Delta x = -p - C\,\Delta y \in \mathrm{IR}^n$.

Chan's algorithm is just Keller's when using specific approximations of the vector p and the matrix C. Since

$$\begin{aligned} x^c - p &= x^c - f_x^{-1}(x^c,y^c)\, f(x^c,y^c) \\ &= \text{Newton approximation of the solution of } f(x,y^c) = 0 \text{ at } x^c \text{ for fixed } y^c \end{aligned}$$

is a (quadratically) good approximation of the solution $\hat{x}$, i.e.

$$x^c - p \approx \hat{x} \quad ,$$

certainly vice versa

$$x^c - \hat{x} \approx p$$

delivers a (quadratically) good approximation of p. Though $\hat{x}$ is not available it can be approximated (with quadratically small error) by the given iteration

$$x^{j+1} := F(x^j, y^c), \quad j = 0, 1, \ldots, \quad x^0 := x^c \quad , \tag{2.11}$$

such that

$$p \approx x^c - x^j, \; j \geq 1 \quad . \tag{2.12}$$

Chan uses $j = 1$ in his version. We shall prefer to determine an appropriate j by requiring the residual decrease to be quadratic. Thus we retain the quadratic approximation quality of the Newton step.

As an approximation of C in the second step of the Block Elimination algorithm Chan uses

$$Ce^i \approx -\frac{1}{h_i}\left[F(x^c, y^c + h_i e^i) - F(x^c, y^c)\right], \quad i = 1, \ldots, m \quad . \tag{2.13}$$

This approximation is exact at the solution if the iteration $x^{j+1} = F(x^j, y)$ in (2.11) is Newton's iteration, and it is at least a relatively good approximation if the F–iteration has small contraction number ([2]).

Using (2.12) and (2.13) one step of Chan's method now reads:

Algorithm ANM (Approximate Newton Method) [2]

1. Compute $p = x^c - F(x^c, y^c)$.
2. For $i = 1, \ldots, m$ compute

$$c_i = -\frac{1}{h_i}\left[F(x^c, y^c + h_i e^i) - F(x^c, y^c)\right] \quad ,$$

where c_i denotes the i–th column of C, h_i denotes a small finite difference parameter and e^i denotes the i–th unit vector.

3.–6. Execute the same steps as in the Block Elimination algorithm.

Note that the Block Elimination algorithm as well as the the ANM algorithm both require $(m+1)$ calls of F in each step and the storage of the $(n \times m)$–matrix C. Therefore Chan applies his algorithm to the case where g has small dimension m compared to f. Typical applications of his from bifurcation calculations would have $m = 1$ or $m = 2$.

2.3 Derivation of the new algorithm

In contrast to the applications of Chan within the coupled system (1.5), (1.6) one has $m \approx n$ and both a large. Thus a modification of Chan's approach has to take into account that

- the dimensions of f and g are approximately of the same size, and it should aim at
- avoiding the computation and storage of the derivatives of f and g.

The basic idea to accomplish this is to solve the Schur complement equation in (2.10) iteratively by means of the Bi–CGSTAB method. In this approach the Schur complement S is only required for matrix–vector operations Sw. This entity — as well as several others — can then be approximated by adequate differencing procedures using the two independent solvers F and G (see (2.2), (2.3)).

Approximation to p
As pointed out above $p = f_x^{-1}(x^c, y^c)\, f(x^c, y^c)$ can be approximated by $p \approx x^c - x^j, j \geq 1$, with x^j from (2.11).

Approximation to Cw
Like S the matrix $C = f_x^{-1} f_y$ is only needed in matrix–vector operations Cw. Approximating f_y by finite differencing we get

$$\frac{1}{h} C(hw) \approx \frac{1}{h} f_x^{-1}(x^c, y^c) \underbrace{\left[f(x^c, y^c + hw) - f(x^c, y^c)\right]}_{\approx f_y(x^c, y^c)\,(hw)} \quad . \tag{2.14}$$

The right hand side in (2.14) is just a Newton step in x–direction to solve $0 = f(x^c, y^c + hw) - f(x, y^c)$ in x at position (x^c, y^c). Using (2.2) and the identity (2.4) this zero can be found by the iteration

$$x^{j+1} := F(x^j, y^c) + f(x^c, y^c + hw), \quad j = 0, 1, \dots, \quad x^0 := x^c \quad , \tag{2.15}$$

and we get the approximation

$$Cw \approx h^{-1}(x^j - x^c), \; j \geq 1 \quad .$$

Approximation to R
The modified right hand side $R = g - g_x p$ can be computed by applying G (see (2.3)) once, since Taylor expansion results in

$$g(x^c - p, y^c) \;\approx\; g(x^c, y^c) - g_x(x^c, y^c)\, p = R(x^c, y^c) \quad .$$

Approximation to Sw
Similar to R, the matrix–vector product Sw can be computed by applying G twice, since for each vector w and scalar h Taylor expansion leads to

$$\begin{aligned} & g(x^c - C(hw), y^c + hw) - g(x^c, y^c) \\ & \approx -g_x(x^c, y^c)\, C(hw) + g_y(x^c, y^c)(hw) \;=\; S(hw) \quad . \end{aligned} \tag{2.16}$$

Approximation to Δx
Having computed Δy in the fifth step of the Block Elimination algorithm Δx may be computed from the equation

$$f(x^c + \Delta x, y^c + \Delta y) = 0 \quad .$$

If we approximate Δx by a Newton step similarly to the computation of p we can use the iteration

$$x^{j+1} := F(x^j, y^c + \Delta y), \quad j = 0, 1, \dots, \quad x^0 := x^c \quad . \tag{2.17}$$

This leads to the approximation

$$\Delta x \approx x^c - x^j, \; j \geq 1 \quad .$$

Finally a suitable size of the parameter h in the approximations to the derivatives (2.14), (2.16) has to be chosen and appropriate stopping criteria for the iterations (2.11), (2.15) and (2.17) have to be set. The latter can be done for example by choosing a maximum iteration number or the prescription that the error in the corresponding f–equations is reduced at least quadratically. For details the reader is referred to [1], as well as to [5] for a description of the Bi–CGSTAB algorithm. We want to complete the derivation of the new algorithm with the description of one approximate Newton step of the method, starting at position (x^c, y^c).

Algorithm IANM (Iterative Approximate Newton Method)

1. Compute $p = x^c - x^j,\ j \geq 1,\ x^j$ from (2.11).
2. Compute the modified right hand side $R = g(x^c - p, y^c)$.
3. Solve $S\Delta y = -R$ using Bi–CGSTAB; for each matrix–vector product Sw within the algorithm do

    3.1 Compute $C(hw) = x^j - x^c,\ j \geq 1,\ x^j$ from (2.15).

    3.2 Compute $Sw = h^{-1}\left[g(x^c - C(hw), y^c + hw) - g(x^c, y^c)\right]$.

4. Compute $\Delta x = x^c - x^j,\ j \geq 1,\ x^j$ from (2.17).

3 Numerical Examples

In the final section we want to present numerical examples which deal with the simulation of the PFBC plant.

Example 1 At first we want to compare the convergence behaviour of the Jacobi, Gauss–Seidel and IANM algorithms (see sections 2.1, 2.3). Whereas the Jacobi coupling (2.7) even in a damped version failed to converge in our case, the results for the (damped variant of the) Gauss–Seidel coupling (2.8) and the IANM method are to be found in tab. 1 and 2 respectively. The computations were made on a HP9000 S700 workstation, the CPU time is related to this machine type.

Table 1: numerical results using the GAUSS–SEIDEL iteration (2.8)

# unkn.	# it. steps eq. (2.8)	# F–It. eq. (2.2)	# G–It. eq. (2.3)	CPU [min]
153	929	2354	2354	10:33
561	1085	2790	2790	84:28

Table 2: numerical results using the IANM algorithm (new approach)

# unkn.	# Newton steps	# F–It. eq. (2.2)	# G–It. eq. (2.3)	CPU [min]
153	9	83	47	00:27
561	3	28	16	00:57
2145	2	19	11	05:21
8385	1	10	6	23:51

Since it is difficult to find an appropriate initial guess for the iterations (1.8), (1.9), as initial guess on a fine mesh an interpolant of the computed solution on a coarse mesh has been used. This is to explain the decreasing number of Newton steps in the IANM method when refining the mesh.

Note that within the IANM method the number of F– and G–iterations differs. This is advantageous if there is a difference in the effort to solve the f– and g–equation (2.2), (2.3) independently. In our example the discretization of (1.8) leads to a symmetric linear system of equations whereas the discretization of (1.9) leads to a non–symmetric one.

Example 2 In fig. 2 we present results from the simulation of the PFBC reactor model (1.1)–(1.3) derived by the new IANM algorithm.

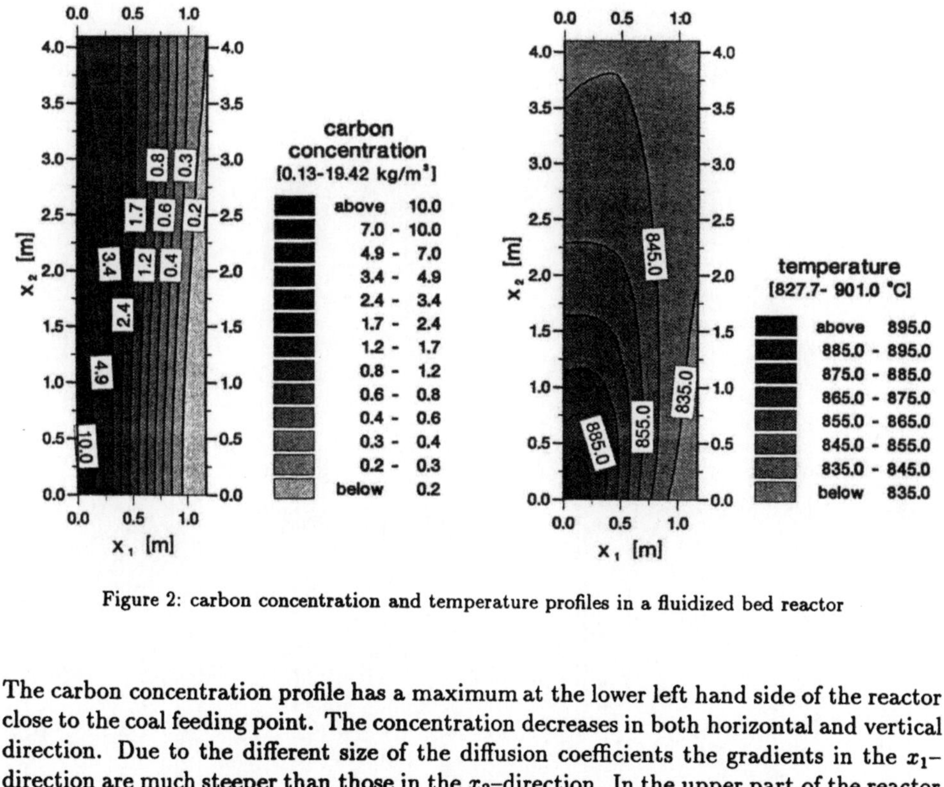

Figure 2: carbon concentration and temperature profiles in a fluidized bed reactor

The carbon concentration profile has a maximum at the lower left hand side of the reactor close to the coal feeding point. The concentration decreases in both horizontal and vertical direction. Due to the different size of the diffusion coefficients the gradients in the x_1–direction are much steeper than those in the x_2–direction. In the upper part of the reactor there are almost no vertical carbon concentration gradients since most of the oxygen for the combustion reaction is used in the lower part such that it is not available in the upper region.

The temperature profile has a similar shape as the carbon concentration profile. This can be explained by the reaction mechanism: In regions with high carbon and oxygen concentration a large amount of heat of reaction is set free compared to regions where one of the concentrations is low. On the other hand the heat of reaction is almost evenly carried off through the heat exchangers.

References

[1] **S. Artlich, W. Mackens**: *Newton–Coupling of Fixed point Iterations or Tiebreaking in Energy Pingpong*, Hamburger Beiträge zur Angewandten Mathematik, Reihe E, Scientific Computing, in preparation.

[2] **T. F. Chan**: *An Approximate Newton Method for Coupled nonlinear Systems*, Research Report YALEU/DCS/RR–300, Yale University, Feb. 1984.

[3] **H. Groenewald**: *Zum Einfluß von Wärmetauscherbündeln auf die Temperaturhomogenität druckaufgeladener Wirbelschichtfeuerungen*, VDI-Verlag, Reihe 6, Nr. 244, Düsseldorf, 1990.

[4] **H. B. Keller**: *Numerical solution of Bifurcation and Nonlinear Eigenvalue problems*, in: P. Rabinowitz (ed.): Applications of Bifurcation Theory, Academic Press, New York, 1977, 359–384.

[5] **H. A. van der Vorst**: *Bi–CGSTAB: A Fast and Smoothly Converging Variant of Bi–CG for the Solution of Nonsymmetric Linear Systems*, SIAM J. Sci. Stat. Comput. **13** (1992), 631–644.

COUPLED PROBLEMS IN MICROSYSTEM TECHNOLOGY

Hans-Joachim Bungartz [1], Stefan Schulte [2]

[1] Institut für Informatik der Technischen Universität München
D-80290 München, Germany
bungartz@informatik.tu-muenchen.de
[2] Siemens AG, Zentralabteilung Forschung und Entwicklung
D-91050 Erlangen, Germany
stefan.schulte@zfe.siemens.de

SUMMARY

In microsystem technology, the numerical simulation of coupled problems is one of the principal challenges. We present a classification of the most important occurring couplings, and we give a survey of existing solution techniques with emphasis on the so-called *partitioned solution*. Here, there is no joint model, neither continuous nor discrete, but the coupled problem is solved by an outer iteration for the coupling and by arbitrary inner solution processes for each single problem. The coupling is done via changed boundary conditions, geometries, or parameters after each step of iteration. This approach seems to be advantageous, since its modularity allows the use of existing and efficient codes for each sub-problem. Thus, only the outer iteration has to be organized with some kind of interface for the coupling. Furthermore, this technique is perfectly suited for parallelization, especially for the use of (heterogeneous) workstation clusters. For the simulation of a micro-miniaturized two-valve membrane pump, first numerical results are presented.

INTRODUCTION

In the last years, the advances in semiconductor technology concerning both the numerical simulation and the production of microchips have led to growing endeavours to profit from the experiences and progresses there in other fields of industrial relevance, too. Especially the efforts for micro-miniaturized sensors like acceleration sensors in modern airbag systems and for microactuators like microvalves or micropumps in medicine, e. g., finally resulted in what we call microsystem technology today. Since a realistic prototyping and the following production are often very costly, the numerical simulation of such microsystems turns out to be imperative.

When we try to tackle the numerical simulation of a microsystem, we learn immediately that we can't avoid to deal with the numerical treatment of coupled problems. There are three main reasons for the fact that, here, the coupling of different physical effects (structural dynamics, fluid dynamics, heat transfer, or electromagnetics, e.g.) and the coupling of the physical level with the system level (integration of a device simulation into the simulation of the surrounding electric circuit, e. g.) are more frequently encountered than in the macro world: First, aspects of scaling often lead to a dominance of surface effects over volume dependent effects. Second, especially in sensors a lot of different physical phenomena are used for signal conversion. Finally, in some microsystems different physical effects may have an influence on each other, which can result in so-called cross talk effects between different closely related conductors, e. g.

Furthermore, the variety of physical effects possibly involved in a microsystem and the resulting great variety of possible couplings show the disadvantages of the standard approach of defining a model for one particular coupled problem or one class of coupled problems and of finding a way for its numerical treatment by adding new features to existing programs. The development of such complex models usually takes a lot of time, and the resulting codes can often be neither applied easily to other problems nor extended in a simple manner if growing accuracy requirements demand a more precise model, e. g. This is the reason why modular solution techniques based on the connection of different existing codes for different sub-problems seem to be a very promising approach for an efficient numerical simulation of coupled problems.

MICROSYSTEMS

As mentioned above, a microsystem could be defined as the integration of microelectronic components like integrated circuits and non-microelectronic components like micromechanical sensors or actuators into one complete system. Typically, the overall dimensions of such systems are in the range of a few thousand μm. The principal structure of a microsystem is illustrated in the following figure (1).

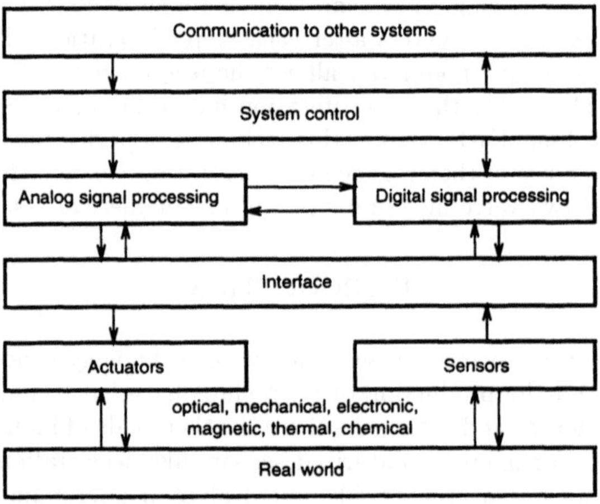

Fig. 1 Principal structure of a microsystem.

An example for a microsystem which is in the development stage now is the future generation of airbag systems. These systems must combine sensors for measuring the acceleration and electric circuits for processing the measured data and comparing them with the reference data in order to decide whether the airbag has to be blown up or not. Finally, if a crash is indicated, the actuators must inflame the explosive. In such complex systems, some components like the so-called force balance accelerometers [13] can be microsystems themselves (sensors, actuators, or electronic control units). Since a microsystem contains components from different disciplines like microelectronics or micromechanics, the term *microelectromechanical system* to describe such devices is often used.

Starting out from the above interpretation of a microsystem, we can think of microsystem technology as a combination of different microtechnologies by the application of system technologies like packaging, interconnections, and testing. Moreover, especially in the context of chemo- and bio-sensitive materials or piezo technologies, material sciences are in the centre of interest, too. Finally, one of the main tasks of system technologies is the development of methods and tools for the computer based design of the whole microsystem (cf. MEMCAD [22]).

The numerical simulation of the behaviour of single components and of the whole system is very important during the design process, because a development based on experiments only would require the fabrication of (very expensive) prototypes and the measuring of the critical parameters. Since the sensors used for that are usually bigger than the whole system to be measured itself, this is often not feasible. Generally, the numerical simulation of the component behaviour means the solution of the partial differential equations which model the most relevant physical effects of this special device, e. g. the mechanical deformation of a cantilever. This kind of simulation is called simulation on the physical level. To describe the behaviour of the whole system, it is necessary to make a more abstract model which can, usually, mathematically be described by systems of ordinary differential equations. An example for this is the simulation of electric circuits. In the future, couplings between different levels like the physical and the system level will have to be taken into account to an increasing degree, especially because of growing requirements concerning accuracy and reliability.

COUPLED PROBLEMS

In the macro world, coupled problems and their numerical treatment have a rather long tradition that goes back to the late sixties. The main emphasis has always been put on topics like fluid-structure interaction, thermo-elasticity, or soil-pore interaction. For fluid-structure interaction, e. g., the examples range from aero-elasticity in the design of aircraft wings to reactor safety [33] and from the earthquake-proof construction of dams [19] to examinations concerning the vibration of bridges — a problem that attained a lot of popularity after the spectacular collapse of the Tacoma bridge in 1940.

This variety of examples for coupled problems from all kinds of areas of application led to a very broad interpretation of the term *coupled problems* and to the development of very different techniques for their numerical treatment [11, 16]. A first kind of a unifying approach was done by Zienkiewicz [32]. There, a coupled problem is defined as a set of n bidirectionally combined sub-problems $\mathcal{P}_i$, $1 \leq i \leq n$,

$$[L_i(x_1, ..., x_{i-1}, x_{i+1}, ..., x_n)]\,(x_i) = f_i(x_i; x_1, ..., x_{i-1}, x_{i+1}, ..., x_n), \qquad (1)$$

with operators L_i, right hand sides f_i, unknowns x_i, and parameters $x_1, ..., x_{i-1}, x_{i+1}, ..., x_n$. Here, the principal requirement is the bidirectional coupling, i. e. the fact that there is neither any possibility for an explicit elimination of parameters which link the different sub-problems nor any way of an independent solution of any $\mathcal{P}_i$.

Following this definition of a coupled problem, Zienkiewicz in [32] also presents a first classification of coupled problems. He distinguishes between two classes of problems: those with totally or partially overlapping domains and those with different domains. In the first case, the coupling takes place via the differential equations that, usually, refer to different underlying physical phenomena. The second class, however, is characterized by a coupling that occurs on the domain interfaces, i. e. via the boundary conditions.

Here, a further distinction is made between problems with different underlying physics or different variables (fluid-structure interaction or structure-electrostatics interaction, e. g.) and problems with identical physics and variables (structure-structure interaction, e. g.).

In contrast to the above definition of a coupled problem by Zienkiewicz, the terms *coupled problems* or *coupled methods* are often used in a broader sense. Then, these also include the so-called mixed problems, which allow both a reducible coupled and an irreducible single problem formulation, but also domain decomposition methods, where all $\mathcal{P}_i$ describe identical physical phenomena on different sub-domains, or the coupling of different discretization schemes like finite element and boundary element methods. Furthermore, even systems of linear or non-linear algebraic equations or ordinary differential equations, to some extent, can be seen as coupled problems.

Finally, it must not be forgotten that the topic of coupled problems has been tackled by researchers from other disciplines than numerical simulation. For instance, the field of multidisciplinary design optimization that is concerned with how to analyse efficiently and design optimally a system governed by multiple coupled disciplines or made up of coupled components works on coupled problems, too [1]. Since the approaches discussed there come from the theory of systems, they are much more generally orientated and do not start from a specific problem.

When we are looking for efficient modular strategies to deal with the numerical solution of coupled problems in a flexible manner that is independent of a specific problem, those various approaches from other disciplines, but also methods known from the iterative solution of large systems of linear equations [10], e. g., can be very helpful. Furthermore, since we want to solve coupled problems from microsystem technology, we have to take into account the special conditions and requirements of the micro world.

COUPLED PROBLEMS IN MICROSYSTEM TECHNOLOGY

The importance of cross coupled effects in micro technology results from three different facts which are closely related to what characterizes a microsystem.

Most of the couplings observed on the physical level can be seen as a consequence of the scaling behaviour of the basic quantities length, surface, and volume. When the dimensions of a system represented by the characteristic length L tend towards zero, the surface and the volume tend towards zero like L^2 and L^3, respectively. Because of that, for decreasing L, there is a growing importance of all terms proportional to the surface. Consequently, surface forces like pressure or electrostatic loads dominate volume forces, for example magnetic or mass dependent loads like inertial forces. Coupled problems resulting from these scaling properties naturally belong to the second class in the scheme of Zienkiewicz, because the interaction occurs only on the domain interfaces.

Especially in sensor design, a lot of different physical phenomena have often to be used for the conversion of the quantity to be measured into a signal which is well suited for further data processing steps. This characteristic transformation of an input signal into an (electric) output signal can be used for a definition of sensors [8]. Here, a unidirectional coupling of different effects is the basic principle of a sensor. Since some of the phenomena used for signal conversion are closely connected, a bidirectional coupling can not always be avoided.

The third reason for the importance of couplings in the micro world stems from the high integration density caused by the small overall dimensions. For example, in so-called smart sensors where the sensor device is combined with electric circuits for the signal conversion and for the compensation of secondary effects, the heat emission of the electronic device can cause thermal strains in the mechanical part of the sensor, if the distances are too small. Furthermore, the field of the so-called electromagnetic compatibilities (EMC) deals with mutual influences or cross talks between electric conductors. Especially problems of this type could require the combination of the simulation on the system level (circuit behaviour) and the simulation on the physical level.

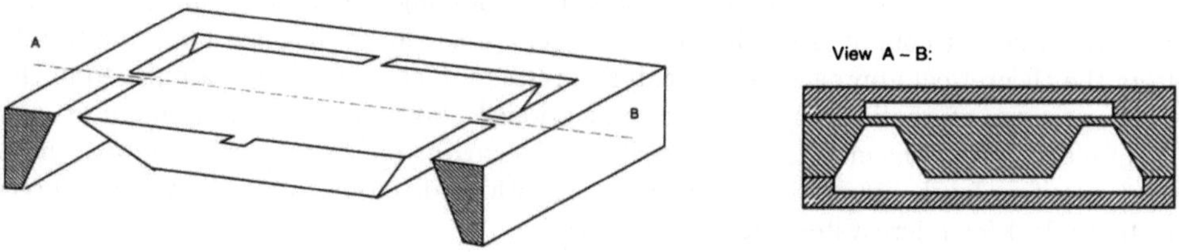

Fig. 2 Piezoresistive accelerometer with seismic mass.

Now, let us look at some micromechanical components whose behaviour is dominated by couplings between different phenomena on the physical level. Figure (2) shows the basic structure of most of today's accelerometers. Here, the deflection of a movable mass caused by the acceleration forces and determined by different possible techniques (capacitive principle, piezoresistive films) is used as a measure for the actual acceleration. In fact, each sensor of this type is a resonator with its main degree of freedom in the direction in which the acceleration shall be measured. The damping behaviour of such a device is strongly influenced by the pressure and the viscosity of the surrounding gas. In [27], the influences of the different parameters such as the distance of the movable mass from the ground plate or the viscosity of the surrounding fluid are analysed, and it is shown that there is a close connection between the fluid flow and the structural dynamic problem. Furthermore, a mathematical model to handle this problem like a squeeze film damper is suggested. The basic idea of this approach, the modeling of the fluid flow with the Reynolds equation from lubrication theory [24], goes back to the sixties [9]. Note that, for lubrication problems, a lot of interesting (and modular) solution approaches have been developed [11, 12, 15].

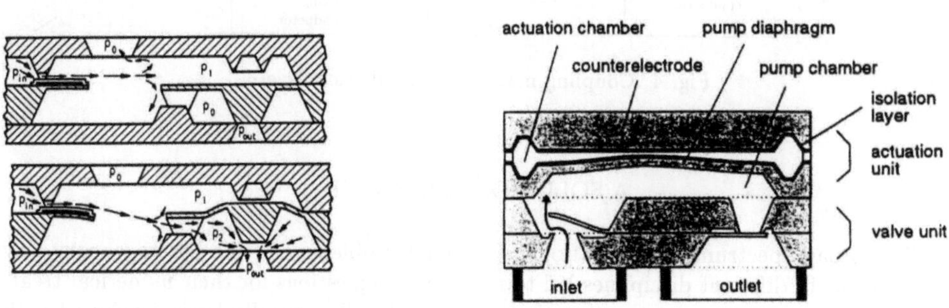

Fig. 3 Hydraulically actuated valve (left, taken from [26]) and two-valve membrane micropump (right, taken from [31]). The micropump has outer dimensions of $14.5 \times 8.5 \times 1.43\,\text{mm}^3$. It is designed for flow rates of about 75 ml/min and a system pressure level of about 6 bar.

Other examples for problems with fluid-structure interaction in micro technology are shown in figure (3). Here, the fluid flow has to be modeled by the complete Navier-Stokes equations. Micropumps like the one of figure (3) will have a broad range of applications in the future (medical drug infiltration systems [23], microfluidic systems for chemical analysis, or environmental applications, e. g.).

To get an overview of the cross coupling problems in microsystem technology, a systematic analysis including both an application based and a theoretical approach with the derivation of the mathematical models of the different considered single field effects in a unique notation has been done. As a result, the 'coupling matrix' in figure (4) summarizes the different couplings on the physical level. In the lower diagonal part, the results from the theoretical approach are indicated, and the upper part shows examples from the different areas of application. As already mentioned above, some types of microsystems require to consider the interaction of phenomena occurring on the physical level with those on the system level. Simulation problems with both kinds of phenomena have turned out to be typical for microsystems (see figure (5)).

components / kind of coupling	structural dynamics	heat conduction	fluid dynamics	electromagnetics
structural dynamics		- resonators (thermally induced high frequency vibrations) - thermal actuators	- accelerometers - gyroscopes - valves, pumpes	- sensors with capacitive detection - electrostatic and piezoelectric actuators
heat conduction	thermo-elasticity $h \rightarrow s$: material law $s \rightarrow h$: energy dissipation		- microcooler - flow- and thermal sensors	- thermopiles (electrical heating with integrated resistors) - thermal actuators (pump actuation, resonators)
fluid dynamics	$f \rightarrow s$: surface forces $s \rightarrow f$: - geometry - boundary cond.	coupled by the energy equation		- sensors and pumps working with the electrohydrodynamic principle
electromagnetics	$e \rightarrow s$: surface and volume forces $s \rightarrow e$: - geometry - moving conductor	$e \rightarrow h$: electromagnetic loss $h \rightarrow e$: material properties	electrohydrodynamics $e \rightarrow s$: surface and volume forces $s \rightarrow e$: - geometry - moving conductor	

Fig. 4 Coupling matrix on the physical level.

A SOLUTION APPROACH

The broad spectrum of applications of coupled problems and the heterogeneous way they appear in different disciplines led to numerous suggestions for their numerical treatment [21]. Among these approaches that have nearly all originally been developed with one special problem in mind and have afterwards been generalized, it is possible to distinguish two classes (see figure (6)).

First, the connection of the different mathematical models via their coupling condition

kind of coupling \ components	circuit simulation	multibody dynamics	structural dynamics	heat conduction	fluid dynamics	electro-magnetics
circuit simulation	\	mechatronic systems - airbag - control units of combustion engines	- smart sensors - force balance accelerometer	thermal design of electronic circuits with respect to placement		electromagnetic compatibility (EMC) on circuit level
multibody dynamics	c→m: control theory m→c: system response	\	elasticity of components in multibody systems, e.g. robot arms		simplified model of the acceleration sensor	multibody motion in an electrostatic field, e.g. motors, generators
structural dynamics	coupling via additional effects, e.g. electrostatics	boundary conditions	\	Cross coupling on the physical level		
heat conduction	h→c: temperature dependent components c→h: heat source			\		
fluid dynamics		similar to fluid-structure interaction			\	
electro-magnetics	current flow ↕ field around conductor	e→m: surface and volume forces m→e: moving conductor				\

Fig. 5 Coupling matrix on the system level.

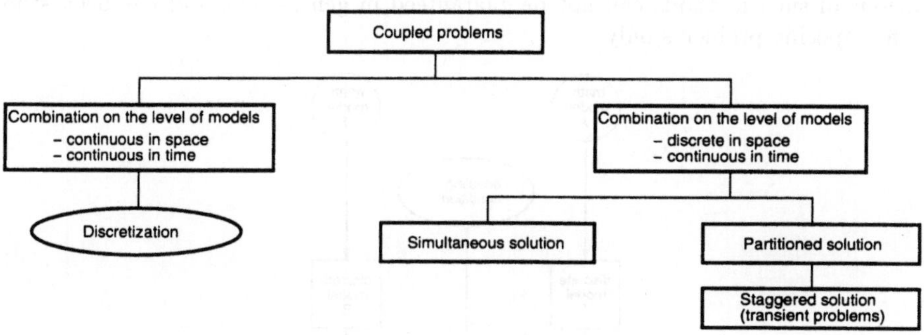

Fig. 6 Classification of the solution approaches.

is the most natural way and results in a system of partial differential equations with appropriate boundary and initial conditions. An example for that is thermo-elasticity [2]. Here, the equations from structural dynamics which regard thermally induced strains and temperature dependent material properties have to be coupled with the heat equation, in which the source term is given by a time dependent change of the strain tensor. An alternative way of deriving combined mathematical models is that of considering the energy functional [6], which finally leads to bond graph techniques [20, 25]. Used especially for the simulation of problems in thermo-electricity, the so-called *tailored modeling* [28] starts from a mathematical model that contains each relevant effect of the considered class of problems. Thus, the couplings are taken into account automatically. In order to reduce the model, only the important variables are considered in the further modeling steps, i. e., the model is tailored according to the given problem.

In contrast to the realization of the coupling on the continuous level, we can also couple the different semi-discrete models (cf. figure (6)) which have the form of algebraic equations if the problems do not depend on time, and which are ordinary differential equations otherwise. The resulting block-structured systems can be handled in two different ways.

In the first method called *simultaneous solution*, standard methods like Gauss elimination or block Gauss elimination in the case of linear equations are used to solve the equations of the combined model. A general disadvantage of the simultaneous solution is the fact that symmetry and sparsity of the resulting matrix may get lost due to the coupling terms even if each single problem is symmetric and sparse. Furthermore, in the case of time dependent problems, the system of ordinary differential equations which results from the combination of two problems which show different time behaviour could be stiff.

To avoid these difficulties and to get a modular approach, the so-called *partitioned solution* shown in figure (7) has been developed [4, 5]. Here, there is no joint model, neither continuous nor discrete, but the coupled problem is solved by an outer iteration for the coupling and by arbitrary inner solution processes for each single problem. The coupling is done via changed boundary conditions, geometries, or parameters after each step of iteration. Concerning the organization of the outer iteration, Jacobi-, Gauss-Seidel-, or SOR-type methods are the state of the art. In the latter case, the relaxation parameters are normally fixed by experiment and strongly problem dependent. The convergence behaviour of such methods can not be guaranteed in general [29] and has been studied for a few specific problems only.

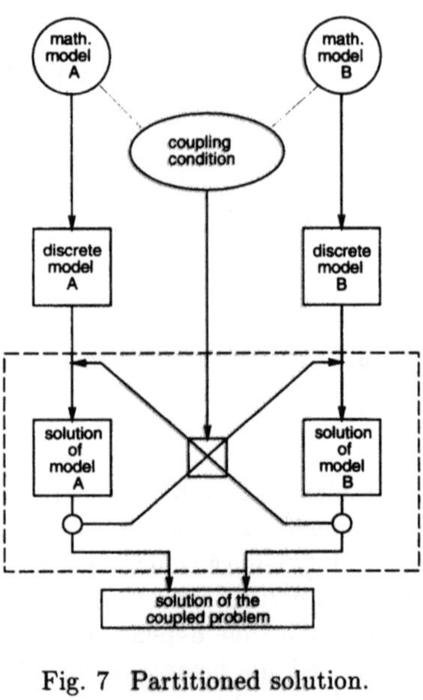

Fig. 7 **Partitioned solution.**

Whereas most of the techniques introduced so far have been developed starting from a specific problem, the partitioned solution leads to a more general and methodical point of view. Following this idea, we suggest an approach in which the coupled solution shows the same structure as the coupled problem to be solved. We start from a description of the coupled problem by its different underlying single field phenomena $\mathcal{P}_i$, represented by the operators L_i of their mathematical models (cf. (1)), and the corresponding coupling operators C_{ij} for the couplings from $\mathcal{P}_i$ to $\mathcal{P}_j$. For the numerical treatment of the whole coupled problem, each $\mathcal{P}_i$ gets its own solver S_i attached, and C_{ij} and C_{ji} are taken into account by an appropriate iterative strategy It_{ij} (see figure (8)).

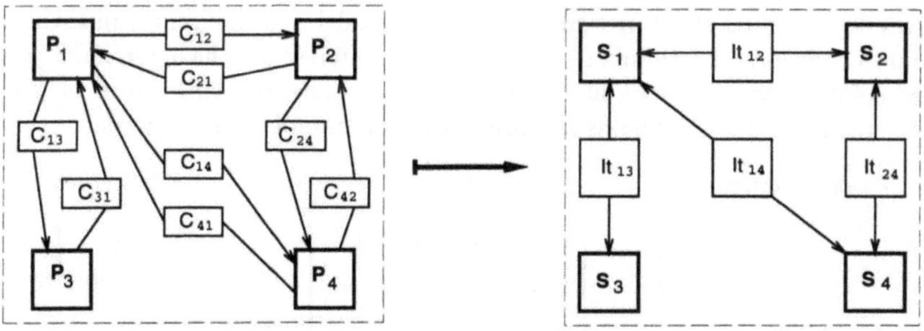

Fig. 8 Coupled problem (left) and its coupled solution (right).

The realization of such an algorithm can be seen as a typical task of scientific computing, because aspects from numerical analysis, computer science, and the respective area of application (microsystem design, e. g.) have to be considered.

From the point of view of numerical analysis, the construction of outer iteration schemes for both stationary and time dependent problems as well as the development of strategies for the variable transformation at the problem interfaces (non-matching discretizations with self-adaptive sub-problem solvers, e. g.) are in the centre of interest. Furthermore, we need error indicators and estimators for the overall procedure in order to be able to decide which sub-problem has to be solved more exactly.

Considering the aspects of computer science, one has to organize the communication between the different solvers at the interfaces as well as the control of the overall procedure. Moreover, there is the question of how to implement the coupled solution process in a heterogeneous computer environment when sub-problem solvers with licences restricted to special nodes have to be used.

Finally, the main task on the part of the applications is the preparation of appropriate mathematical models for the single effects and for the couplings between them. Especially for the derivation of the coupling operators C_{ij}, there are often several alternatives, and the choice of the most suitable one may depend on the respective outer iteration strategy.

The main objective of our work in that context is to analyse the connections between the different components like outer iteration scheme, single field solvers, coupling and interface operators, and so on, in order to optimize the overall procedure introduced in figure (8). As a result of such an analysis, we want to get some kind of list of requirements which should be fulfilled by each single field solver to facilitate its efficient use in coupled solution processes. On the other hand, if a problem and the corresponding available single field solvers are given, we want to provide criteria for the proper decision which solution strategy should be used.

In micropumps like the one shown in figure (3), different kinds of couplings can be observed. The development of one single model for the whole device would require the consideration of the electrostatics-structure interaction used in the actuation unit and the fluid-structure interaction in the valve unit. Furthermore, the actuation unit is coupled to an electric circuit in order to control the flow rate. For a first model, however, we restrict ourselves to the fluid-structure interaction and model the actuation unit via velocity boundary conditions. Furthermore, since our main interest are the algorithmic aspects of coupled solution, we only look at the 2D case, first. The following figure (9) shows the physical model used and explains the principal structure of such a device.

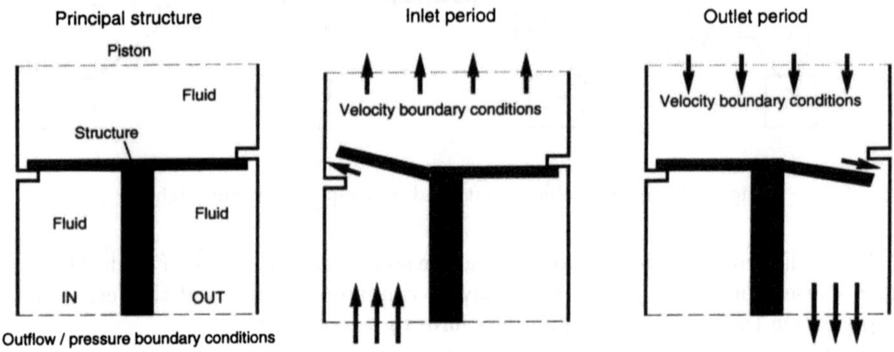

Fig. 9 Principal structure and functional description of the model.

First, we concentrate on the stationary form of the problem: The fluid flow influences the structure via pressure loads at the wetted surface of the structure (interface). As a result, the corresponding coupling operator is given as the identity restricted to the interface nodes. On the other hand, the deformation of the structure changes the geometry of the fluid domain. This leads to a non-linear form of the coupled problem even in the case of a linear model for the fluid flow like a potential flow, because changes in domain geometry result in a changed discretized form of the fluid flow operator.

For modeling the fluid problem, the incompressible form of the stationary Navier-Stokes equations is taken. Based on measured flow rates and the underlying geometry of the flaps [30], the Reynolds number was estimated to take values of about 300. Laminar flow conditions are supposed for the whole domain and for each position of the valve. Note that this might be unrealistic, especially around the valves, but on the other hand, the small dimensions and an appropriate choice of the geometry like in figure (3) should help to prevent turbulence. Anyway, it is a crucial task to find an accurate physical model for the flow around an opening flap, because the flow conditions change form squeeze film flow for very small slots (Reynolds equation from lubrication theory) to the one which can be described by the full Navier-Stokes equations only.

For the flow simulation, a Navier-Stokes solver based on the MAC-technique and implemented by Griebel et. al. [3] was used. The main advantage of this code from our point of view is that it makes possible a very simple and flexible handling of complicated geometries and moving boundaries. For a detailed description of the code and its main properties, see [3].

To calculate the deflection of an arbitrary elastic structure in 3D caused by a pressure distribution on the surface, the equations of linear elasticity (Navier equation) have to be used. Due to the special geometry (high aspect ratio of the flap length to its thickness), the use of the simpler plate equation is possible. For 2D problems, this results in an ordinary differential equation known from the bending of beams. Thus, for the description of the mechanical part of the valve behaviour, the model is based on the one of a cantilever fixed at one end and loaded by a continuous pressure distribution on its surface.

To get an idea of the strength of the coupling and the general flow situation inside such a device, the modeling was done in three main steps. First, the flaps were taken as non-flexible plates opened and closed in order to control the flow inside the pump (see figure (9)). In the second step, the deformation of the valves was calculated using the plate equation, which can be solved analytically (see figure (10)).

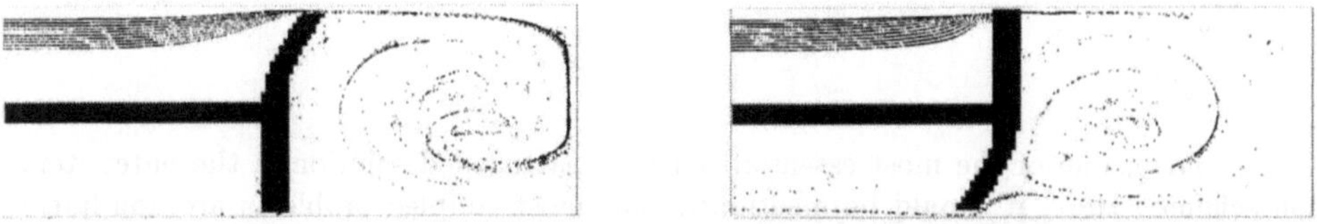

Fig. 10 Streaklines during inlet (left) and outlet (right) period.

Finally, we modeled the deformation of the flaps assuming a plain strain problem. For this simulation task from structural mechanics, we used the commercial finite element package ADINA. Now, the coupled solution was realized with a partitioned solution scheme like the one of figure (7). For the first tests, each iteration step started with the creation of a new input file based on the results of the previous run. Note that this is not necessary, because in the structural problem only the right hand side (load vector) has to be modified, whereas the factorized stiffness matrix can still be used. First of all, Jacobi- and Gauss-Seidel-type iterations were tested. As in [29], a strongly problem dependent convergence behaviour could be observed. For the coupled stationary simulation of the inlet valve, table (1) shows the influence of the stiffness of the valve itself (Young's modulus) and the influence of the viscosity of the surrounding fluid (Reynolds number) on the number of (outer) iteration steps that are necessary till the flap has reached its final (open) position or a state where it alternates between two or three neighbouring positions.

CONCLUDING REMARKS

In this paper, we gave a short survey of both the relevance and the numerical treatment of coupled problems in microsystem technology, and we presented a modular strategy based on the partitioned solution approach as well as first numerical results concerning the simulation of a two-valve membrane pump. We can summarize that the coupled problems which have been observed in the micro world up to now are very heterogeneous. It depends strongly on the specific problem whether the coupling has to be considered or not and which mathematical model has to be used for an efficient numerical treatment. Another interesting problem which shall not be discussed in detail here is the question of the validity of the classical models from the macro world. Especially for flow problems with low Knudsen numbers, e. g., a modification of the models is necessary [17].

Table 1 Dependence of the convergence behaviour on problem parameters.

Outer iteration	**Problem parameters**		**Number of iterations**
	Young's modulus	Reynolds number	*: alternating states
Gauss-Seidel	$2.0 \cdot 10^{11}$	100	7
		200	10
		333	8*
	$1.5 \cdot 10^{11}$	100	9*
		200	13*

Of course, one of the most essential parts in partitioned solution is the outer iteration scheme. Here, it should be mentioned that most coupled problems are non-linear, even if all of the underlying sub-problems are linear (electrostatics-structure interaction, e. g.). Starting on the semi-discrete level and formulating the coupled problem as a block-structured system of algebraic equations or ordinary differential equations, its solution is in principle the same as the treatment of systems of non-linear equations [18], and the techniques mentioned above can be seen as generalized linear methods described in [18]. Furthermore, the application of modern minimization techniques for the solution of non-linear equations to the iteration procedures used for solving coupled problems may lead to an acceleration of those procedures [14]. Finally, time dependent coupled problems described by block-structured systems of ordinary differential equations or by differential algebraic ones can be treated with techniques developed for the parallel solution of systems of ordinary differential equations [7]. Those aspects and the integration of multilevel approaches into our coupled solution context will be in the centre of our future work.

REFERENCES

[1] R. J. Balling and J. Sobieszczanski-Sobieski, *Optimization of coupled problems: A critical overview of approaches*, Technical Report, ICASE: Institute for Computer Applications in Science and Engineering, NASA Langley Research Center, Hampton, VA, USA, 1994.

[2] A. Boley and H. Weiner, *Theory of thermal stresses*, Wiley, New York, 1962.

[3] T. Dornseifer, M. Griebel, and T. Neunhoeffer, *Numerische Simulation — Eine praxisorientierte Einführung am Beispiel der Strömungsmechanik*, Vieweg Verlag, to appear.

[4] C. A. Felippa and T. L. Geers, *Partitioned analysis for coupled mechanical systems*, Eng. Comput., 7 (1988), pp. 331–342.

[5] C. A. Felippa and K. C. Park, *Staggered transient analysis procedures for coupled mechanical systems*, Computer Methods in Applied Mechanics and Engineering, 24 (1980), pp. 61–111.

[6] H. Fujita and T. Ikoma, *Numerical determination of the electromechanical field for a micro servosystem*, Sensors and Actuators, A21-A23 (1990), pp. 215–218.

[7] C. Gear, *The potential for parallelism in ordinary differential equations*, tech. report, University of Illinois at Urbana-Champaign, 1986.

[8] W. Göpel, J. Hesse, and J. Zemel, *Sensors, A Comprehensive Survey*, vol. 1: Fundamentals and General Aspects, VCH Verlagsgesellschaft mbH, Weinheim, 1989.

[9] W. S. Griffin, H. H. Richardson, and S. Yamanami, *A study of fluid squeeze-film damping*, Transactions of the ASME, Journal of Basic Engineering, (1966), pp. 451–456.

[10] W. Hackbusch, *Iterative Lösung großer schwachbesetzter Gleichungssysteme*, Teubner, Stuttgart, 1991.

[11] E. Hinton, P. Bettess, and R. W. Lewis, eds., *Numerical Methods for Coupled Problems — Proceedings of the International Conference Held at the University College, Swansea, 7.-11. 9. 1981*, Pineridge Press, 1981.

[12] J. Itoh, T. Ebihara, Y. Kubota, and A. Takahashi, *Simulation of tape floating behavior in VTR*, Trans. ASME, Journal of Tribology, 115 (1993), pp. 561–565.

[13] P. Kaufmann, *Force balance accelerometer for cost-sensitive application*, Sensors, (1988), p. 19.

[14] T. Kerkhoven and Y. Saad, *On acceleration methods for coupled nonlinear elliptic systems*, Numerische Mathematik, (1992), pp. 525–548.

[15] K. Kohno, S. Takahashi, and K. Saki, *Elasto-hydrodynamic lubrication analysis of journal bearings with combined use of boundary elements and finite elements*, Engineering Analysis with Boundary Elements, 13 (1994), pp. 273–281.

[16] R. W. Lewis, P. Bettess, and E. Hinton, eds., *Numerical Methods in Coupled Systems*, John Wiley & Sons, 1984.

[17] Y. Mitsuya, *Modified Reynolds equation for ultra-thin film gas lubrication using 1.5-order slip-flow model and considering surface accommodation coefficient*, ASME Journal of Tribology, 115 (1993), pp. 289–294.

[18] J. Ortega and W. Rheinboldt, *Iterative Solution of Nonlinear Equations in Several Variables*, Academic Press, Inc., 1970.

[19] D. K. Paul, O. C. Zienkiewicz, and E. Hinton, *Transient dynamic analysis of reservoir-dam interaction using staggered solution schemes*, in Numerical Methods for Coupled Problems, Pineridge Press, 1981.

[20] R. Rosenberg and D. Karnopp, *Introduction to Physical Dynamics*, McGraw-Hill, New York, 1983.

[21] S. Schulte, *Simulation of cross coupled effects in physical sensors*, vde-verlag gmbh Berlin, Offenbach, 1994, pp. 833–842.

[22] S. Senturia, R. Harris, B. Johnson, S. Kim, and K. Nabors, *A computer-aided design system for microelectromechanical systems (MEMCAD)*, Journal of Microelectromechanical Systems, 1 (1992), pp. 3–13.

[23] W. SPENCER, W. CORBETT, L. DOMINGUEZ, AND B. SHAFER, *An electronically controlled piezoelectric insulin pump and valves*, IEEE Transactions On Sonics And Ultrasonics, SU-25 (1978), pp. 153–156.

[24] J. B. STARR, *Squeeze-film damping in solid-state accelerometers*, in Tech. Digest of IEEE Solid-State Sensor and Actuator Workshop, Hilton Head Island, USA, 1990, pp. 44–47.

[25] J. THOMA, *Simulation by Bondgraphs. Introduction to a Graphical Method*, Springer-Verlag, Berlin-Heidelberg, 1990.

[26] H.-P. TRAH, H. BAUMANN, C. DÖRING, H. GOEBEL, T. GRAUER, AND M. METTNER, *Micromachined valve with hydraulically actuated membrane subsequent to a thermoelectrically controlled bimorph cantilever*, Sensors and Actuators A, 39 (1993), pp. 169–176.

[27] T. TSCHAN, *Simulation, Design and Characterization of a Silicon Piezoresistive Accelerometer, Fabricated by a Bipolar-Compatible Industrial Process*, PhD thesis, University of Neuchatel, Switzerland, 1992.

[28] G. WACHUTKA, *Tailored modeling of miniaturized electrothermomechanical systems using thermodynamic methods*, in Micromechanical Systems, D. Cho, J. Peterson, A. Pisano, and C. Friedrich, eds., vol. 40, 1992.

[29] H. YIE, X. CAI, AND J. WHITE, *Convergence properties of relaxation versus the surface-Newton generalized-conjugate residual algorithm for self-consistent electromechanical analysis of 3-d micro-electro-mechanical structures*, in International Workshop on Numerical Modeling of Processes and Devices for Integrated Circuits: NUPAD V (Cat. No.94TH0640-3), Honolulu, IEEE 0 7803 1867 6, June 1994, pp. 137–140.

[30] R. ZENGERLE, *Mikro-Membranpumpen als Komponenten für Mikro-Fluidsysteme*, PhD thesis, Universität der Bundeswehr München, 1994.

[31] R. ZENGERLE, S. KLUGE, M. RICHTER, AND A. RICHTER, *A bidirectional silicon micropump*. To be published, 1995.

[32] O. C. ZIENKIEWICZ, *Coupled problems and their numerical solution*, in Numerical Methods in Coupled Systems, R. W. Lewis, P. Bettess, and E. Hinton, eds., John Wiley & Sons, 1984.

[33] O. C. ZIENKIEWICZ AND R. E. NEWTON, *Coupled vibrations of a structure submerged in a compressible fluid*, in Proc. Int. Symp. on Finite Element Techniques, Stuttgart, 1.–15. 5. 1969.

An a posteriori error estimate for the unsymmetric coupling of FEM and BEM

Carsten Carstensen
Department of Mathematics, Heriot–Watt University, Edinburgh EH14 4AS, UK.

Darius Zarrabi
Fachbereich Mathematik, TH Darmstadt, D-64277 Darmstadt, FRG

Abstract

The coupling of the finite element method (FEM) and boundary element method (BEM) can be performed in at least two different ways leading to a symmetric or an unsymmetric discrete problem. While the symmetric coupling is proved to converge in a large number of situations, this is known for the unsymmetric coupling only in case of a smooth boundary or under some improper restrictions on the mesh–sizes. Since the unsymmetric coupling is less expensive, it is used by engineers and seemingly gives satisfactory numerical results. Besides the lack of a priori error control one might ask for the justification of the actual numerical output after its computation. In this note we firstly present such a tool and prove a posteriori error estimates for the unsymmetric coupling. Since the computable upper error bound is local, we may use this information to steer the automatic mesh–refinement. The resulting adaptive algorithm is illustrated in a numerical example.

1 Introduction

The coupling of the finite element method (FEM) and boundary element method (BEM) is useful for the numerical analysis of interface problems where, for example, a nonlinear problem is under consideration in some bounded domain Ω and a simple linear problem with constant coefficients in the unbounded exterior domain Ω_c. In this paper we treat the following model problem with a Laplace problem in $\Omega_c := \mathbb{R}^2 \setminus \overline{\Omega}$ and a nonlinear problem in the two–dimensional Lipschitz domain Ω with boundary $\Gamma = \partial\Omega$: We look for a function $u \in H^1_{loc}(\Omega \cup \Omega_c)$ and a real constant b satisfying

$$-\operatorname{div}A(Du) = f \quad \text{in } \Omega \tag{1}$$

$$\Delta u = 0 \quad \text{in } \Omega_c \tag{2}$$

$$u(x) = \frac{b}{2\pi} \cdot \log(x) + o(1) \quad \text{as } |x| \to \infty \tag{3}$$

$$u|_\Omega = u|_{\Omega_c} \quad \text{on } \Gamma \tag{4}$$

$$A(Du|_\Omega) \cdot n = Du|_{\Omega_c} \cdot n \quad \text{on } \Gamma. \tag{5}$$

Here we have a possibly nonlinear mapping $A : L^2(\Omega)^2 \to L^2(\Omega)^2$ which is assumed to be uniformly monotone and Lipschitz, i.e., there exist positive constants c_A and C_A with

$$c_A \| \sigma - \tau \|^2_{L^2(\Omega)^2} \leq \int_\Omega (A(\sigma) - A(\tau))(\sigma - \tau)\, dx \tag{6}$$

$$\| A(\sigma) - A(\tau) \|_{L^2(\Omega)^2} \leq C_A \| \sigma - \tau \|_{L^2(\Omega)^2} \quad (\sigma, \tau \in L^2(\Omega)^2). \tag{7}$$

Furthermore, $f \in L^2(\Omega)$ is a given right hand side; D denotes the gradient and Δ denotes the Laplacian; $o(1)$ is the Landau symbol with $\lim_{|x|\to\infty} o(1) = 0$; n is the unit normal on Γ pointing into Ω_c.

It is stressed that the model situation could be modified to other operators, e.g., to linear elasticity, or other dimensions (with other radiation conditions (3)), or further boundary conditions.

In order to rewrite the exterior problem, we use the representation formula for $x \in \Omega^c$

$$u(x) = \frac{1}{2\pi}\int_\Gamma \phi \cdot \log|x-z|\,ds_z - \frac{1}{2\pi}\int_\Gamma w(z)\cdot \frac{\partial}{\partial n_z}\log|x-z|\,ds_z. \tag{8}$$

As it is well-known, (8) yields a solution of (2) and (3) for all values of ϕ and w which may be choosen as $w = u|_\Gamma$ and $\phi = \frac{\partial u}{\partial n}$ where n is the exterior normal along Γ.

Given $v \in H^{1/2}(\Gamma)$ and $\phi \in H^{-1/2}(\Gamma)$ we define boundary integral operators for $z \in \Gamma$

$$\begin{aligned}(V\phi)(z) &:= -\frac{1}{\pi}\int_\Gamma \phi(\zeta)\log|z-\zeta|\,ds_\zeta\\ (Kv)(z) &:= -\frac{1}{\pi}\int_\Gamma v(\zeta)\frac{\partial}{\partial n_\zeta}\log|z-\zeta|\,ds_\zeta.\end{aligned}$$

This defines the single layer potential V and the double layer potential K which are linear and bounded mappings between the following Sobolev–spaces [13], for $s \in [-1/2, 1/2]$,

$$V : H^{s-1/2}(\Gamma) \to H^{s+1/2}(\Gamma), \qquad K : H^{s+1/2}(\Gamma) \to H^{s+1/2}(\Gamma).$$

The single layer potential V is symmetric and positive definite (in the above spaces with $s = 0$) [13] provided the capacity of Γ is smaller than one; we refer, e.g., to [9, 13, 28, 26, 27] for proofs and more details: sufficient is that Ω lies in a ball with radius less than 1 so that this condition on Γ can always be achieved by scaling [28].

As a carefull analysis involving jump relations shows, letting x tend towards the boundary in (8) we obtain a condition for v and ϕ, namely,

$$(K-1)w - V\phi = 0, \qquad w = u|_\Gamma. \tag{9}$$

Using this and standard arguments one gains the weak form of the interface problem (1)–(5): *Find $u \in H^1(\Omega)$ satisfying (9) and*

$$\int_\Omega A(Du)\cdot D\eta\,dx - \int_\Gamma \phi\cdot\eta|_\Gamma\,ds = \int_\Omega f\cdot\eta\,dx \quad (\eta \in H^1(\Omega))\,. \tag{10}$$

The problems (1)–(5) and (9), (10) are equivalent in the following sense: If $u \in H^1_{loc}(\Omega\cup\Omega_c)$ is a solution of (1)–(5), then $u|_\Omega$ solves (10). Conversely, if $u \in H^1(\Omega)$ is a solution of (10), then u can be extended by using the representation formula (8) to a function $u \in H^1_{loc}(\Omega\cup\Omega_c)$ which solves (1)–(5) (see, e.g., [9, 2, 16, 21, 22] for details).

From standard results in the theory of monotone operators, we gain existence and uniqueness of solutions in our model problem.

2 Discretization

For simplicity, we assume that Ω is a polygon and consider finite partitions $\mathcal{T}_h$ of Ω,

$$\overline{\Omega} = \bigcup_{T\in\mathcal{T}_h}\overline{T}$$

such that $T \in \mathcal{T}_h$ is (the interior of) a triangle with angles greater than a global constant $c_\theta > 0$ and diameter $h_T > 0$. We assume that two non–identical triangles or so-called finite elements in $\mathcal{T}_h$ share at most a common edge or a common vertex.

With any partition $\mathcal{T}_h$ we associate the piecewise linears $S^1_h(\Omega)$, i.e., continuous functions which are affine on each triangle T in $\mathcal{T}_h$. The element sides

$$\mathcal{E}_h := \{E : E \text{ open edge of } T \in \mathcal{T}_h\}$$

consist of interior elements sides

$$\mathcal{E}_h(\Omega) := \{E \in \mathcal{E}_h : E \subset \Omega\}$$

or of boundary sides

$$\mathcal{E}_h(\Gamma) := \{E \in \mathcal{E}_h : E \subset \Gamma\}.$$

For each $E \in \mathcal{E}_h = \mathcal{E}_h(\Omega) \cup \mathcal{E}_h(\Gamma)$ let h_E denote the length of E. Let n_E be the exterior unit normal vector for $E \in \mathcal{E}_h(\Gamma)$ and chose some fixed unit normal n_E for $E \in \mathcal{E}_h(\Omega)$.

In order to discretize the boundary integral operators, we associate the discrete function spaces of piecewise linears $S^1_h(\Gamma)$ and piecewise constants $S^0_h(\Gamma)$ to the partition $\mathcal{E}_h(\Gamma)$ of the boundary, i.e. $S^1_h(\Gamma)$ (resp. $S^0_h(\Gamma)$) consists of continuous (resp., in general, discontinuous) functions on Γ which are affine (resp. constant) on each so–called boundary element E in $\mathcal{E}_h(\Gamma)$.

The discrete problem reads: *Find* $u_h \in S^1_h(\Omega)$ *and* $\phi_h \in S^0_h(\Gamma)$ *satisfying*

$$\int_\Omega A(Du_h)\cdot D\eta_h\,dx + \int_\Gamma \phi_h \cdot \eta_h|_\Gamma\,ds \;=\; \int_\Omega f\cdot\eta_h\,dx \in S^1_h(\Omega) \qquad (11)$$

$$\int_\Gamma (V\phi_h - (K-1)u_h|_\Gamma)\cdot\eta|_\Gamma\,ds \;=\; 0 \qquad (12)$$

for all $\eta_h \in S^1_h(\Omega)$.

Assuming that a solution of the discrete problem exists let

$$\sigma_h := A(Du_h) \quad\text{and}\quad \phi_h := V_h^{-1}(K_h - j_h^* i_h)u_h|_\Gamma \qquad (13)$$

be discrete analogs of

$$\sigma := A(Du) \quad\text{and}\quad \phi := V^{-1}(K-1)u|_\Gamma.$$

Note that Du_h and ϕ_h are piecewise constant and, as a condition on A, we assume that σ_h is piecewise constant as well. On $E \in \mathcal{E}_h(\Gamma)$, $[\sigma_h n_E]$ is defined as the jump of the discrete tractions $\sigma_h n_E$ across the common element edge E. Let $\frac{\partial}{\partial s}$ denote the derivative with respect to the arc–length (at least in the distributional sense). Then, for each triangle T we may define a non–negative real number $\eta_h(T)$ by

$$\begin{aligned}\eta_h(T)^2 &:= h_T^2\cdot\|f\|^2_{L^2(T)}\\ &+ \frac{1}{2}\sum_{E\subset\partial T,\,E\in\mathcal{E}_h(\Omega)} h_E\cdot\|[\sigma_h n_E]\|^2_{L^2(E)}\\ &+ \sum_{E\subset\partial T,\,E\in\mathcal{E}_h(\Gamma)} h_E\cdot\|\phi_h - \sigma_h n_E\|^2_{L^2(E)}\\ &+ \sum_{E\subset\partial T,\,E\in\mathcal{E}_h(\Gamma)} h_E\cdot\Big\|\frac{\partial}{\partial s}\big(V\phi_h - (K-1)u_h|_\Gamma\big)\Big\|^2_{L^2(E)}.\end{aligned}$$

The first two terms in the definition of $\eta_h(T)$ are well established in residual based a posteriori error analysis of the FEM while the remaining two contributions are related to equilibrium on Γ and to (12).

3 The a posteriori error estimate

For the unsymmetric coupling we present the following a posteriori error estimate which is the analog of [6] in the symmetric case.

Theorem 1 *There exists a constant $c > 0$, which depends only on Ω and c_θ, c_A and C_A, such that*

$$\| u - u_h \|_{H^1(\Omega)} \leq c \cdot \Big(\sum_{T \in \mathcal{T}_h} \eta_h(T)^2 \Big)^{1/2}. \tag{14}$$

We emphasize that $\eta_h(T)$ can be computed — at least approximated numerically — once the discrete solution u_h is known; see [9, 10] for details in the implementation.

Based on Theorem 1 we may define an adaptive algorithm for automatic mesh-refinements following the literature. The heuristic idea is first to regard $\eta_h(T)$ as an (approximation to the unknown) local error related to T and secondly to refine T if $\eta_h(T)$ is comparably large. Here, the refinement is steered by a parameter θ with $0 \leq \theta \leq 1$.

ALGORITHM (A_θ)

(a) *Start with a coarse initial mesh $\mathcal{T}_{h_0}$. Put $k = 0$.*

(b) *Solve the discrete problem u_{h_k} with respect to the actual mesh $\mathcal{T}_{h_k}$.*

(c) *Compute $\eta_{h_k}(T)$ for each T in $\mathcal{T}_{h_k}$.*

(d) *Compute the upper bound (14) and decide to stop (then terminate computation) or to refine*

(e) *Refine (e.g., halve the largest edge of) $T \in \mathcal{T}_{h_k}$ provided*

$$\eta_{h_k}(T) \geq \theta \cdot \max_{T' \in \mathcal{T}_{h_k}} \eta_{h_k}(T'). \tag{15}$$

(f) *Refine further triangles to avoid hanging nodes. This defines a new mesh $\mathcal{T}_{h_{k+1}}$. Replace k by $k+1$ and go to (b).*

Adaptive algorithms similar to Algorithm (A_θ) are considered in [6, 9, 10] for the symmetric coupling.

4 Proof of the a posteriori error estimate

We use the following result from adaptive BEM.

Theorem 2 ([6]) *Assume $f \in H^1(\Gamma)$ has at least one zero in each of the elements in $\mathcal{E}_h(\Gamma)$. Then,*

$$\| f \|_{H^\alpha(\Gamma)} \leq c(\log(1+\kappa))^{1/2} \Big(\sum_{E \in \mathcal{E}_h(\Gamma)} h_E \cdot \| \frac{\partial}{\partial s} f \|^2_{L^2(E)} \Big)^{1/2}$$

where the constant $c > 0$ dependent on c_theta only. □

Proof of Theorem 1. Because of (6) and (10), elementwise integration yields

$$\begin{aligned}
c_A \cdot \|Du - Du_h\|^2_{L_2(\Omega)} &\le \sum_T \int_T (\operatorname{div} \sigma_h + f) \cdot (u - v_h)\, dx \\
&+ \frac{1}{2} \sum_E \int_E [\sigma_h \cdot n] \cdot (u - v_h)|_\Gamma\, ds \\
&+ \int_\Gamma (\phi - \phi_h) \cdot (u - u_h)\, ds \\
&+ \int_\Gamma (\phi_h - \sigma_h \cdot n) \cdot (u - v_h)\, ds\ .
\end{aligned}$$

Following [11], there exists a family of interpolation operators $\{I_h\}_h$ such that

$$\begin{aligned}
\|u - u_h - I_h(u - u_h)\|_{L_2(T)} &\le c_1 \cdot h \cdot |u - u_h|_{H^1(N_T)} \\
\|u - u_h - I_h(u - u_h)\|_{L_2(E)} &\le c_2 \cdot \sqrt{h} \cdot |u - u_h|_{H^1(N_E)}
\end{aligned}$$

with some constants $c_1, c_2 > 0$. Here, N_T denotes the union of all neighbours of T. Since v_h above may be choosen arbitrarily, let $v_h = u_h - I_h(u - u_h)$. Then, there exists some $c > 0$ such that

$$\begin{aligned}
c_A/c \cdot \|Du - Du_h\|^2_{L_2(\Omega)} &\le \sum_T h \cdot \|\operatorname{div} \sigma_h + f\|_{L_2(T)} \cdot |u - u_h|_{H^1(N_T)} \\
&+ \frac{1}{2} \sum_E \sqrt{h} \cdot \|[\sigma_h \cdot n]\|_{L_2(E)} \cdot |u - u_h|_{H^1(N_E)} \\
&+ \sum_j \sqrt{h} \cdot \|\phi_h - \sigma_h \cdot n\|_{L_2(\Gamma_j)} \cdot |u - u_h|_{H^1(N_{\Gamma_j})} \\
&+ \int_\Gamma (\phi - \phi_h) \cdot (u - u_h)\, ds\ .
\end{aligned}$$

The term $\int_\Gamma (\phi - \phi_h) \cdot (u - u_h)\, ds$ is estimated with $\phi_h^* := V^{-1}(K - 1)u_h$ as

$$\begin{aligned}
\int_\Gamma (\phi - \phi_h) \cdot (u - u_h)\, ds &= - \int_\Gamma V^{-1}(K - 1)(u - u_h) \cdot (u - u_h)\, ds \\
&\quad + \int_\Gamma (\phi_h^* - \phi_h) \cdot (u - u_h)\, ds \\
&\le - c \cdot \|u - u_h\|^2_{H^{1/2}} + \|u - u_h\|_{H^{1/2}} \cdot \|\phi_h^* - \phi_h\|_{H^{-1/2}} \\
&\le -\frac{c}{2} \|u - u_h\|^2_{H^{1/2}} + \frac{1}{2c} \|\phi_h^* - \phi_h\|^2_{H^{-1/2}}\ .
\end{aligned}$$

By Theorem 2

$$\begin{aligned}
&\alpha \cdot (\|Du - D(u_h)\|^2_{L_2(\Omega)} + \|u - u_h\|^2_{H^{1/2}}) \\
&\le \sum_T h \|\operatorname{div} \sigma_h + f\|_{L_2(T)} \cdot |u - u_h|_{H^1(N_T)} \\
&+ \frac{1}{2} \sum_E \sqrt{h} \|[\sigma_h \cdot n]\|_{L_2(E)} \cdot |u - u_h|_{H^1(N_E)} \\
&+ \sum_j \sqrt{h} \|\phi_h - \sigma_h \cdot n\|_{L_2(\Gamma_j)} \cdot |u - u_h|_{H^1(N_{\Gamma_j})} \\
&+ \sum_j h \| \frac{\partial}{\partial s} ((K - 1/2)u_h - V\phi_h)\, \|^2_{L_2(\Gamma_j)}.
\end{aligned}$$

The first summand on the right hand side of the last estimate ist estimated as follows where we apply Cauchy's, Korn's and Young's inequality succesively.

$$
\begin{aligned}
&\sum_T h\|\operatorname{div}\sigma_h + f\|_{L_2(T)} \cdot |u - u_h|_{H^1(N_T)} \\
&\leq c_3 \cdot \left(\sum_T h^2 \|\operatorname{div}\sigma_h + f\|^2_{L_2(T)}\right)^{1/2} \cdot \|D(u-u_h)\|_{L_2(\Omega)} \\
&\leq a \cdot \left(\sum_T h^2 \|\operatorname{div}\sigma_h + f\|^2_{L_2(T)}\right) + \frac{1}{a}\|D(u-u_h)\|^2_{L_2(\Omega)}.
\end{aligned}
$$

Then, we estimate the second and the third summand in the same manner. Choosing the constant a in the three terms sufficiently small we conclude the proof with a final application of Korn's inequality. □

5 Numerical results

In this section we report on some numerical examples to compare the symmetric with the unsymmetric coupling of FEM and BEM. We include jumps on the interface such that our problem reads: *Find* $(u,\phi) \in H^1(\Omega) \times H^{-1/2}(\Gamma)$ *such that (10) holds with* $A = Id$ *and* $f = 0$ *and*

$$
\langle \psi, V\phi + (1-K)\gamma u \rangle = \langle \psi, Vt_0 + (1-K)u_0 \rangle
$$

for all $(v,\psi) \in H^1(\Omega) \times H^{-1/2}(\Gamma)$. We precribe jumps of the solution and its normal derivative on the boundary, namely, u_0 and t_0, such that the exact solution is

$$
\begin{aligned}
u(x,y) &= \Im(x+iy)^{2/3} \\
u_c(x,y) &= \log|(x+1/2, y-1/2)|\,.
\end{aligned}
$$

In the following b_N denotes the value of the right hand side in the appearing a posteriori error estimates and

$$
e_N := \|u - u_h\|_{H^1(\Omega)} + \left\|\frac{\partial u}{\partial n} - \phi_h\right\|_{H^{-1/2}(\Gamma)}
$$

is the error in the energy norm.

Table 1: Symmetric coupling for the Laplace problem, $\theta = 0$

N	e_N	b_N	$\frac{e_N}{b_N}$
16	0.63636 D+00	0.24960 D+01	0.25495 D+00
19	0.60685 D+00	0.20679 D+01	0.29346 D+00
37	0.42442 D+00	0.16991 D+01	0.24979 D+00
49	0.38701 D+00	0.13376 D+01	0.28933 D+00
97	0.27670 D+00	0.11036 D+01	0.25072 D+00
145	0.24931 D+00	0.86370 D+00	0.28865 D+00
298	0.17789 D+00	0.70868 D+00	0.25102 D+00
481	0.15919 D+00	0.55277 D+00	0.28799 D+00

Table 2: Unsymmetric coupling for the Laplace problem, $\theta = 0$

N	e_N	b_N	$\frac{e_N}{b_N}$
16	0.65712 D+00	0.16154 D+01	0.40678 D+00
19	0.64725 D+00	0.13745 D+01	0.47090 D+00
37	0.43289 D+00	0.11768 D+01	0.38731 D+00
49	0.39993 D+00	0.94589 D+00	0.42281 D+00
97	0.28226 D+00	0.80371 D+00	0.35120 D+00
145	0.25764 D+00	0.63165 D+00	0.40788 D+00
289	0.18133 D+00	0.52752 D+00	0.34374 D+00
481	0.16436 D+00	0.41144 D+00	0.39948 D+00

For uniform meshes ($\theta = 0$), the results of our experiments are given in Table 1 and 2. We observe that even the unsymmetric method seems to converge which is not justified a priori but which is clearly indicated by the decrease of b_N. Of certain interest is the quotient e_N/b_N which is bounded from above in agreement with Theorem 1. Moreover, these quotients seem to be bounded below as well which clearly indicates the sharpness of Theorem 1. We stress that this efficiency can be proved as in [5, 6] for uniform meshes in model situations like here.

Table 3: Symmetric coupling for the Laplace problem, $\theta = 0.5$

N	e_N	b_N	$\frac{e_N}{b_N}$
16	0.63636 D+00	0.24960 D+01	0.25495 D+00
19	0.60685 D+00	0.20679 D+01	0.29346 D+00
23	0.47838 D+00	0.18632 D+01	0.25675 D+00
27	0.43429 D+00	0.15545 D+01	0.27938 D+00
32	0.34192 D+00	0.13723 D+01	0.24916 D+00
36	0.32110 D+00	0.12134 D+01	0.26463 D+00
45	0.26198 D+00	0.10724 D+01	0.24429 D+00
57	0.23503 D+00	0.94526 D+00	0.24864 D+00
66	0.19794 D+00	0.85006 D+00	0.23285 D+00

Table 4: Unsymmetric coupling for the Laplace problem, $\theta = 0.5$

N	e_N	b_N	$\frac{e_N}{b_N}$
16	0.65712 D+00	0.16154 D+01	0.40678 D+00
19	0.64725 D+00	0.13745 D+01	0.47090 D+00
29	0.43995 D+00	0.11986 D+01	0.39286 D+00
34	0.41417 D+00	0.10202 D+01	0.40597 D+00
46	0.31066 D+00	0.94310 D+00	0.32940 D+00
57	0.28763 D+00	0.80443 D+00	0.35756 D+00
71	0.21476 D+00	0.69569 D+00	0.30870 D+00

To illustrate the adaptive scheme we applied Algorithm ($A_{1/2}$) and obtained meshes as shown in Fig. 1 and Fig. 2. In both cases we see some refinments towards the origin where the exact solution has a singularity so that this is expected to improve the approximation. Indeed, as seen in Tab. 3 and 4, e.g., for $e_N < 0.2$ we need $N > 200$ in the uniform case but $N < 66$ in the adaptive case. This indicates efficiency of the adaptive scheme in this example.

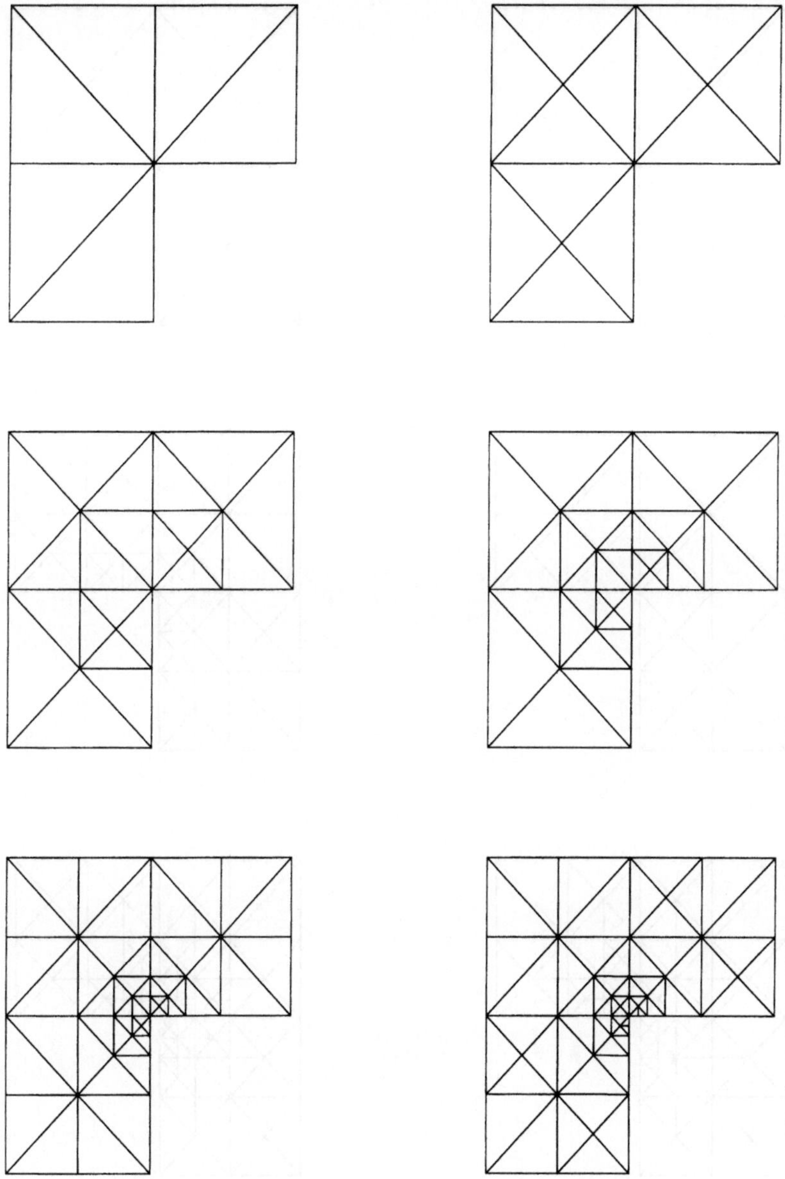

Figure 1: Adapted meshes (symmetric coupling)
Adapted meshes after step 0 (initial mesh) ,1,3,5 and 7
Third row, last figure: Mesh after step 8

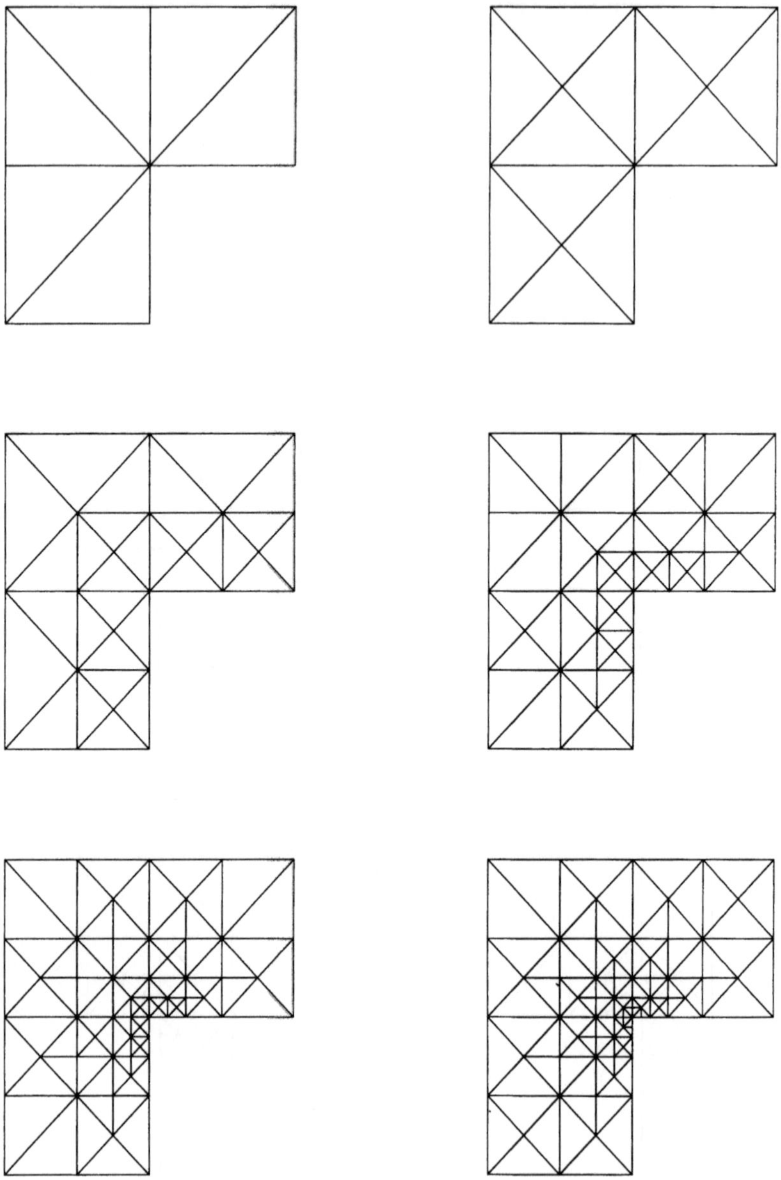

Figure 2: Adapted meshes (unsymmetric coupling)
Adapted meshes after step 0 (initial mesh) ,1,3,5 and 7
Third row, last figure: Mesh after step 8

References

[1] I Babuška, A. Miller: A feedback finite element method with a posteriori error estimation. *Comp. Meth. Appl. Mech. Eng.* **61** (1987) 1–40.

[2] C. Carstensen: Interface problem in holonomic elastoplasticity. *Math. Meth. in the Appl. Sci.* **16** (1993) 819—835.

[3] C. Carstensen: Adaptive boundary element methods and adaptive finite element and boundary element coupling in M. Çostabel, M. Dauge, S. Nicaise (eds.): Proceedings *Boudary Value Problems and Integral Equations on Non-Smooth Domains* (in print 1994).

[4] C. Carstensen: Efficiency of a posteriori error estimates for boundary element methods on uniform meshes. *Math. Comp.* (1994); accepted for publication.

[5] C. Carstensen: An a posteriori error estimate for a first kind integral equation. (1995) in preparation.

[6] C. Carstensen: A posteriori error estimates for the symmetric coupling of finite elements and boundary elements. (1995) in preparation.

[7] C. Carstensen, E.P.Stephan: A posteriori error estimates for boundary element methods. *Math. Comp.* (1994); in press.

[8] C. Carstensen, E.P. Stephan: Adaptive boundary element methods for some first kind integral equations. Preprint Institut für Angewandte Mathematik, Universität Hannover (1993).

[9] C. Carstensen, E.P. Stephan: Adaptive coupling of boundary elements and finite elements. Preprint Institut für Angewandte Mathematik, Universität Hannover (1993).

[10] C. Carstensen, S. Funken & E.P. Stephan: On the adaptive coupling of FEM and BEM in 2–d–elasticity. Preprint Institut für Angewandte Mathematik, Universität Hannover (1993).

[11] P. Clement: Approximation by finite element functions using local regularization. *RAIRO Sér. Rouge Anal. Numér.* **R-2** 77-84 (1975).

[12] M. Costabel: Symmetric methods for the coupling of finite elements and boundary elements. In: C.A. Brebia et al. (Eds.), Boundary Elements IX, Vol. 1, 411–420, Springer–Verlag, Berlin 1987.

[13] M. Costabel: Boundary integral operators on Lipschitz domains: Elementary results. *SIAM J. Math. Anal.* **19** (1988) 613–626.

[14] M. Costabel, E.P. Stephan: Boundary integral equations for mixed boundary value problems in polygonal domains and Galerkin approximation. *Banach Center Publ.* **15** (1985) 175–251.

[15] M. Costabel, E.P. Stephan: A direct boundary integral equation method for transmission problems. *J. Math. Anal. Appl.* **106** (1985) 367–413.

[16] M. Costabel, E.P. Stephan: Coupling of finite and boundary element methods for an elastoplastic interface problem. *SIAM J. Numer. Anal.* **27** (1990) 1212–1226.

[17] K. Eriksson, C. Johnson: An adaptive finite element method for linear elliptic problems. *Math. Comp.* **50** (1988) 361—3883.

[18] K. Eriksson, C. Johnson: Adaptive finite element methods for parabolic problems I. A linear model problem *SIAM J. Numer. Anal.* **28** (1991) 43—77.

[19] B. Faehrmann: *Lokale a-posteriori-Fehlerschätzer bei der Diskretisierung von Randintegralgleichungen.* PhD–thesis, University of Kiel, FRG (1993).

[20] G.N. Gatica, G.C. Hsiao: On a class of variational formulations for some nonlinear interface problems. *Rendiconti di Mathematica Ser. VII* **10**(1990) 681-715.

[21] G.N. Gatica, G.C. Hsiao: On the coupled BEM and FEM for a nonlinear exterior Dirichlet problem in R^2. *Numer. Math.* **61** (1992) 171-214.

[22] H. Han: A new class of variational formulations for the coupling of finite and boundary element methods. *J. Comput. Math.* **8** (1990) 223–232.

[23] N. Heuer: hp–Versionen der Randelementemethode. PhD Dissertation, University of Hannover, FRG (1992).

[24] C. Johnson, P. Hansbo: Adaptive finite element methods in computational mechanics. *Comput. Meth. Appl. Mech. Engin.* **101** (1992) 143–181.

[25] J.L Lions, E. Magenes: *Non-homogeneous boundary value problems and applications, Vol. I.* Berlin–Heidelberg–New York: Springer 1972.

[26] J. Saranen and W.L. Wendland: Local residual–type error estimates for adaptive boundary element methods on closed curves. *Applicable Analysis* **48** (1993) 37—50.

[27] E.P. Stephan, W.L. Wendland, G.C. Hsiao: On the intregal equation method for the plane mixed boundary value problem of the Laplacian. *Math. Meth. Appl. Sci.* **1** (1979) 265—321.

[28] I.H. Sloan, A. Spence: The Galerkin Method for Integral Equations of the first kind with Logarithmic Kernel: Theory. *IMA J. Numer. Anal.* **8** (1988) 105—122.

[29] R. Verfürth: A review of a posteriori error estimation and adaptive mesh–refinement techniques. Preprint. (1993).

[30] W.L. Wendland: On Asymptotic Error Estimates for Combined BEM and FEM. in *Finite and boundary element techniques from mathematical and engineering point of view.* CISM Courses **301** E. Stein, W.L. Wendland, eds., Springer-Verlag New York, 1988, 273–331.

On Numerical Treatment of Coupled BEM and FEM for Nonlinear Exterior Problems

M. Feistauer
Charles University Prague
Faculty of Mathematics and Physics
Malostranské nám 25, 118 00 Praha 1
Czech Republic

G. C. Hsiao and R. E. Kleinman
University of Delaware
Dept. of Mathematical Sciences
Newark, DE 19716, USA

Summary

This paper presents a brief survey on the investigation of the coupled BEM and FEM applied to a nonlinear exterior boundary value problem. The aim is to find the solution of a nonlinear partial differential equation considered in an annular bounded domain and the Laplace equation outside. These equations are bound together by transmission conditions and are equipped with boundary conditions. The nonlinear problem in the interior domain is combined with an integral equation obtained with the aid of potential theory and then the whole problem is reformulated in a weak sense. The discretization is carried out by the coupled finite element - boundary element method. We discuss the existence and uniqueness of the solution of the discrete as well continuous problem, the convergence of approximate solutions to an exact one, provided the size of the grid tends to zero, and the iterative solution of the nonlinear discrete problem. Proofs are the subject of more detailed papers (see, e.g., [5]).

1 Continuous Problem

Let $\Omega^- \subset \mathbb{R}^2$ be a bounded domain with a Lipschitz continuous boundary $\partial\Omega^- = \Gamma_0 \cup \Gamma$. Here, Γ_0, Γ are simple closed curves, Γ_0 is piecewise C^3, $\Gamma_0 \subset \operatorname{Int}\Gamma$, $\Gamma \in C^\infty$, $\operatorname{Int}\Gamma$ is convex. We set $\Omega^+ = \operatorname{Ext}\Gamma$.

Problem (P). Find $u : \overline{\Omega}^- \cup \overline{\Omega}^+ \longrightarrow \mathbb{R}$ (sufficiently smooth) and $\beta \in \mathbb{R}$ such that

$$-\sum_{i=1}^{2} \frac{\partial}{\partial x_i} a_i(\cdot, u, \nabla u) + a_0(\cdot, u, \nabla u) = f \quad \text{in } \Omega^-, \tag{1.1}$$

$$u|\Gamma_0 = 0, \tag{1.2}$$

$$\Delta u = 0 \quad \text{in } \Omega^+, \tag{1.3}$$

$$\text{a)} \quad u^- = u^+, \qquad \text{b)} \quad \sum_{i=1}^{2} a_i(\cdot, u, \nabla u)\, n_i|^- = \frac{\partial u}{\partial n}|^+ \quad \text{on } \Gamma, \tag{1.4}$$

$$\lim_{|x|\to\infty} (u(x) - u_\infty(x) - \frac{\alpha}{2\pi} \ln |x|) = \beta. \tag{1.5}$$

By $+$ and $-$ we denote the limits at points on Γ approaching them from Ω^+ and Ω^-, respectively, $n = (n_1, n_2)$ is the unit outer normal to $\partial\Omega^-$. The function u_∞ and the constant α are given.

We consider the following *assumptions on data*: There exists a domain $\Omega^* \subset I\!R^2$ such that $\overline{\Omega}^- \subset \Omega^*$, $f \in L^2(\Omega^*)$, the functions $a_i = a_i(x,\xi)$, $x \in \Omega^*$, $\xi = (\xi_0, \xi_1, \xi_2) \in I\!R^3$, are continuous, the derivatives $\partial a_i/\partial \xi_j$ are continuous and bounded and

$$|a_i(x,\xi)| \le c_0 \left(1 + \sum_{j=0}^{2} |\xi_j|\right), \quad x \in \Omega^*,\ \xi \in I\!R^3,\ i = 0,1,2, \tag{1.6}$$

$$\sum_{i=0}^{2} a_i(x,\xi)\,\xi_i \ge c_1(\xi_1^2 + \xi_2^2) - c_2 \left(1 + \sum_{j=0}^{2} |\xi_j|\right), \quad x \in \Omega^*,\ \xi \in I\!R^3, \tag{1.7}$$

$$\sum_{i,j=1}^{2} \frac{\partial a_i}{\partial \xi_j}(x,\xi)\,\theta_i\,\theta_j \ge \alpha(\theta_1^2 + \theta_2^2),\ x \in \Omega^*,\ \xi \in I\!R^3,\ \theta = (\theta_1, \theta_2) \in I\!R^2, \tag{1.8}$$

where the constants c_0, c_1, c_2, $\alpha > 0$ are independent of x, ξ, θ. Further, let $u_\infty \in C^2(I\!R^2)$ and $\Delta u_\infty = 0$. Hence, $\Delta(u - u_\infty) = 0$ in Ω^+.

Using the fundamental solution $E(x,y) = -(\ln|x-y|)/(2\pi)$ of the Laplace equation and basic results from the potential theory, we obtain the representation

$$\begin{aligned}(u - u_\infty)(x) &= \int_\Gamma \frac{\partial}{\partial n(y)} E(x,y)\,(u - u_\infty)^+(y)\,dS_y - \\ &- \int_\Gamma E(x,y) \left(\frac{\partial(u-u_\infty)}{\partial n}\right)^+ (y)\,dS_y - \lambda, \quad x \in \Omega^+,\end{aligned} \tag{1.9}$$

and the condition

$$\int_\Gamma \left(\frac{\partial(u-u_\infty)}{\partial n}\right)^+ dS = \alpha, \tag{1.10}$$

where $\lambda = \text{const}$. The passage to limit in (1.9) as $x \in \Omega^+$ tends to points on Γ and the differentiation with respect to n yields the identities

$$\begin{aligned}\text{a)}\quad (u-u_\infty)^+ &= \left(\frac{1}{2}I + K\right)(u-u_\infty)^+ - \\ &- V\left(\frac{\partial(u-u_\infty)}{\partial n}\right)^+ - \lambda \quad \text{on } \Gamma, \\ \text{b)}\quad \left(\frac{\partial(u-u_\infty)}{\partial n}\right)^+ &= -W(u-u_\infty)^+ + \\ &+ \left(\frac{1}{2}I - K'\right)\left(\frac{\partial(u-u_\infty)}{\partial n}\right)^+ \quad \text{on } \Gamma,\end{aligned} \tag{1.11}$$

where

$$\begin{aligned}(V\xi)(x) &= \int_\Gamma E(x,y)\,\xi(y)\,dS_y, \\ K(\xi)(x) &= \int_\Gamma \left\{\frac{\partial}{\partial n(y)} E(x,y)\right\} \xi(y)\,dS_y, \\ (K'\xi)(x) &= \int_\Gamma \left\{\frac{\partial}{\partial n(x)} E(x,y)\right\} \xi(y)\,dS_y, \\ (W\xi)(x) &= -\frac{\partial}{\partial n(x)} \int_\Gamma \left\{\frac{\partial}{\partial n(y)} E(x,y)\right\} \xi(y)\,dS_y.\end{aligned} \tag{1.12}$$

Cf., e.g., [12, 13].

By $W^{k,p}(\Omega^-)$ and $H^k(\Omega^-) = W^{k,2}(\Omega^-)$ we denote the well-known Sobolev spaces, $H^{1/2}(\Gamma)$ is the space of traces on Γ of all $v \in H^1(\Omega^-)$ and $H^{-1/2}(\Gamma)$ is the dual of $H^{1/2}(\Gamma)$. For $\xi \in H^{-1/2}(\Gamma)$, $\phi \in H^{1/2}(\Gamma)$ we denote by $\langle \xi, \phi \rangle$ the duality pairing between $H^{-1/2}(\Gamma)$ and $H^{1/2}(\Gamma)$ such that

$$\langle \xi, \phi \rangle = \int_\Gamma \xi \, \phi \, dS \quad \text{for } \xi \in L^2(\Gamma), \ \phi \in H^{1/2}(\Gamma). \tag{1.13}$$

In [4] it was shown that the mappings

$$\begin{aligned} V &: H^{-1/2}(\Gamma) \to H^{1/2}(\Gamma), & K &: H^{1/2}(\Gamma) \to H^{1/2}(\Gamma), \\ K' &: H^{-1/2}(\Gamma) \to H^{-1/2}(\Gamma), & W &: H^{1/2}(\Gamma) \to H^{-1/2}(\Gamma) \end{aligned} \tag{1.14}$$

are continuous.

Further, we define the following spaces:

$$\begin{aligned} &\text{a)} \quad \mathcal{V} = \{v \in C^\infty(\overline{\Omega}^-); \ \operatorname{supp} v \subset \Omega^- \cup \Gamma\}, \\ &\text{b)} \quad V = \overline{\mathcal{V}}^{H^1(\Omega^-)} = \{v \in H^1(\Omega^-); v|\Gamma_0 = 0\}, \\ &\text{c)} \quad H_0^{-1/2}(\Gamma) = \{\delta \in H^{-1/2}(\Gamma); \langle \delta, 1 \rangle = 0\}, \\ &\text{d)} \quad W = H^1(\Omega^-) \times H^{-1/2}(\Gamma), \\ &\text{e)} \quad H = V \times H_0^{-1/2}(\Gamma) \subset W. \end{aligned} \tag{1.15}$$

The function

$$\|(v,\sigma)\|_W = \left(\|v\|^2_{H^1(\Omega^-)} + \|\sigma\|_{H^{-1/2}(\Gamma)}\right)^{1/2}, \quad (v,\sigma) \in W \tag{1.16}$$

represents the norm in the spaces in W and H. By the Friedrichs inequality, there exists a constant $\hat{c}_1 > 0$ such that

$$|v|_{H^1(\Omega^-)} := \left(\int_{\Omega^-} |\nabla v|^2 \, dx\right)^{1/2} \le \|v\|_{H^1(\Omega^-)} \le \hat{c}_1 |v|_{H^1(\Omega^-)}, \quad v \in V. \tag{1.17}$$

Putting $\sigma = (\partial u/\partial n)^+$, using the transmission conditions (1.4, a–b) and identities (1.10), (1.11, a–b), we find that the unknown functions $u : \overline{\Omega}^- \to \mathbb{R}$, $\sigma : \Gamma \to \mathbb{R}$ and constant $\lambda \in \mathbb{R}$ satisfy (1.1), (1.2) and

$$\begin{aligned} &\sum_{i=1}^2 a_i(\cdot, u, \nabla u)\, n_i|^- = \\ &= -W(u-u_\infty)^- + \left(\frac{1}{2}I - K'\right)\left(\sigma - \frac{\partial u_\infty}{\partial n}\right) + \frac{\partial u_\infty}{\partial n} \ \text{on } \Gamma, \end{aligned} \tag{1.18}$$

$$\left(\frac{1}{2}I - K\right)(u-u_\infty)^- + V\left(\sigma - \frac{\partial u_\infty}{\partial n}\right) + \lambda = 0 \quad \text{on } \Gamma, \tag{1.19}$$

$$\int_\Gamma \left(\sigma - \frac{\partial u_\infty}{\partial n}\right) dS = \alpha. \tag{1.20}$$

The above relations lead us to the weak formulation of the problem obtained in such a way that (1.1) is multiplied by an arbitrary $v \in V$, integrated over Ω^-, transformed by Green's theorem, condition (1.18) is used and to the resulting identity the relation (1.19) multiplied by any $\delta \in H_0^{-1/2}(\Gamma)$ and "integrated" over Γ is added. Putting $\sigma^* = \left(\int_\Gamma \frac{\partial u_\infty}{\partial n} dS + \alpha\right)/|\Gamma|$, we arrive at the *weak formulation* of Problem (P):

Problem (WP). Find $(u, \mu) \in H$ such that

$$a(u,v) + B((u,\mu),\,(v,\delta)) = F(v,\delta) \quad \forall\,(v,\delta) \in H, \tag{1.21}$$

where the forms a, B and F are defined for $u,\ v \in H^1(\Omega^-)$ and $\mu,\ \delta \in H^{-1/2}(\Gamma)$ in the following way:

$$a(u,v) = \int_{\Omega^-} \left[\sum_{i=1}^{2} a_i(\cdot,u,\nabla u)\,\frac{\partial v}{\partial x_i} + a_0(\cdot,u,\nabla u)\,v\right] dx, \tag{1.22}$$

$$B((u,\mu),\,(v,\delta)) = \tag{1.23}$$
$$= \left\langle W\,u^- - \left(\frac{1}{2}I - K'\right)\,\mu,\,v^-\right\rangle + \left\langle \delta,\,V\,\mu + \left(\frac{1}{2}I - K\right)u^-\right\rangle,$$

$$F(v,\delta) = \int_{\Omega^-} f v\,dx - B\left(\left(u_\infty,\,\frac{\partial u_\infty}{\partial n} - \sigma^*\right),(v,\delta)\right) + \left\langle\frac{\partial u_\infty}{\partial n},v\right\rangle, \tag{1.24}$$

In [9, 10] it was shown that

$$B((u,\mu),\,(v,\delta)) = b\left(\frac{du}{ds},\,\frac{dv}{ds}\right) + b(\mu,\delta) + \tag{1.25}$$
$$+d(v,\mu) - d(u,\delta) + \langle\delta,u\rangle - \langle\mu,v\rangle,\ \ u,\,v\in V,\ \ \mu,\delta\in H_0^{-1/2}(\Gamma),$$

where

$$b(\xi,\delta) = \frac{1}{2\pi}\int_\Gamma\int_\Gamma \ln\frac{d(\Gamma)}{|x-y|}\,\xi(y)\,\delta(x)\,dS_x dS_y, \tag{1.26}$$
$$\xi,\ \delta \in H^{-1/2}(\Gamma),\ \ d(\Gamma) = 2\,\text{diameter of } \Gamma,$$
$$d(\phi,\delta) = \frac{1}{2\pi}\int_\Gamma\int_\Gamma \frac{(x-y)\cdot n(y)}{|x-y|^2}(\phi(y)-\phi(x))\,\delta(x)\,dS_x dS_y,$$
$$\phi \in H^{1/2}(\Gamma),\ \ \delta \in H^{-1/2}(\Gamma).$$

d/ds means the derivative along Γ. It is necessary to note that the forms b and d are defined by (1.26) for sufficiently regular functions and extended by continuity on the mentioned spaces. Moreover, there exists a constant $c_b > 0$ such that

$$B((u,\mu),(u,\mu)) \geq c_b\|\mu\|^2_{H^{-1/2}(\Gamma)}, \quad (u,\mu) \in H. \tag{1.27}$$

It is possible to define the mapping $A : H \to H^*$ by

$$\langle A(u,\mu),(v,\delta)\rangle_{H^*H} = a(u,v) + B((u,\mu),(v,\delta)),\ (u,\mu),(v,\delta)\in H. \tag{1.28}$$

Here $\langle\cdot,\cdot\rangle_{H^*H}$ denotes the duality between the dual H^* to H and the space H.

In a similar way as in [8] we can prove the following properties of the operator A:

Theorem 1.1. a) *The operator A is Lipschitz-continuous and maps any bounded set into a bounded set.*
b) *The operator A satisfies the condition* (S): *If*

$$(u_n,\mu_n) \to (u,\mu) \quad \textit{weakly in } H \textit{ as } n\to\infty \tag{1.29}$$

and

$$\langle A(u_n, \mu_n) - A(u, \mu), (u_n, \mu_n) - (u, \mu)\rangle_{H^* H} \to 0, \tag{1.30}$$

then

$$(u_n, \mu_n) \to (u, \mu) \quad \textit{strongly in } H. \tag{1.31}$$

Provided the condition

$$\sum_{i,j=0}^{2} \frac{\partial a_i}{\partial \xi_j}(x, \xi)\, \theta_i \theta_j \geq \alpha(\theta_1^2 + \theta_2^2), \;\; x \in \Omega^*, \;\; \xi, \; \theta = (\theta_0, \theta_1, \theta_2) \in I\!R^3 \tag{1.32}$$

is satisfied with $\alpha > 0$ independent of x, ξ, θ, the operator A is strongly monotone: there exists $c^* > 0$ such that

$$\begin{aligned} &\langle A(u, \mu) - A(v, \delta), (u, \mu) - (v, \delta)\rangle_{H^* H} \geq \\ &\geq c^* \|(u - v, \; \mu - \delta)\|_W^2 \quad (u, \mu), \; (v, \delta) \in H. \end{aligned} \tag{1.33}$$

It is suitable to note that condition (S) is a consequence of assumption (1.8) and properties of the form B. It represents a pseudomonotonicity property of the problem and plays an important role in the investigation of the convergence of approximate solutions to an exact one.

2 Discrete Problem

Let us consider a system $\{\Omega_h^-\}_{h \in (0,h_0)}$, $h_0 > 0$, of polygonal approximations of the domain Ω^-. Let $\mathcal{T}_h$ be a triangulation of $\overline{\Omega_h^-}$ formed by a finite number of closed triangles with usual properties. See, e. g., [3]. By $\sigma_h = \{P_1, \ldots, P_{N_h}\}$ we denote the set of all vertices of all $T \in \mathcal{T}_h$ and assume that

a) $\sigma_h \subset \overline{\Omega_h^-}$, $\sigma_h \cap \partial\Omega_h^- \subset \partial\Omega^-$,

b) σ_h contains the points of $\partial\Omega^-$ where the condition of C^3-regularity of Γ_0 is not satisfied.

By h_T and ϑ_T we denote the length of the largest side of $T \in \mathcal{T}_h$ and the magnitude of the smallest angle of $T \in \mathcal{T}_h$' respectively, and put

$$h = \max_{T \in \mathcal{T}_h} h_T, \qquad \vartheta_h = \min_{T \in T} \vartheta_T.$$

Moreover, we consider only such triangulations $\mathcal{T}_h$ that each $T \in \mathcal{T}_h$ has at most two vertices lying on $\partial\Omega^-$. The triangles with two vertices on $\partial\Omega^-$ will be called *boundary triangles.* By Γ_{0h} and Γ_h we denote the components of $\partial\Omega_h^-$ approximating Γ_0 and Γ, respectively. Let $\mathcal{G}_h$ denote the set of all sides $S \subset \Gamma_h$ of $T \in \mathcal{T}_h$ adjacent to Γ_h. By $|T|$ and $|S|$ we denote the area a triangle $T \in \mathcal{T}_h$ and the length of a side $S \in \mathcal{G}_h$, respectively.

We assume that the system $\{\mathcal{T}_h\}_{h \in (0,h_0)}$ is *regular*, i. e.,

$$\vartheta_h \geq \vartheta_0 > 0 \quad \forall\, h \in (0, h_0) \tag{2.1}$$

and the *inverse inequality* is satisfied on Γ_h:

$$\max_{S \in \mathcal{G}_h} |S| \leq \nu |S| \quad \forall\, S \in \mathcal{G}_h, \; \forall\, h \in (0, h_0). \tag{2.2}$$

ϑ_0 and ν are constants independent of T, S and h. In virtue of (2.1), there exists $\sigma > 0$ such that

$$\sigma h_T^2 \leq |T| \quad \forall T \in \mathcal{T}_h,\ \forall h \in (0, h_0). \tag{2.3}$$

In the finite element – boundary element discretization of Problem (WP) we use the following finite dimensional spaces:

$$\begin{aligned} X_h &= \left\{ v_h \in C(\overline{\Omega^-}_h); v_h \text{ is linear on each } T \in \mathcal{T}_h \right\}, \\ V_h &= \{ v_h \in X_h; v_h | \Gamma_{0h} = 0 \}, \\ H_h^{-1/2} &= \left\{ \xi_h \in L^2(\Gamma_h); \xi_h \text{ is constant on each } S \in \mathcal{G}_h \right\}, \\ H_{0h}^{-1/2} &= \left\{ \xi_h \in H_h^{-1/2}; \int_{\Gamma_h} \xi_h \, dS = 0 \right\}, \\ H_h &= V_h \times H_{0h}^{-1/2} \end{aligned} \tag{2.4}$$

and equip H_h with the norm

$$\|(v, \delta)\|_h = \left(\|v\|^2_{H^1(\Omega_h^-)} + \|\delta\|^2_{H^{-1/2}(\Gamma_h)} \right)^{1/2}, \quad (v, \delta) \in H_h. \tag{2.5}$$

Obviously,

$$\begin{aligned} &V_h \subset X_h \subset H^1(\Omega_h^-), \quad H_h^{-1/2} \subset L^2(\Gamma_h) \subset H^{-1/2}(\Gamma_h), \\ &H_{0h}^{-1/2} \subset H_0^{-1/2}(\Gamma_h) = \left\{ \xi \in H^{-1/2}(\Gamma_h); \langle \xi_h, 1 \rangle_h = 0 \right\}, \end{aligned} \tag{2.6}$$

where $\langle \cdot, \cdot \rangle_h$ denotes the duality pairing between $H^{-1/2}(\Gamma_h)$ and $H^{1/2}(\Gamma_h)$ induced by $L^2(\Gamma_h)$ – scalar product.

Let us put

$$\begin{aligned} \tilde{a}_h(u, v) &= \int_{\Omega_h^-} \left[\sum_{i=1}^{2} a_i(\cdot, u, \nabla u) \frac{\partial v}{\partial x_i} + a_0(\cdot, u, \nabla u)\, v \right] dx, \\ &\qquad u,\ v \in H^1(\Omega_h^-), \\ \tilde{L}_h(v) &= \int_{\Omega_h^-} f\, v \, dx, \quad v \in H^1(\Omega_h^-), \\ \tilde{b}_h(\xi, \delta) &= \frac{1}{2\pi} \int_{\Gamma_h} \int_{\Gamma_h} \ln \frac{d(\Gamma)}{|x - y|} \xi(y)\, \delta(x) \, dS_x dS_y, \\ &\qquad \xi,\ \delta \in H^{-1/2}(\Gamma_h), \\ \tilde{d}_h(\phi, \delta) &= \frac{1}{2\pi} \int_{\Gamma_h} \int_{\Gamma_h} \frac{(x - y) \cdot n_h(y)}{|x - y|^2} (\phi(y) - \phi(x))\, \delta(x) \, dS_x dS_y, \\ &\qquad \phi \in H^{1/2}(\Gamma_h),\ \delta \in H^{-1/2}(\Gamma_h). \end{aligned} \tag{2.7}$$

Here $n_h(y)$ can be defined in two possible ways: a) $n_h(y)$ is the unit outer normal to $\partial\Omega_h^-$ at $y \in \Gamma_h$, b) $n_h(y)$ is the piecewise linear interpolation of the unit outer normal n to Γ with interpolation points from the set $\sigma_h \cap \partial\Omega$. Further, we set

$$\begin{aligned} &\tilde{B}_h((u, \mu), (v, \delta)) = \tilde{b}_h \left(\frac{du}{ds_h}, \frac{dv}{ds_h} \right) + \tilde{b}_h(\mu, \delta) + \\ &+ \tilde{d}_h(v, \mu) - d_h(u, \delta) + \langle \delta, u \rangle_h - \langle \mu, v \rangle_h, \\ &\qquad u,\ v \in H^1(\Omega_h^-),\ \mu,\ \delta \in H^{-1/2}(\Gamma_h), \\ &\tilde{F}_h(v, \delta) = \tilde{L}_h(v) - \tilde{B}_h \left(\left(u_\infty, \frac{\partial u_\infty}{\partial n_h} - \sigma^* \right), (v, \delta) \right) + \left\langle \frac{\partial u_\infty}{\partial n_h}, v \right\rangle_h, \\ &\qquad v \in H^1(\Omega_h^-),\ \delta \in H^{-1/2}(\Gamma_h), \end{aligned} \tag{2.8}$$

where d/ds_h is the derivative along Γ_h. The discrete problem can be formulated in the following way:

Problem (DP). Find $(u_h, \mu_h) \in H_h$ such that

$$\tilde{a}_h(u_h, v_h) + \tilde{B}_h((u_h, \mu_h), (v_h, \delta_h)) = \tilde{F}_h(v_h, \delta_h) \ \forall (v_h, \delta_h) \in H_h. \tag{2.9}$$

The solvability of Problem (DP) is established in the following theorem.

Theorem 2.1. *Problem (DP) has at least one solution. Under condition (1.32) the solution of (DP) is unique.*

3 Convergence

The convergence of approximate solutions (u_h, μ_h) satisfying (2.9) to the solution (u, μ) of the continuous problem (WP) as $h \to 0$ is connected with several difficulties caused by the nonlinearity of the problem and by different domains of definition of approximate and exact solutions. This causes that $V_h \not\subset V$, $H_{0h}^{-1/2} \not\subset H_0^{-1/2}(\Gamma)$, $H_h \not\subset H$ and the finite element – boundary element discretization is *nonconforming*. To overcome this problem, we define modifications of elements $u_h \in V_h$ and $\mu_h \in H_{0h}^{-1/2}$ which allow us to compare approximate and exact solutions.

Let us define the mapping $\psi_h : \Gamma_h \xrightarrow{\text{onto}} \Gamma$ so that for $x \in \Gamma_h$, $y = \psi_h(x)$ is the point on Γ closest to x. For h sufficiently small, ψ_h is one-to-one and ψ_h as well as its inverse ψ_h^{-1} are Lipschitz-continuous and piecewise smooth. For any $\xi_h \in H_{0h}^{-1/2}$ we define its *modification* $\hat{\xi}_h$ by

$$\hat{\xi}_h(y) = \xi_h(\psi_h^{-1}(y))\, J_h(y), \quad y \in \Gamma, \tag{3.1}$$

where

$$J_h(y(s)) = \left| \frac{d}{ds}\, \psi_h^{-1}(y(s)) \right| \tag{3.2}$$

is the one-dimensional Jacobian and $y = y(s)$ is the parametrization of Γ with respect to the arclength s measured on Γ. Obviously, $\hat{\xi}_h \in L^2(\Gamma)$ and

$$\int_\Gamma \hat{\xi}_h \, dS = \int_{\Gamma_h} \xi_h \, dS = 0 \quad \text{for } \xi_h \in H_{0h}^{-1/2}. \tag{3.3}$$

Hence, $\hat{\xi}_h \in H_0^{-1/2}(\Gamma) \cap L^2(\Gamma)$.

If $T \in \mathcal{T}_h$ is a boundary triangle with vertices $P_1^T \in \Omega_h^-$ and P_2^T, $P_3^T \in \partial\Omega$ and $\Sigma_T \subset \partial\Omega$ is the arc approximated by the side $P_2^T P_3^T$ of T, then the curved triangle $T^{id} \subset \Omega^-$ with two straight sides $P_1^T P_2^T$, $P_1^T P_3^T$ and curved side Σ_T is called the *ideal triangle* associated with T.

In the same way as in [6 – 8] we define the *natural extension* $\overline{v}_h$ of $v_h \in V$. This means that $\overline{v}_h|T = v_h|T$ for each $T \in \mathcal{T}_h$ and, if T is a boundary triangle, $v_h|T = p|T$ where p is a linear polynomial, then $\overline{v}_h|T^{id} = p|T^{id}$. Furthermore, for any $v_h \in V_h$ we define its *modification* $\hat{v}_h \in V$:

$$\begin{aligned} &\text{a)}\ \ \hat{v}_h|T = v_h|T, \ \text{if } T \in \mathcal{T}_h \text{ has at most one vertex lying on } \partial\Omega_h, \\ &\text{b)}\ \ \hat{v}_h|T^{id} = I_h^{id}\, v_h|T^{id} = \text{Zlámal's ideal interpolation (cf. [16, 8])}. \end{aligned} \tag{3.4}$$

c) On $\overline{\text{Int}\Gamma} - \text{Int}\Gamma_h$ we set $\hat{v}_h(x) = v_h(y)$
for all x from the straight segment joining $y \in \Gamma_h$
with $\psi_h(y) \in \Gamma$.

Now we say that approximate solutions $(u_h, \mu_h) \in H_h$ of Problem (DP) converge to a couple $(u, \mu) \in H$ as $h \to 0$, if

$$\lim_{h\to 0} \|(\hat{u}_h, \hat{\mu}_h) - (u, \mu)\|_W = 0. \tag{3.5}$$

The detailed analysis can be summarized in the following theorems:

Theorem 3.1. *Let $\{(u_h, \mu_h)\}_{h\in(0,h_0)}$ be the system of approximate solutions satisfying* (2.9). *Then there exist a sequence $h_n \to 0$ and $(u, \mu) \in H$ such that*

$$\lim_{n\to\infty} \|(\hat{u}_{h_n}, \hat{\mu}_{h_n}) - (u, \mu)\|_W = 0. \tag{3.6}$$

The couple (u, μ) is a solution of Problem (WP).

Theorem 3.2. *If condition* (1.32) *is satisfied, then the solution (u, μ) of* (WP) *is unique and* (3.5) *holds. Moreover, if $u \in H^2(\Omega^-)$ and $\mu \in H^{1/2}(\Gamma)$, then there exist constants $h_1,\ c > 0$ such that*

$$\|(\hat{u}_h, \hat{\mu}_h) - (u, \mu)\|_W \le c\, h^\alpha, \quad h \in (0, h_1), \tag{3.7}$$

where $\alpha = 1/2$ or $\alpha = 1$, provided n_h is the unit outer normal to $\partial\Omega_h$ or n_h is a piecewise linear approximation of the unit outer normal to $\partial\Omega$, respectively (see Section 2).

4 Iterative Procedures for the Solution of the Discrete Problem

Due to the nonlinearity of the problem, it is necessary to apply a suitable iterative process to the solution of the discrete problem (DP). We propose two possible methods which can be treated even within the frame of the continuous problem (WP).

Let $G : H \times H \to \mathbb{R}$ be a symmetric H-elliptic continuous bilinear form.

a) *Strongly monotone case.* It follows from (1.27) and the Friedrichs inequality that $H = V \times H_0^{-1/2}(\Gamma)$ can be considered as a Hilbert space with the scalar product

$$((w, \vartheta), (v, \delta)) = \int_{\Omega^-} \nabla w \cdot \nabla v \, ds + b(\vartheta, \delta), \quad (w, \vartheta), (v, \delta) \in H. \tag{4.1}$$

If (1.32) holds and, hence, the operator A is strongly monotone, then it is suitable to use the following algorithm:

$$\begin{aligned}
&\text{a)} \quad (u^0, \mu^0) \in H \quad \text{is arbitrary,} \\
&\text{b)} \quad (\theta^{k+1/2}, \vartheta^{k+1/2}) \in H \quad \text{is a solution of the linear problem} \\
&\qquad G((\theta^{k+1/2}, \vartheta^{k+1/2}), (v, \delta)) = \\
&\qquad a(u^k, v) + B((u^k, \mu^k), (v, \delta)) - F(v, \delta), \quad (v, \delta) \in H, \\
&\text{c)} \quad (u^{k+1}, \mu^{k+1}) = (u^k, \mu^k) - \omega(\theta^{k+1/2}, \vartheta^{k+1/2}), \quad k \ge 0.
\end{aligned} \tag{4.2}$$

It is possible to prove the existence of $\tilde{\omega} > 0$ such that provided $\omega \in (0, \tilde{\omega})$, process (4.2, a–c) yields the sequence (u^k, μ^k) converging to the solution (u, μ) of Problem (WP).

b) *Nonmonotone case.* If the problem is not strongly monotone, the situation is more complicated and the iterative method (4.2, a-c) can fail. As one possibility which seems to be sufficiently robust, we can use the *method of least squares and conjugate gradients* developed in [1, 2, 11] for the solution of strongly nonlinear nonmonotone problems.

Let G have the same properties as above. For arbitrary $(u, \mu) \in H$ we define $(\theta, \vartheta) = (\theta(u, \mu), \vartheta(u, \mu))$ as a solution of the problem

$$\begin{aligned} &\text{a)} \quad (\theta, \vartheta) \in H, \\ &\text{b)} \quad G((\theta, \vartheta), (v, \delta)) = a(u, v) + B((u, \mu), (v, \delta)) - F(v, \delta), \ (v, \delta) \in H, \end{aligned} \tag{4.3}$$

and introduce the functional $J : H \to I\!R$ by

$$J(u, \mu) = \frac{1}{2} G((\theta(u, \mu), \vartheta(u, \mu)), (\theta(u, \mu), \vartheta(u, \mu))). \tag{4.4}$$

It is evident that Problem (WP) is equivalent to the following minimization problem: Find $(u, \mu) \in H$ such that

$$J(u, \mu) = \arg \min_{(z, \gamma) \in H} J(z, \gamma) \ \ \& \ \ J(u, \mu) = 0. \tag{4.5}$$

This is an optimal control problem with the cost functional J defined via the solution of the state equation represented by (4.3, a–b).

The minimization of J can be carried out by the conjugate gradient method with preconditioning given by the form G.

The *choice of the preconditioning form G* plays an important role in both cases a) and b). It turns out suitable to choose

$$G((w, \vartheta), (v, \delta)) = ((w, \vartheta), (v, \delta)) \tag{4.6}$$

(see (4.1)), which has the required properties. Then equations (4.2, b), (4.3, b) and the equation for the gradient of J split into a separate Poisson boundary value problem in Ω^- and an integral equation of the first kind over Γ. The above preconditioning form G was successfully used in [5] for the solution of nonlinear monotone problem of potential subsonic gas flow.

In the discretization of algorithms (4.2) and (4.3) – (4.5) we proceed in a similar way as in Section 2. The forms a, B, F and the space H are replaced by $\tilde{a}_h$, $\tilde{B}_h$, $\tilde{F}_h$ and H_h, respectively, and the preconditioner G defined with the aid of (4.1), (4.6) is approximated by

$$\begin{aligned} &G_h((w_h, \vartheta_h), (v_h, \delta_h)) = \beta_h(v_h, w_h) + \tilde{b}_h(\vartheta_h, \delta_h), \\ &\beta_h(v_h, w_h) = \int_{\Omega_h^-} \nabla v_h \cdot \nabla w_h \, dx, \quad (w_h, \vartheta_h), (v_h, \delta_h) \in H_h. \end{aligned} \tag{4.7}$$

Acknowledgements: The research of M. Feistauer has been supported by the Grant No. 201/94/1067 of the Czech Grant Agency. The research of G. C. Hsiao and R. E. Kleinman has been supported under ONR Grant No. N00014 – 91 – J – 1700.

References

[1] H. Berger, G. Warnecke, W. Wendland: On the coupling of the finite elements and boundary elements for transonic potential flows. *Numer. Meth. Part. Diff. Eqns.* **6** (1990), 17–42.

[2] M. O. Bristeau, R. Glowinski, J. Periaux, P. Perrier, O. Pironneau, G. Poirier: Application of optimal control and finite element methods to the calculation of transonic flows and incompressible viscous flows. In: B. Hunt (ed.): *Numerical Methods in Applied Fluid Dynamics*. Academic Press, New York – San Francisco – London, 1980, 203–312.

[3] P. G. Ciarlet: *The Finite Element Method for Elliptic Problems*. North–Holland, Amsterdam, 1978.

[4] M. Costabel: Boundary integral operators on Lipschitz domains: Elementary results. *SIAM J. Math. Anal.* **19** (1988), 613–626.

[5] M. Feistauer, G. C. Hsiao, R. E. Kleinman: Analysis and numerical realization of coupled BEM and FEM for nonlinear exterior problems. In: R. E. Kleinman, R. Kress, E. Martensen (eds.): *Inverse Scattering and Potential Problems in Mathematical Physics*, Methoden und Verfahren der Mathematischen Physik, Band 40, P. Lang, Frankfurt am Main, etc., 1995, 47–73.

[6] M. Feistauer, V. Sobotíková: Finite element approximation of nonlinear elliptic problems with discontinuous coefficients. *M^2AN* **24** (1990), 457–500.

[7] M. Feistauer, A. Ženíšek: Finite element solution of nonlinear elliptic problems. *Numer. Math.* **50** (1987), 457–475.

[8] M. Feistauer, A. Ženíšek: Compactness method in the finite element theory of nonlinear elliptic problems. *Numer. Math.* **52** (1988), 147–163.

[9] G. N. Gatica: On the coupling of boundary integral and finite element methods for nonlinear boundary value problems. Ph. D. Dissertation, University of Delaware, 1989.

[10] G. N. Gatica, G. C. Hsiao: The coupling of boundary element and finite element methods for a nonlinear exterior boundary value problem. *Z. Analysis Anwend.* (ZAA) **8** (1989), 377–387.

[11] R. Glowinski: *Numerical Methods for Nonlinear Variational Problems*. Springer–Verlag, New York, etc., 1984.

[12] G. C. Hsiao: The coupling of BEM and FEM – a brief review. In: C. A. Brebbia et al. (eds): *Boundary elements X*, Vol. I, Springer-Verlag, Berlin 1988, 431–445.

[13] G. C. Hsiao: The coupling of boundary element and finite elements methods. *ZAMM* **70** (1990), 493–503.

[14] G. C. Hsiao, R. E. Kleinman: Feasible error estimates in boundary element methods. In: C. A. Brebbia et al. (eds): *Boundary Element Techniques VII*, Elsevier Applied Sciences, London – New York, 1992, 875–886.

[15] G. C: Hsiao, W.L. Wendland: A finite element method for an integral equation of the first kind. *J. Math. Anal. Appl.* **58** (1977), 449–481.

[16] M. Zlámal: Curved elements in the finite element method I. *SIAM J. Numer. Anal.* **10** (1973), 229–240.

Coupling of a global coarse discretization and local fine discretizations

P.J.J. Ferket

Eindhoven University of Technology,
Department of Mathematics and Computing Science,
P.O. Box 513, 5600 MB Eindhoven, The Netherlands.

Abstract

In this paper we discuss a discretization approach for elliptic boundary value problems on composite grids. In this approach global approximations of the continuous solution are used to define artificial Dirichlet boundary conditions on the interfaces between the global coarse grid and the local fine grids. Local approximations of the continuous solution are used to improve a basic global discretization by a local defect correction. To analyze the effect of the local defect correction step the local discretization error is discussed. If the defect correction step is performed at the correct places then the accuracy of the resulting approximation is in agreement with the resolution of the composite grid. Numerical examples illustrate the theoretical insights.

1 Introduction

Many boundary value problems produce solutions that possess *highly localized properties*. In this paper we consider two-dimensional elliptic boundary value problems with one or a few small regions with high activity. In these regions the solution varies much more rapidly than in the remaining part of the domain. We are mainly interested in problems in which this behaviour is due to the source term (e.g. a strong well). In general, from the point of view of efficiency, it is not attractive to use a uniform grid for discretizing such a problem. Often the use of *local grid refinement techniques* will be advantageous.

In this paper we study a local grid refinement technique based on the combination of several uniform grids with different grid sizes that cover different parts of the domain. The continuous solution is then approximated on the *composite grid* which is the union of the uniform subgrids. Methods based on such a technique have been addressed by several authors. The finite volume element method used in McCormick's fast adaptive composite grid method is of this type and an analysis of this composite grid discretization is given in [1, 6, 7]. This finite volume type of method uses vertex-centered approximations. A finite volume method for composite grids using special cell-centered approximations is analyzed in [2, 5]. The local defect correction method introduced in [4] is a very general approach which can be used for discretization on a composite grid too.

In this paper we analyze a simple discretization method for elliptic boundary value problems on composite grids. In our approach discrete problems on the global coarse grid and on the local fine subgrids are coupled. Global approximations of the continuous solution are used to define *artificial Dirichlet boundary conditions* on the interfaces between the coarse grid and the local fine grids. Local approximations of the continuous solution are used to improve a basic global discretization by a *local defect correction*. On the uniform grids finite differences are used. Discrete problems on the uniform grids are solved exactly. The method considered in this paper is a special case of the local defect correction method by Hackbusch [4].

We will discuss the effect of a local defect correction step. Therefore, we also consider a discretization approach on composite grids without local defect correction. In this *basic approach* information from the global coarse grid is used on the local fine grids, but not the other way around. Comparison of the discretization errors of both methods shows that the local defect correction method yields better approximations than the basic approach.

With respect to the accuracy of the local defect correction approximation it appears to be important at which grid points in the local regions the basic discretization is updated by a defect correction. The region of defect correction is determined by a parameter *dist* in the local defect correction method. An analysis of the discretization error of the local defect correction method indicates how to choose this parameter *dist*. For a suitable choice of *dist* the local defect correction approximation is, in general, in agreement with the resolution of the composite grid.

The remainder of this paper is organized as follows. In Section 2 a model situation is described and both the basic approach and the local defect correction method are introduced. In Section 3 the discretization error vectors are analyzed and compared. Further the role of the parameter *dist* is considered. Finally, in Section 4 some numerical examples are given that confirm the theoretical insights obtained in Section 3.

2 Description of the local defect correction method

In this section we describe the *local defect correction method* for a model situation. We consider a Dirichlet boundary value problem in the unit square $\Omega = (0,1) \times (0,1)$:

$$\begin{aligned} \mathcal{L}U &= f \quad \text{in } \Omega, \\ U &= g \quad \text{on } \partial\Omega, \end{aligned} \tag{2.1}$$

with $\mathcal{L}$ a scalar linear elliptic second-order differential operator. We assume that the right hand sides f and g are such that there is one region in which the solution U of (2.1) varies much more rapidly than in the remainder of the domain. This so called *high activity region* is contained in a rectangular local region $\Omega_l \subset \Omega$. The *interface* between Ω_l and Ω is denoted by Γ. We assume that U has relatively slow variation near this interface.

In the local defect correction method a numerical approximation of U is computed using a global grid and a local grid. The *global coarse grid* Ω^H is a uniform grid with grid size H that covers the domain Ω. Grid functions on Ω^H are denoted by a superscript H, e.g. w^H. Similar notation is used for grids that are introduced in the remainder of this section. The *local fine grid* Ω_l^h is a uniform grid with grid size $h < H$ that covers the local region Ω_l. We assume that the *interface* Γ between Ω_l and Ω coincides with grid lines of Ω^H and that $H/h \in \mathbb{N}$. We introduce the *coarse interface grid* $\Gamma^H = \Gamma \cap \Omega^H$ and its fine grid analogue Γ^h, the *fine interface grid*. The *composite grid* $\Omega^{H,h}$ is the union of the global coarse grid and the local fine grid ($\Omega^{H,h} = \Omega^H \cup \Omega_l^h$). In Figure 1 examples of the grids Ω^H, Ω_l^h and $\Omega^{H,h}$ are shown.

In the local defect correction method we also use a *local coarse subgrid* Ω_s^H. This is a uniform grid with grid size H that covers a subregion $\Omega_s \subseteq \Omega_l$. The interface between

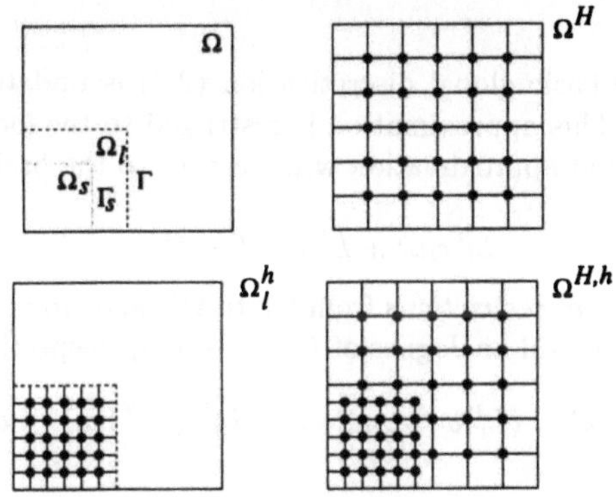

Figure 1: Local regions Ω_l and Ω_s and grids Ω^H, Ω_l^h and $\Omega^{H,h}$.

Ω_s and Ω is denoted by Γ_s (see Figure 1). We assume that Γ_s coincides with grid lines of Ω^H. Further we assume that a constant *dist* exists such that

$$\forall \mathbf{x} \in \Gamma_s \;:\; \min\{|\mathbf{x} - \mathbf{y}| \mid \mathbf{y} \in \Gamma\} = dist. \tag{2.2}$$

If $dist = 0$ then $\Omega_s = \Omega_l$, $\Gamma_s = \Gamma$, and $\Omega_s^H = \Omega_l^H$. The *local coarse grid* Ω_l^H is a uniform grid with grid size H that covers the local region Ω_l.

The local defect correction method is a discretization method for boundary value problem (2.1) on the composite grid $\Omega^{H,h}$. A basic global discretization is improved by a local discretization defined in the subdomain Ω_l. Thus (2.1) is discretized on uniform grids only. For discretization on uniform grids *standard finite differences* are used.

Finite difference discretization of boundary value problem (2.1) on the global coarse grid Ω^H yields the *basic global discretization*:

$$L^H \tilde{u}^H = f^H. \tag{2.3}$$

L^H is a finite difference operator and $\tilde{u}^H$, f^H are grid functions on Ω^H. In the following we identify finite difference operators and matrices as well as grid functions and vectors. The Dirichlet boundary conditions on $\partial\Omega$ are incorporated in f^H.

Next (2.1) is discretized with respect to the local fine grid Ω_l^h. Dirichlet boundary conditions on the interface Γ result from interpolation between the values $\tilde{u}^H(\mathbf{x})$ with $\mathbf{x} \in \Gamma^H$ and, if $\partial\Omega_l \cap \partial\Omega \neq \emptyset$, the values of $g(\mathbf{x})$ with $\mathbf{x} \in \Gamma \cap \partial\Omega$. Then a system of linear equations on the local fine grid results:

$$L_l^h \tilde{u}_l^h = f_l^h - L_\Gamma^h p_\Gamma \tilde{u}^H. \tag{2.4}$$

L_l^h is a finite difference operator with respect to Ω_l^h. *Natural Dirichlet boundary conditions* on $\partial\Omega_l \cap \partial\Omega$ are incorporated in f_l^h. *Artificial Dirichlet boundary conditions* on Γ appear explicitly in (2.4) by the term $-L_\Gamma^h p_\Gamma \tilde{u}^H$. In this term p_Γ is an interpolation operator on the interface (we only consider linear and quadratic interpolation). L_Γ^h is a finite difference operator that deals with Dirichlet boundary conditions on Γ. If (2.3) and (2.4) are solved exactly the following *basic approximation* of U results on the composite grid:

$$\tilde{u}^{H,h}(\mathbf{x}) := \begin{cases} \tilde{u}_l^h(\mathbf{x}) & \mathbf{x} \in \Omega_l^h \\ \tilde{u}^H(\mathbf{x}) & \mathbf{x} \in \Omega^{H,h} \backslash \Omega_l^h \end{cases}. \tag{2.5}$$

In the next step the basic global discretization (2.3) is updated using the local fine grid approximation $\tilde{u}_l^h$. This approximation is restricted to the local coarse grid Ω_l^H and the defect of this restricted approximation with respect to the basic global discretization is computed:

$$L_l^H r_l \tilde{u}_l^h + L_\Gamma^H r_\Gamma \tilde{u}^H - f_l^H. \tag{2.6}$$

In (2.6) r_l and r_Γ are *trivial restrictions* from Ω_l^h to Ω_l^H and from Ω^H to Γ^H respectively. L_l^H, L_Γ^H and f_l^H are coarse grid analogues of L_l^h, L_Γ^h and f_l^h respectively and we have that:

$$(L^H w^H)(\mathbf{x}) = (L_l^H w^H{}_{|\Omega_l^H})(\mathbf{x}) + (L_\Gamma^H r_\Gamma w^H)(\mathbf{x}) \quad \mathbf{x} \in \Omega_l^H \tag{2.7}$$

for all grid functions w^H on Ω^H. So L_l^H, L_Γ^H and f_l^H represent the basic global discretization restricted to the local coarse grid Ω_l^H.
The defect (2.6) is added to the right hand side of the basic global discretization at all grid points of the coarse subgrid Ω_s^H:

$$\bar{f}^H := f^H + \chi(L_l^H r_l \tilde{u}_l^h + L_\Gamma^H r_\Gamma \tilde{u}^H - f_l^H), \tag{2.8}$$

where the characteristic function χ is defined by

$$(\chi w)(\mathbf{x}) := \begin{cases} w(\mathbf{x}) & \mathbf{x} \in \Omega_s^H \\ 0 & \mathbf{x} \in \Omega^H \backslash \Omega_s^H \end{cases}. \tag{2.9}$$

The following updated global discretization results:

$$L^H \bar{u}^H = \bar{f}^H. \tag{2.10}$$

We note that if $\Omega_s = \Omega_l = \Omega$ (i.e. in case of *global* defect correction) then the updated global coarse grid approximation is equal to the global fine grid approximation restricted to the global coarse grid. So in this case the updated global discretization is an improvement on the basic global discretization. In the next sections we will see that this also holds in case of *local* defect correction.

If (2.10) is solved exactly we can update the local fine grid problem (2.4):

$$L_l^h \bar{u}_l^h = f_l^h - L_\Gamma^h p_\Gamma \bar{u}^H. \tag{2.11}$$

The *local defect correction approximation* of U is given by:

$$\bar{u}^{H,h}(\mathbf{x}) := \begin{cases} \bar{u}_l^h(\mathbf{x}) & \mathbf{x} \in \Omega_l^h \\ \bar{u}^H(\mathbf{x}) & \mathbf{x} \in \Omega^{H,h} \backslash \Omega_l^h \end{cases}. \tag{2.12}$$

We note that this approximation results after solving two discrete problems on the global coarse grid and two discrete problems on the local fine grid. The basic approximation $\tilde{u}^{H,h}$ results after solving one discrete problem on the global coarse grid and one discrete problem on the local fine grid. So computing $\bar{u}^{H,h}$ is approximately twice as expensive as computing $\tilde{u}^{H,h}$. We also not that computing $\bar{u}^{H,h}$ is (much) cheaper than computing an approximation on the global fine discretization grid if Ω_l is sufficiently small.

In the next section we discuss the errors of the basic approximation and the local defect correction approximation.

3 Discretization errors

In this section we describe the *local discretization errors* of the discrete problems (2.3), (2.4), (2.10) and (2.11). Further we compare the error vectors $U^{H,h}-\tilde{u}^{H,h}$ and $U^{H,h}-\bar{u}^{H,h}$, where $U^{H,h}$ is the continuous solution U of (2.1) restricted to the composite grid.

The discretization error vector that corresponds with the basic global discretization (2.3) is defined by

$$d^H := L^H U^H - f^H. \tag{3.1}$$

Here U^H is the continuous solution U restricted to the global coarse grid Ω^H. In the remainder we also use U_l^h, U_Γ^h and U_Γ^H that are defined in a similar way.
Due to the relatively large variations of the continuous solution U inside the high activity region, discretization errors, $|d^H(\mathbf{x})|$, will be relatively large at certain grid points of Ω^H inside the high activity region compared with discretization errors at grid points of Ω^H outside the local region Ω_l.

The discretization error vector that corresponds with the discretization process on the local fine grid with *exact Dirichlet boundary conditions on the interface* is defined by

$$d_l^h := L_l^h U_l^h + L_\Gamma^h U_\Gamma^h - f_l^h. \tag{3.2}$$

Since $h < H$ it is reasonable to assume that $|d_l^h(\mathbf{x})| < |d^H(\mathbf{x})|$ at all $\mathbf{x} \in \Omega_l^H$.

The discretization error vector that corresponds with the local fine grid problem (2.4) is defined by

$$\tilde{d}_l^h := L_l^h U_l^h + L_\Gamma^h p_\Gamma \tilde{u}^H - f_l^h. \tag{3.3}$$

Lemma 3.1. The discretization error vector $\tilde{d}_l^h$ satisfies

$$\tilde{d}_l^h = d_l^h - L_\Gamma^h p_\Gamma (U^H - \tilde{u}^H) + L_\Gamma^h (p_\Gamma U^H - U_\Gamma^h). \tag{3.4}$$

Proof. By definition we have

$$\begin{aligned}
\tilde{d}_l^h &= L_l^h U_l^h + L_\Gamma^h p_\Gamma \tilde{u}^H - f_l^h \\
&= L_l^h U_l^h + L_\Gamma^h U_\Gamma^h - f_l^h + L_\Gamma^h (p_\Gamma U^H - U_\Gamma^h) + L_\Gamma^h p_\Gamma (\tilde{u}^H - U^H) \\
&= d_l^h - L_\Gamma^h p_\Gamma (U^H - \tilde{u}^H) + L_\Gamma^h (p_\Gamma U^H - U_\Gamma^h).
\end{aligned}$$

□

The discretization error vector $\tilde{d}_l^h$ consists of three terms. The first term, d_l^h, is the discretization error vector corresponding with boundary value problem (2.1) and the local fine discretization grid Ω_l^h. The second term, $L_\Gamma^h p_\Gamma (U^H - \tilde{u}^H)$, is due to the fact that *artificial* Dirichlet boundary conditions on the interface are used. The third term, $L_\Gamma^h (p_\Gamma U^H - U_\Gamma^h)$, is due to the fact that on the interface interpolation is used. In the following we do not consider the interpolation error on the interface. If the order of interpolation is sufficiently high the contribution of the third term to $\tilde{d}_l^h$ is neglegible. In many cases linear or quadratic interpolation suffices since U has relatively slow variation near the interface.

The accuracy of the basic approximation $\tilde{u}^{H,h}$ from (2.5) depends on d^H and on $\tilde{d}_l^h$.

Theorem 3.2. The error $U^{H,h} - \tilde{u}^{H,h}$ satisfies

$$(U^{H,h} - \tilde{u}^{H,h})(\mathbf{x}) = \begin{cases} ((L_l^h)^{-1} \tilde{d}_l^h)(\mathbf{x}) & \mathbf{x} \in \Omega_l^h \\ ((L^H)^{-1} d^H)(\mathbf{x}) & \mathbf{x} \in \Omega^{H,h} \backslash \Omega_l^h \end{cases}. \tag{3.5}$$

Proof. Combination of (2.3) and (3.1) yields

$$U^H - \tilde{u}^H = (L^H)^{-1} d^H.$$

Combination of (2.4) and (3.3) yields

$$U_l^h - \tilde{u}_l^h = (L_l^h)^{-1} \bar{d}_l^h.$$

From (2.5) and the definition of $U^{H,h}$ it follows that

$$(U^{H,h} - \tilde{u}^{H,h})(\mathbf{x}) = \begin{cases} (U_l^h - \tilde{u}_l^h)(\mathbf{x}) & \mathbf{x} \in \Omega_l^h \\ (U^H - \tilde{u}^H)(\mathbf{x}) & \mathbf{x} \in \Omega^{H,h} \backslash \Omega_l^h \end{cases}.$$

Now (3.5) follows immediately. □

From (3.4) and (3.5) it follows that the accuracy of the basic approximation is restricted by the accuracy of the global coarse grid approximation $\tilde{u}^H$ on the interface. We have that $U^H - \tilde{u}^H = (L^H)^{-1} d^H$. We recall that $|d^H|$ is a very special vector: at several grid points $\mathbf{x}$ inside Ω_l the error $|d^H(\mathbf{x})|$ is much larger than at grid points outside Ω_l. In general, this does not hold for the vector $|(L^H)^{-1} d^H|$. The inverse $(L^H)^{-1}$ *spreads local effects* over a (large) part of the domain. Due to this spreading, the accuracy of the basic approximation at grid points outside Ω_l is mainly determined by components of d^H at grid points inside the high activity region. It is clear that the accuracy of the basic approximation can only be in agreement with the resolution of the composite grid inside Ω_l (i.e. $\|(L_l^h)^{-1} \bar{d}_l^h\| \approx \|(L_l^h)^{-1} d_l^h\|$) if the components of $|d^H|$ at grid points inside the high activity region are not too large compared with the components of $|d_l^h|$ at corresponding grid points and if local effects are not spreaded too much by $(L^H)^{-1}$. In the next section several numerical examples are given in which the accuracy of the basic approximation is not in agreement with the resolution of the composite grid.

The discretization error vector that corresponds with the updated global discretization is defined by

$$\bar{d}^H := L^H U^H - \bar{f}^H. \tag{3.6}$$

Lemma 3.3. The discretization error vector $\bar{d}^H$ satisfies

$$\begin{aligned} \bar{d}^H \;=\;& (1-\chi) d^H + \\ & \chi L_l^H r_l (L_l^h)^{-1} d_l^h + \\ & \chi L_l^H r_l (L_l^h)^{-1} L_\Gamma^h (p_\Gamma U^H - U_\Gamma^h) + \\ & \chi (L_\Gamma^H r_\Gamma - L_l^H r_l (L_l^h)^{-1} L_\Gamma^h p_\Gamma)(U^H - \tilde{u}^H). \end{aligned} \tag{3.7}$$

Proof. By definition we have

$$\begin{aligned} \bar{d}^H &= L^H U^H - f^H - \chi(L_l^H r_l \tilde{u}_l^h + L_\Gamma^H r_\Gamma \tilde{u}^H - f_l^H) \\ &= (1-\chi)(L^H U^H - f^H) + \chi(L^H U^H - L_l^H r_l \tilde{u}_l^h - L_\Gamma^H r_\Gamma \tilde{u}^H) \\ &= (1-\chi) d^H + \chi(L^H U^H - L_l^H r_l \tilde{u}_l^h - L_\Gamma^H r_\Gamma \tilde{u}^H) \end{aligned}$$

Using (2.7) we get

$$\bar{d}^H = (1-\chi) d^H + \chi(L_l^H r_l (U_l^h - \tilde{u}_l^h) + L_\Gamma^H r_\Gamma (U^H - \tilde{u}^H)).$$

With $U_l^h - \tilde{u}_l^h = (L_l^h)^{-1}\bar{d}_l^h$ (3.7) follows using the result of Lemma 3.1. □

The discretization error vector that corresponds with the updated local fine grid problem (2.11) is defined by

$$\bar{d}_l^h := L_l^h U_l^h + L_\Gamma^h p_\Gamma \bar{u}^H - f_l^h. \tag{3.8}$$

Lemma 3.4. The discretization error vector $\bar{d}_l^h$ satisfies

$$\bar{d}_l^h = d_l^h - L_\Gamma^h p_\Gamma (U^H - \bar{u}^H) + L_\Gamma^h (p_\Gamma U^H - U_\Gamma^h). \tag{3.9}$$

Proof. Analogous to the proof of Lemma 3.1. □

The accuracy of the local defect correction approximation $\bar{u}^{H,h}$ from (2.12) depends on $\bar{d}^H$ and on $\bar{d}_l^h$.

Theorem 3.5. The error $U^{H,h} - \bar{u}^{H,h}$ satisfies

$$(U^{H,h} - \bar{u}^{H,h})(\mathbf{x}) = \begin{cases} ((L_l^h)^{-1}\bar{d}_l^h)(\mathbf{x}) & \mathbf{x} \in \Omega_l^h \\ ((L^H)^{-1}\bar{d}^H)(\mathbf{x}) & \mathbf{x} \in \Omega^{H,h} \backslash \Omega_l^h \end{cases} . \tag{3.10}$$

Proof. Analogous to the proof of Theorem 3.2. □

The accuracy of the local defect correction approximation is restricted by the accuracy of the global coarse grid approximation $\bar{u}^H$ on the interface. There is a significant difference between the discretization error vectors d^H and $\bar{d}^H$. As we have seen above, possible poor accuracy of the basic approximation is due to large components of $|d^H|$ at grid points inside the high activity region. At the right hand side of (3.7) only components of d^H at grid points outside Ω_s appear, according to the definition of χ in (2.9). If *dist* in (2.2) is small enough, e.g. $dist = 0$, then all large components of $|d^H|$ are eliminated at the right hand side of (3.7). In that case the term $(1-\chi)d^H$ is in agreement with the resolution of the composite grid (*coarse* grid size H *outside* Ω_l).

At the right hand side of (3.7) three other terms appear. The first one, $\chi L_l^H r_l (L_l^h)^{-1} d_l^h$, involves the discretization error vector that corresponds with the discretization process on the local fine grid with exact Dirichlet boundary conditions on the interface. This discretization error vector is also in agreement with the resolution of the composite grid (*fine* grid size h *inside* Ω_l).

The second one, $\chi L_l^H r_l (L_l^h)^{-1} L_\Gamma^h (p_\Gamma U^H - U_\Gamma^h)$, involves an interpolation error on the interface. The third one, $\chi(L_\Gamma^H r_\Gamma - L_l^H r_l (L_l^h)^{-1} L_\Gamma^h p_\Gamma)(U^H - \tilde{u}^H)$, involves the error of the basic global coarse grid approximation $\tilde{u}^H$ on the interface. As before the interpolation error on the interface is neglected with respect to the error of the basic approximation on the interface.

Since $U^H - \tilde{u}^H = (L^H)^{-1} d^H$ the discretization error vector $\bar{d}^H$ (and thus $U^{H,h} - \bar{u}^{H,h}$) still depends on components of d^H that correspond with grid points inside Ω_l. In the following we analyze the term $|\chi(L_\Gamma^H r_\Gamma - L_l^H r_l (L_l^h)^{-1} L_\Gamma^h p_\Gamma)(U^H - \tilde{u}^H)|$. In order to accomplish this we rewrite this term as follows.

Consider the Dirichlet boundary value problem

$$\begin{aligned} &\mathcal{L}W = 0 && \text{in } \Omega_l, \\ &W(\mathbf{x}) = 0 && \mathbf{x} \in \partial\Omega_l \backslash \Gamma, \\ &W(\mathbf{x}) = (p_\Gamma(U^H - \tilde{u}^H))(\mathbf{x}) && \mathbf{x} \in \Gamma. \end{aligned} \tag{3.11}$$

Here we have assumed that the result of the interpolation p_Γ is a function on Γ (note that this leaves some freedom).
Discretization of (3.11) with respect to Ω_l^H yields

$$L_l^H w_l^H + L_\Gamma^H r_\Gamma(U^H - \tilde{u}^H) = 0, \tag{3.12}$$

with L_l^H, L_Γ^H and r_Γ as in (2.7).
The discretization error vector that corresponds with (3.12) is defined by

$$\tau_l^H := L_l^H W_l^H + L_\Gamma^H r_\Gamma(U^H - \tilde{u}^H). \tag{3.13}$$

Here W_l^H is the continuous solution W restricted to the local coarse grid Ω_l^H.
Discretization of (3.11) with respect to Ω_l^h yields

$$L_l^h w_l^h + L_\Gamma^h p_\Gamma(U^H - \tilde{u}^H) = 0, \tag{3.14}$$

with L_l^h, L_Γ^h and p_Γ as in (2.4).
The discretization error vector that corresponds with (3.14) is defined by

$$\tau_l^h := L_l^h W_l^h + L_\Gamma^h p_\Gamma(U^H - \tilde{u}^H). \tag{3.15}$$

Here W_l^h is the continuous solution W restricted to the local fine grid Ω_l^h.
The fourth term at the right hand side of (3.7) is related to these discretization error vectors.

Lemma 3.6. The fourth term at the right hand side of (3.7) satisfies

$$\chi(L_\Gamma^H r_\Gamma - L_l^H r_l (L_l^h)^{-1} L_\Gamma^h p_\Gamma)(U^H - \tilde{u}^H) = \chi(\tau_l^H - L_l^H r_l (L_l^h)^{-1} \tau_l^h) \tag{3.16}$$

Proof. Since

$$\begin{aligned} L_\Gamma^H r_\Gamma(U^H - \tilde{u}^H) &= -L_l^H w_l^H, \\ L_\Gamma^h p_\Gamma(U^H - \tilde{u}^H) &= -L_l^h w_l^h \end{aligned}$$

we have

$$(L_\Gamma^H r_\Gamma - L_l^H r_l (L_l^h)^{-1} L_\Gamma^h p_\Gamma)(U^H - \tilde{u}^H) = L_l^H r_l w_l^h - L_l^H w_l^H.$$

Since r_l is a trivial injection we have

$$L_l^H r_l w_l^h - L_l^H w_l^H = L_l^H (W_l^H - w_l^H) - L_l^H r_l (W_l^h - w_l^h).$$

Combination of (3.12), (3.13) and (3.14), (3.15) yields

$$\begin{aligned} (W_l^H - w_l^H) &= (L_l^H)^{-1} \tau_l^H, \\ (W_l^h - w_l^h) &= (L_l^h)^{-1} \tau_l^h. \end{aligned}$$

Combination of the results above yields (3.16). □

The fourth term at the right hand side of (3.7) consists of a discretization error vector corresponding to a discretization process on the local coarse grid and one corresponding to a discretization process on the local fine grid. Both discretization processes deal with boundary value problem (3.11). Now note that the right hand sides f and g, which induce the large variations of U from (2.1) are absent in (3.11). Therefore W from (3.11) has in general much less variation inside the high activity region than U from (2.1). As a result discretization errors for (3.11) are (much) smaller than corresponding discretization errors for (2.1) at grid points inside the high activity region. If $\|\tau_l^H - L_l^h r_l (L_l^h)^{-1}\tau_l^H\| \lesssim \max\{\|(1-\chi)d^H\|, \|L_l^H r_l (L_l^h)^{-1} d_l^h\|\}$, then the accuracy of the local defect correction approximation is in agreement with the resolution of the composite grid. In the next section several numerical examples are given to illustrate this.

For some problems (e.g. *convection dominated* problems) W from (3.11) has large derivatives near the interface. Then $|\tau_l^H(\mathbf{x})|$ is large at grid points near the interface. These large components are eliminated from the right hand side of (3.7) if *dist* is large enough. For such problems an optimal choice of *dist* with respect to the first term at the right hand side of (3.7) is in conflict with an optimal choice of *dist* with respect to the fourth term at the right hand side of (3.7). Often a small *dist* > 0 (e.g. *dist* = H) is a good compromise (see Example 4.2 in the next section).

4 Numerical examples

In this section we compare the basic approximation and the local defect correction approximation for some model problems. In the first two examples one dimensional boundary value problems are considered. In the one dimensional case interpolation on the interface is not needed and both the third term at the right hand side of (3.7) and the third term at the right hand sides of (3.4) and (3.9) vanish.

Example 4.1. Consider the one dimensional diffusion problem

$$-\frac{d^2U}{dx^2} = f(x), \qquad 0 < x < 1.$$

Let the right hand side f and the Dirichlet boundary conditions be such that the continuous solution is given by

$$U(x) = (\tanh(A(x-z)) + 1)/2. \tag{4.1}$$

If $A \gg 1$ this solution has a high activity region near z. Let Ω_l be given by $\Omega_l = (x_L, x_R)$ with $0 < x_L < z < x_R < 1$. Standard central differences are used.
In this case the corresponding boundary value problem (3.11) is given by

$$\begin{array}{l} \frac{d^2W}{dx^2} = 0, \qquad x_L < x < x_R, \\ W(x_L) = U(x_L) - \tilde{u}^H(x_L), \\ W(x_R) = U(x_R) - \tilde{u}^H(x_R). \end{array}$$

The discretization error vectors τ_l^H and τ_l^h from (3.13) and (3.15) depend on fourth derivatives of W. Since W is a linear function these discretization error vectors are equal to zero and the fourth term at the right hand side of (3.7) vanishes. The accuracy of $\bar{u}^{H,h}$ only depends on $(1-\chi)d^H$ and d_l^h and is therefore in agreement with the resolution of the composite grid if *dist* is small enough. Since W is a linear function it is optimal to choose *dist* = 0.
In Table 1 the errors (with respect to the maximum norm) of the basic approximation, the local defect correction approximation and the approximation that results after *global* refinement are given for several values of the fine grid size h. The coarse grid size is kept fixed: $H = 1/32$. We have choosen $A = 100$, $z = 0.4$, $\Omega_l = (10/32, 16/32)$ and *dist* = 0.

Table 1: Results for 1D diffusion problem.

h	$\|U^{H,h} - \tilde{u}^{H,h}\|_\infty$	$\|U^{H,h} - \bar{u}^{H,h}\|_\infty$	$\|U^{h,h} - u^{h,h}\|_\infty$
1/64	$2.2E+01$	$2.7E+00$	$2.7E+00$
1/128	$2.2E+01$	$4.7E-02$	$4.7E-02$
1/256	$2.2E+01$	$5.1E-03$	$5.0E-03$
1/512	$2.2E+01$	$1.3E-03$	$1.2E-03$
1/1024	$2.2E+01$	$3.7E-04$	$3.1E-04$
1/2048	$2.2E+01$	$1.4E-04$	$7.6E-05$

The numbers in the second column show that the basic approximation does not improve when h decreases. The accuracy of the basic approximation is comparable to the accuracy of the global coarse grid approximation $\tilde{u}^H$ ($\|U^H - \tilde{u}^H\|_\infty = 2.7e+01$). The accuracy of the local defect correction approximation is for most values of h comparable with the accuracy of the *global fine grid approximation* $u^{h,h}$. Of course this does not hold anymore if h becomes too small. For these values of h global fine grid accuracy can be obtained if H is decreased sufficiently.

Similar results as in Example 4.1 have been obtained for convection diffusion problems. If the problem is convection dominated then the parameter *dist* in the local defect correction method has to be chosen larger than zero in order to obtain satisfactory results.

Example 4.2. Consider the one dimensional convection diffusion problem

$$-\epsilon\frac{d^2U}{dx^2} + \frac{dU}{dx} = f(x), \qquad 0 < x < 1.$$

Let the right hand side f and the Dirichlet boundary conditions be such that the continuous solution is given by

$$U(x) = (\tanh((x-z)/\epsilon) + 1)/2. \tag{4.2}$$

If $\epsilon \ll 1$ this solution has a high activity region near z. Let Ω_l be given by $\Omega_l = (x_L, x_R)$ with $0 < x_L < z < x_R < 1$. The second derivative is discretized using standard central differences. For the first derivative upwind differences are used.
In this case the corresponding boundary value problem (3.11) is given by

$$\begin{aligned} &-\epsilon\tfrac{d^2W}{dx^2} + \tfrac{dW}{dx} = 0, \qquad x_L < x < x_R,\\ &W(x_L) = U(x_L) - \tilde{u}^H(x_L),\\ &W(x_R) = U(x_R) - \tilde{u}^H(x_R). \end{aligned}$$

For $\epsilon \ll 1$ the solution W varies rapidly near x_R if $b \neq a$ (due to a *boundary layer*). Then the discretization error vector $|\tau_l^H|$ has one (or a few) large components near x_R.
In Table 2 the errors of the basic approximation, the local defect correction approximation and the approximation that results after *global* refinement are given for several values of h and *dist*. The coarse grid size is kept fixed: $H = 1/32$. We have chosen $\epsilon = 0.01$, $z = 0.4$ and $\Omega_l = (10/32, 16/32)$. The basic approximation becomes worse if h decreases. (Note that $\|U^H - \tilde{u}^H\|_\infty = 1.5e+00$). The local defect correction method yields, for all values of *dist* considered here, a better approximation than the basic approximation. Due to the large variations of W near x_R, the local defect correction method with *dist* $= 0$

Table 2: **Results for 1D convection dominated problem.**

h	$\|U^{H,h} - \tilde{u}^{H,h}\|_\infty$	$\|U^{H,h} - \bar{u}^{H,h}\|_\infty$ $dist = 0$	$dist = H$	$dist = 2H$	$\|U^{h,h} - u^{h,h}\|_\infty$
1/64	$5.5E-01$	$2.2E-01$	$2.2E-01$	$2.3E-01$	$2.2E-01$
1/128	$7.2E-01$	$2.3E-01$	$1.1E-01$	$1.2E-01$	$1.1E-01$
1/256	$9.1E-01$	$2.0E-01$	$6.3E-02$	$7.2E-02$	$6.1E-02$
1/512	$1.1E+00$	$3.0E-01$	$3.2E-02$	$7.6E-02$	$3.2E-02$

does not yield satisfactory results. The best approximations are obtained by the local defect correction method with $dist = H$. For all values of h in Table 2 the accuracy of the approximation that results from the local defect correction method with $dist = H$ is comparable to the accuracy of the approximation that results after global refinement.

In our final example a two dimensional model boundary value problem is considered. In this case interpolation on the interface is required.

Example 4.3. Consider the two dimensional Poisson problem

$$-(\frac{d^2U}{dx^2} + \frac{d^2U}{dy^2}) = f(x,y), \qquad (x,y) \in (0,1) \times (0,1).$$

Let the right hand side f and the Dirichlet boundary conditions be such that the continuous solution is given by

$$U(x) = (\tanh(A(x + y - 0.125)) + 1)/2. \tag{4.3}$$

If $A \gg 1$ this solution has a high activity region near the line segment $x + y = 0.125$ in the unit square. In this example $A = 25$ and $\Omega_l = (0,1/4) \times (0,1/4)$. Standard central differences and quadratic interpolation on the interface are used. The reason for using quadratic interpolation is given in [3].

The corresponding boundary value problem (3.11) is a Laplace problem on $(0,1/4) \times (0,1/4)$ with homogeneous Dirichlet boundary conditions on the left and lower boundary and inhomogeneous Dirichlet boundary conditions on the right and upper boundary. Therefore the solution W is a smooth function.

In Table 3 the errors of the basic approximation, the local defect correction approximation (obtained for $dist = 0$) and the approximation that results after *global* refinement are given for several values of h. The coarse grid size is kept fixed: $H = 1/16$. The results are similar to those in Example 4.1. The local defect correction approximation is (much) better than the basic approximation (note that $\|U^H - \tilde{u}^H\|_\infty = 6.08e-02$) and its accuracy is in agreement with the resolution of the composite grid. Approximately the same

Table 3: **Results for 2D diffusion problem.**

h	$\|U^{H,h} - \tilde{u}^{H,h}\|_\infty$	$\|U^{H,h} - \bar{u}^{H,h}\|_\infty$	$\|U^{h,h} - u^{h,h}\|_\infty$
1/32	$2.29E-02$	$2.29E-02$	$2.30E-02$
1/64	$2.29E-02$	$5.58E-03$	$5.63E-03$
1/128	$2.29E-02$	$1.39E-03$	$1.44E-03$
1/256	$2.29E-02$	$6.70E-04$	$3.57E-04$

numbers as in Table 3 are obtained if linear interpolation is used on the interface. This indicates that in this example interpolation errors are indeed neglegible. For $h = 1/256$ global fine grid accuracy of the local defect correction approximation is obtained in case $H = 1/32$.

Acknowledgement. The author wishes to thank Arnold Reusken for critically reading the manuscript.

References

[1] Cai, Z., McCormick, S.F.: *On the accuracy of the finite volume element method for diffusion equations on composite grids*, SIAM J. Numer. Anal. 27, 636-655 (1990).

[2] Ewing, R.E., Lazarov, R.D., Vassilevski, P.S.: *Local refinement techniques for elliptic problems on cell-centered grids. I: error analysis*, Math. Comp. 56, 437-461 (1991).

[3] Ferket, P.J.J., Reusken, A.A.: *A finite difference discretization method for elliptic problems on composite grids*, RANA 95-01, Department of Mathematics and Computing Science, Eindhoven University of Technology (1995) (submitted).

[4] Hackbusch, W.: *Local defect correction method and domain decomposition techniques*, Computing, Suppl 5, 89-113 (1984).

[5] Lazarov, R.D., Mishev, I.D., Vassilevski, P.S.: *Finite volume methods with local refinement for convection-diffusion problems*, Computing 53, 33-58 (1994).

[6] Mc Cormick, S., Thomas, J.: *The fast adaptive composite grid (FAC) method for elliptic equations*, Math. Comp., 46 (1986), pp 439-456.

[7] McCormick, S.F.: *Multilevel adaptive methods for partial differential equations*, SIAM, Philadelphia (1989).

The Fourier–finite–element method for elliptic problems in axisymmetric domains

Bernd Heinrich
Technische Universität Chemnitz-Zwickau, Fachbereich Mathematik,
PSF 964, D–09107 Chemnitz, Germany

Abstract

The paper deals with the Fourier–finite–element method for solving elliptic problems in three–dimensional axisymmetric domains and gives some survey of this approach to Poisson-like equations. The method combines the approximating Fourier method with the finite–element method and reduces the solution of a problem in 3D to the solution of coupled or even decoupled systems of elliptic problems in 2D. Algorithmic aspects of this method, its parallelization and basic results of the numerical analysis, particularly for the Poisson equation in domains with edges, are discussed. The results are illustrated by some numerical example, where the rate of convergence of the Fourier-finite-element approximation and the run time for partial computational steps are given.

1 Introduction

The so-called Fourier-finite-element method (FFEM) is a method which combines the approximate Fourier method and the finite-element method. The FFEM is often applied to three-dimensional boundary value problems (BVPs) in physics and engineering, if the domain is axisymmetric or prismatic. The remaining data as well as the solution of the BVP may vary with all coordinates. More precisely, on the one hand the approximating Fourier method is used, which involves trigonometric polynomials of degree $\leq N$ at least in one space direction, here with respect to the rotational angle φ; see e.g. [2, 3, 13]. On the other hand, the finite-element-method (cf. e.g. [4]) is taken for the approximate calculation of Fourier coefficients of the solution on triangulations with mesh size h of the plane meridian domain $\Omega_a \subset R^2_+$ generating the axisymmetric (or prismatic) domain. It should be noted that the FFEM has several advantages in comparison with other discretization methods. Thus, by means of the FFEM the approximate solution of the elliptic problem in 3D can be reduced to the approximate solution of a finite set of elliptic problems in 2D yielding approximations of the Fourier coefficients u_k of the solution $\hat{u}$ in 3D. Moreover, for basic problems involving Laplace- or Lamé-like operators, the BVPs in 2D are not coupled and can be solved in parallel. For highly oscillating data on the right-hand side, especially with prescribed frequencies, it is not hard to solve the systems of equations for the corresponding Fourier coefficients, since the condition numbers of the stiffness matrices are diminishing if the frequency is growing. For the analysis, the algorithm, software, and applications of the FFEM, we refer to [1, 10, 12, 15, 19, 20, 21, 24, 25, 26].

Error estimates of the FFEM are given in [3, 8, 11, 14, 23].
At the Faculty of Mathematics in Chemnitz, the package FEFOP (Fischer/Steidten/ Weber/ Lang) was developed, cf. [5, 10], where the finite-element package FEMGP (Jung/ Queck/Steidten) is an essential part of it. By means of FEFOP, elliptic problems in axisymmetric domains with Laplace-like operators can be solved using FFEM. Utilizing a parallel version of the Fast Fourier Transform (FFT), a parallel implementation of the FFEM was created, see [16, 17]. Recently, some experiments have been carried out by means of this package on the parallel computer Parsytec GC/PowerPlus-128. For example, Fourier-finite-element systems with $4 \cdot 10^6$ unknown nodal values are solved by means of 128 parallel working processors (cf. [9]), which leads to considerable shortening of the run time in comparison with the sequential computation. This shows that parallelized Fourier-finite-element techniques can be employed for the efficient solution of elliptic problems on parallel computers.

2 The 3D-BVP and its Fourier decomposition

Let (x_1, x_2, x_3) be the Cartesian coordinates of the point $x \in \mathbf{R}^3$. Furthermore, let $\hat{\Omega}$ denote some bounded domain in $\mathbf{R}^3$, with boundary $\hat{\Gamma} := \partial\hat{\Omega}$. Suppose that $\hat{\Omega}$ is axisymmetric with respect to the x_3–axis and that $\hat{\Omega}\backslash\Gamma_0$ (Γ_0 : part of the x_3–axis contained in $\hat{\Omega}$) is generated by the rotation of a bounded plane meridian domain Ω_a about the x_3-axis . Denote by $\partial\Omega_a$ the boundary of Ω_a and employ $\Gamma_a := \partial\Omega_a\backslash\overline{\Gamma}_0$, cf. Fig. 1.

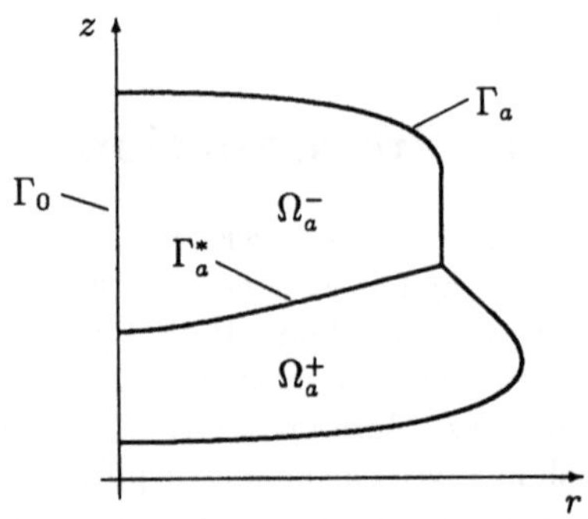

Fig. 1

We consider the Fourier decomposition for some Poisson–like equation under Dirichlet boundary conditions in $\hat{\Omega}$, i.e.

$$- \hat{p}\,\Delta\hat{u} = -\hat{p}\sum_{i=1}^{3}\frac{\partial^2\hat{u}}{\partial x_i^2} = \hat{f} \quad \text{in } \hat{\Omega}^{\pm}, \qquad \hat{u} = \hat{g} \quad \text{on } \hat{\Gamma}, \tag{2.1}$$

with the usual interface conditions at the interface $\hat{\Gamma}^* := \overline{\hat{\Omega}}^- \cap \overline{\hat{\Omega}}^+$. Here, $\hat{p}$ is a piecewise constant coefficient with $\hat{p} = p^{\pm} > 0$ on $\hat{\Omega}^{\pm}$, and $\hat{\Omega}^{\pm}$ are axisymmetric subdomains of $\hat{\Omega}$, with $\overline{\hat{\Omega}}^- \cup \overline{\hat{\Omega}}^+ = \overline{\hat{\Omega}}$, $\hat{\Omega}^- \cap \hat{\Omega}^+ = \emptyset$. The meridian planes of $\hat{\Omega}^{\pm}$ are denoted by $\Omega_a^{\pm}$ (see Fig. 1), and the rotation of the curve $\Gamma_a^* = \overline{\Omega}_a^- \cap \overline{\Omega}_a^+$ about the x_3–axis yields the interface $\hat{\Gamma}^*$. Assume that $\hat{\Gamma}$, $\hat{\Gamma}^* \in C^{0,1} \cap PC^2$ (twice piecewise continously differentiable) and that

$\hat{f} \in L_2(\hat{\Omega})$, $\hat{g} \in H^1(\hat{\Omega})$ hold at least. Let r, φ, z ($x_1 = r\cos\varphi$, $x_2 = r\sin\varphi$, $x_3 = z$) denote the cylindrical coordinates, with $\varphi \in (-\pi, \pi]$. Then, $\hat{\Omega}\backslash\Gamma_0$ is transformed into Ω, $\Omega := \Omega_a \times (-\pi, \pi]$. For each function $\hat{u}(x), x \in \hat{\Omega}\backslash\Gamma_0$, some function u on Ω is defined by

$$u(r,\varphi,z) := \hat{u}(r\cos\varphi, r\sin\varphi, z). \tag{2.2}$$

For functions given on $\hat{\Omega}$ or Ω_a, let H^s ($s \geq 0$, s real, $H^0 = L_2$) denote the usual Sobolev-Slobodetskiĭ -space, with the norm $\|\cdot\|_{H^s}$, moreover, $H^1_g(\hat{\Omega}) := \{\hat{v} \in H^1(\hat{\Omega}) : \hat{v}|_{\hat{\Gamma}} = \hat{g}\}$. The mapping based on (2.2) generates mappings $H^k(\hat{\Omega}\backslash\Gamma_0) \to X^k_{1/2}(\Omega)$, $(k = 0, 1)$, $H_g(\hat{\Omega}) \to V_g(\Omega)$, where

$$\begin{aligned}
X^0_{1/2}(\Omega) &:= \Big\{u = u(r,\varphi,z) : \int_\Omega |u|^2\, r dr d\varphi dz < \infty,\ 2\pi\text{-periodic with respect to } \varphi\Big\},\\
X^1_{1/2}(\Omega) &:= \Big\{u \in X^0_{1/2}(\Omega) : \frac{\partial u}{\partial r}, \frac{\partial u}{\partial z}, \frac{1}{r}\frac{\partial u}{\partial \varphi} \in X^0_{1/2}(\Omega)\Big\}, \qquad (2.3)\\
V_g(\Omega) &:= \Big\{u \in X^1_{1/2}(\Omega) : u|_{\Gamma_a \times (-\pi,\pi]} = g|_{\Gamma_a \times (-\pi,\pi]}\Big\}.
\end{aligned}$$

The generalized formulation of (2.1) is given by: Find $\hat{u} \in H^1_g(\hat{\Omega})$ such that

$$\hat{b}(\hat{u},\hat{v}) := \int_{\hat{\Omega}} \hat{p} \sum_{i=1}^{3} \frac{\partial \hat{u}}{\partial x_i}\frac{\overline{\partial \hat{v}}}{\partial x_i}\, dx = \int_{\hat{\Omega}} \hat{f}\,\overline{\hat{v}}\, dx =: \hat{f}(\hat{v}) \quad \forall \hat{v} \in H^1_0(\hat{\Omega}), \tag{2.4}$$

which can be reformulated using cylindrical coordinates as follows. Find $u \in V_g(\Omega)$ such that

$$\begin{aligned}
b(u,v) &:= \int_\Omega p(r,z)\Big\{\frac{\partial u}{\partial r}\frac{\overline{\partial v}}{\partial r} + \frac{1}{r^2}\frac{\partial u}{\partial \varphi}\frac{\overline{\partial v}}{\partial \varphi} + \frac{\partial u}{\partial z}\frac{\overline{\partial v}}{\partial z}\Big\}\, r dr d\varphi dz\\
&= \int_\Omega f\,\overline{v}\, r dr d\varphi dz =: f(v) \quad \forall v \in V_0(\Omega), \qquad (2.5)
\end{aligned}$$

where, according to the previous assumptions, p depends only on r, z. For functions $v(r,\varphi,z), v \in X^0_{1/2}(\Omega)$, partial Fourier analysis with respect to the rotational angle $\varphi \in (-\pi,\pi]$ is applied. Take the system $\{e^{ik\varphi}\}_{k\in\mathbf{Z}}$ ($i^2 = -1, \mathbf{Z} := \{0, \pm1, \pm2, \ldots\}$), and expand v as a Fourier series, with Fourier coefficients $v_k(r,z), (r,z) \in \Omega_a$:

$$v(r,\varphi,z) = \sum_{k\in\mathbf{Z}} v_k(r,z)\, e^{ik\varphi}, \qquad v_k(r,z) := \frac{1}{2\pi}\int_{-\pi}^{\pi} v(r,\varphi,z)\, e^{-ik\varphi} d\varphi \quad \text{for } k \in \mathbf{Z}. \tag{2.6}$$

Generalized completeness relations lead to the Fourier decomposition of the spaces $X^k_{1/2}(\Omega)$ ($k = 0, 1$) and of the functionals $b(u,v)$, $f(v)$. Thus, the following relations can be proved,

$$\begin{aligned}
b(u,v) &= 2\pi \sum_{k\in\mathbf{Z}} b_k(u_k, v_k), \quad f(v) = 2\pi \sum_{k\in\mathbf{Z}} f_k(v_k),\\
\text{with } b_k(u_k,v_k) &= \int_{\Omega_a} p\Big\{\frac{\partial u_k}{\partial r}\frac{\overline{\partial v_k}}{\partial r} + \frac{\partial u_k}{\partial z}\frac{\overline{\partial v_k}}{\partial z} + \frac{k^2}{r^2} u_k \overline{v}_k\Big\}\, r dr dz, \qquad (2.7)\\
f_k(v_k) &= \int_{\Omega_a} f_k \overline{v}_k\, r dr dz \quad \text{for } k \in \mathbf{Z},
\end{aligned}$$

where u_k, v_k and f_k are the Fourier coefficients of u, v and f. Moreover, it can be shown (cf. [7, 23]) that for $f \in X^0_{1/2}(\Omega)$ there is a unique solution $u \in V_g(\Omega)$ of (2.5) and that the Fourier coefficients u_k are unique solutions of the following variational equations in 2D:

$$\begin{aligned} k=0: \quad \text{find } u_0 \in V^a_{g0}(\Omega_a): \quad b_0(u_0,w) &= f_0(w) \quad \forall\, w \in V^a_0(\Omega_a) \\ k \in \mathbf{Z}\setminus\{0\}: \quad \text{find } u_k \in W^a_{g_k}(\Omega_a): \quad b_k(u_k,w) &= f_k(w) \quad \forall\, w \in W^a_0(\Omega_a), \end{aligned} \tag{2.8}$$

with $b_k(\cdot,\cdot)$ and $f_k(\cdot)$ from (2.7), and with Fourier coefficients g_k of g according to $g(r,\varphi,z) = \sum_{k\in\mathbf{Z}} g_k(r,z)e^{ik\varphi}$ for $(r,z) \in \Gamma_a$. The linear manifolds or subspaces $V^a_\cdot, W^a_\cdot$ are defined by

$$\begin{aligned} V^a_q(\Omega_a) &:= \{v = v(r,z) : r^{1/2}v,\, r^{1/2}\tfrac{\partial v}{\partial r},\, r^{1/2}\tfrac{\partial v}{\partial z} \in L_2(\Omega_a), v|_{\Gamma_a} = q|_{\Gamma_a}\} \\ &\qquad \text{for} \quad q = g_0 \quad \text{or} \quad q = 0, \\ W^a_q(\Omega_a) &:= \{v \in V^a_q(\Omega_a) : r^{-1/2}v \in L_2(\Omega_a)\} \\ &\qquad \text{for} \quad q = g_k \quad (k \in \mathbf{Z}\setminus\{0\}) \quad \text{or} \quad q = 0. \end{aligned} \tag{2.9}$$

3 The Fourier-finite-element method

The solution $u(r,\varphi,z)$ of the BVP (2.5) is now approximated by the truncated Fourier series

$$u_N(r,\varphi,z) = \sum_{|k|\le N} u_k(r,z)\, e^{ik\varphi} \quad \text{for } N > 0. \tag{3.1}$$

Here, $u_k(r,z)$ $(k = 0, \pm 1, \pm 2, \ldots, \pm N)$ denote the first $2N+1$ Fourier coefficients of u, which are the solutions of the 2D-BVPs (2.8) for $|k| \le N$. They shall be calculated approximately by the finite-element method (FEM) applied to the sequence of BVPs (2.8) on the plane meridian Ω_a of $\widehat{\Omega}$. For simplicity, we consider piecewise linear elements and homogeneous boundary conditions, i.e. $g \equiv 0$ on $\widehat{\Gamma}$.
First, approximate Ω_a and $\Omega^\pm_a$ as usual by polygonally bounded domains Ω_{ah} and $\Omega^\pm_{ah}$, respectively, with boundaries $\partial\Omega_{ah}$, $\partial\Omega^\pm_{ah}$, such that $\Omega^+_{ah} \cap \Omega^-_{ah} = \emptyset$, $\overline{\Omega}_{ah} = \overline{\Omega}^+_{ah} \cup \overline{\Omega}^-_{ah}$. Let h be the mesh parameter of the triangulation $\mathcal{T}_h = \{T\}$, which is based on triangles T. Suppose that the usual assumptions on $\mathcal{T}_h$ are satisfied (see e.g. [4]; for details concerning corners on Γ_a and interfaces, cf. [8, 11]). Introduce finite element spaces V^a_{0h} and W^a_{0h} by

$$\begin{aligned} V^a_{0h} &:= \{w = w(r,z) : \; w \in C(\overline{\Omega}_{ah}),\; w \in \mathbf{P}_1(T) \;\; \forall T \in \mathcal{T}_h,\; w = 0 \text{ on } \overline{\Gamma}_{ah}\}, \\ W^a_{0h} &:= \{w = w(r,z) : \; w \in V^a_{0h},\; w = 0 \text{ on } \overline{\Gamma}_0\}, \end{aligned} \tag{3.2}$$

where $\mathbf{P}_1(T)$ denotes the space of all polynomials of degree ≤ 1 on T, $\Gamma_{ah} := \partial\Omega_{ah}\setminus\overline{\Gamma}_0$. For $h \in (0, h_0]$ and integer $N > 0$, the Fourier-finite element space is defined by

$$V_{hN} := \Big\{v : \; v(r,\varphi,z) = \sum_{|k|\le N} v_{kh}(r,z)\, e^{ik\varphi}, \quad v_{0h} \in V^a_{0h},\;\; v_{kh} \in W^a_{0h} \text{ for } k \ne 0\Big\}. \tag{3.3}$$

Moreover, let $\overline{\omega}_{ah}$ denote the set of all nodes of $\mathcal{T}_h$. Define subsets of $\overline{\omega}_{ah}$ by

$$\omega_{ah} := \overline{\omega}_{ah} \cap \Omega_a, \quad \gamma_{ah} := \overline{\omega}_{ah} \cap \overline{\Gamma}_a, \quad \gamma_{0h} := \overline{\omega}_{ah} \cap \Gamma_0 \tag{3.4}$$

and numbers n_i $(i = 1,2,3)$ of nodes by $n_0 = |\omega_{ah} \cup \gamma_{0h}|$, $n_1 = |\omega_{ah}|$ and $n_2 = |\gamma_{ah}|$. For the finite-element spaces we can prove that the following relations hold

$$V_{0h}^a \subset V_0^a(\Omega_{ah}),\ W_{0h}^a \subset W_0^a(\Omega_{ah}),\ V_{hN} \subset V_0(\Omega_h), \tag{3.5}$$

where the subspaces of the functions on Ω_{ah} are defined by analogy to that of Ω_a, $V_0(\Omega_h) := \{v \in X_{1/2}^1(\Omega_h) : v\, |_{\Gamma_{ah}\times(-\pi,\pi]} = 0\}$ with $\Omega_h := \Omega_{ah} \times (-\pi, \pi]$, cf. [8, 11, 23]. The dimensions of these spaces are given by

$$\begin{aligned} &\dim V_{0h}^a = n_0,\ \ \dim W_{0h}^a = n_1,\ \ \dim V_{hN} = n_0 + 2N * n_1, \\ &\text{with} \quad n_i = \mathcal{O}(h^{-2}),\ i = 0, 1. \end{aligned} \tag{3.6}$$

The Fourier-finite-element approximation u_{hN} of the solution u from (2.5) is given as follows. Find $u_{hN} \in V_{hN}$ such that

$$\begin{aligned} b_h(u_{hN}, v_{hN}) &= f_h(v_{hN}) \quad \forall v_{hN} \in V_{hN}, \\ \text{with } b_h(u_{hN}, v_{hN}) &= \int_{\Omega_h} p_h \left\{ \frac{\partial u_{hN}}{\partial r}\frac{\overline{\partial v_{hN}}}{\partial r} + \frac{1}{r^2}\frac{\partial u_{hN}}{\partial \varphi}\frac{\overline{\partial v_{hN}}}{\partial \varphi} + \frac{\partial u_{hN}}{\partial z}\frac{\overline{\partial v_{hN}}}{\partial z} \right\} r dr d\varphi dz, \\ f_h(v_{hN}) &= \int_{\Omega_h} f_h \overline{v_{hN}}\, r dr d\varphi dz, \end{aligned} \tag{3.7}$$

and p_h, f_h are defined by $p_h(r,z) = p^\pm$ for $(r,z) \in \Omega_{ah}^\pm$, $f_h(r,\varphi,z) = f(r,\varphi,z)$ on $\Omega \cap \Omega_h$, $f_h = 0$ otherwise. By analogy to (2.7), (2.8), we get a splitting of (3.7) into a decoupled system of $(2N+1)$ variational equations in 2D, the solutions u_{kh} of which are the finite-element approximations of the Fourier coefficients u_k, $|k| \le N$,

$$\begin{aligned} \text{find } u_{0h} \in V_{0h}^a : &\quad b_{0h}(u_{0h}, w_h) = f_{0h}(w_h) \ \ \forall w_h \in V_{0h}^a, \\ u_{kh} \in W_{0h}^a : &\quad b_{kh}(u_{kh}, w_h) = f_{kh}(w_h) \ \ \forall w_h \in W_{0h}^a,\ 1 \le |k| \le N, \end{aligned} \tag{3.8}$$

$$\begin{aligned} \text{with } b_{kh}(u_{kh}, w_h) &= \int_{\Omega_{ah}} p_h \left\{ \frac{\partial u_{kh}}{\partial r}\frac{\overline{\partial w_h}}{\partial r} + \frac{\partial u_{kh}}{\partial z}\frac{\overline{\partial w_h}}{\partial z} + \frac{k^2}{r^2} u_{kh}\overline{w_h} \right\} r dr dz, \\ f_{kh}(w_h) &= \int_{\Omega_{ah}} f_{kh}\overline{w_h}\, r dr dz, \end{aligned}$$

and f_{kh}, $k \in \mathbb{Z}$, are the Fourier coefficients of f_h from (3.7).

Remark: In the numerical example we have $\Omega_a \cap \Omega_{ah} = \Omega_a$ (here, $\partial\Omega_a$ is straightline near the non-convex parts of Γ_a) such that $\Omega \cap \Omega_h = \Omega$, and therefore $f_h = f$ holds on Ω_h. For simplicity, in the following we shall use f instead of f_h and f_k instead of f_{kh}.

4 Parallel computation of the Fourier-finite-element approximation

The Fourier-finite-element method is often implemented by engineers in sequentially working codes, mainly for elasticity problems in axisymmetric 3D-domains and starting from the Fourier-finite-element approximation of the energy functional, see e.g. [1, 20, 24]. In this section we follow the line of an algorithm proposed in [10], which is implemented in the package FEFOP and extended to a parallel version in [17].
In order to get real systems of simultaneous equations in correspondence to real data $\hat{p}$, $\hat{f}$, $\hat{g}$ and real solution $\hat{u}$ of the BVP, we switch over to the sin-cos-representation of truncated Fourier series, e.g. given for u_{hN} by

$$u_{hN}(r,\varphi,z) = \sum_{|k|\le N} u_{kh}(r,z)\,e^{ik\varphi} = \tfrac{1}{2}u^c_{0h}(r,z) + \sum_{k=1}^{N}\left\{u^c_{kh}(r,z)\cos k\varphi + u^s_{kh}(r,z)\sin k\varphi\right\},$$

$$\text{with}\quad u^c_{kh}(r,z) = \frac{1}{\pi}\int_{-\pi}^{\pi} u_{hN}(r,\varphi,z)\cos(k\varphi)d\varphi \quad \text{for}\quad k=0,1,\ldots,N \tag{4.1}$$

$$\text{and}\quad u^s_{kh}(r,z) = \frac{1}{\pi}\int_{-\pi}^{\pi} u_{hN}(r,\varphi,z)\sin(k\varphi)d\varphi \quad \text{for}\quad k=1,2,\ldots,N,$$

where u^c_{kh}, u^s_{kh} are related with u_{kh} by

$$u^c_{0h} = 2u_{0h},\quad u^c_{kh} = u_{kh} + u_{-kh},\quad u^s_{kh} = i\,\{u_{kh} - u_{-kh}\}\quad \text{for}\quad k=1,2,\ldots,N.$$

In order to realize the Fourier-finite-element algorithm, we need the Fourier coefficients f^c_0, f^c_k and f^s_k ($k = 1,2,\ldots,N$) of the right–hand side f (cf. (3.7), (3.8)). This means, integrals of the type (4.1), but now with f instead of u_{hN}, are to be approximated numerically. For this, we apply the trapezoidal rule with an equidistant subdivision of $(-\pi,\pi]$, i.e. quadrature points $\varphi_j := -\pi + j\Delta\varphi$, $\Delta\varphi := \frac{2\pi}{n}$ ($j = 0,1,\ldots,n-1$; $n = 2^m$ with a positive integer m), and get

$$f^c_k(r,z) \approx \tfrac{2}{n}\sum_{j=0}^{n-1} f(r,\varphi,z)\cos k\varphi_j \quad \text{for}\ k=0,1,\ldots,N\ \text{and}\ (r,z)\in\omega_f. \tag{4.2}$$

Here, ω_f denotes the set of all cubature points assigned to f. Clearly, there is an analogous relation where $f^s_k(r,z)$ and $\sin k\varphi_j$ are involved instead of $f^c_k(r,z)$ and $\cos k\varphi_j$.
The parallel computation of the Fourier-finite-element approximation u_{hN} is based on the splitting (3.8), which is the Fourier decomposition of (3.7). Since the BVPs (3.8) are decoupled, they can be solved simultaneously on several processors of a parallel computer. Moreover, for the Fourier analysis of the right–hand side f and for the Fourier synthesis of the 2D–solutions u^c_{0h}, u^c_{kh} and u^s_{kh} ($k = 1,2,\ldots,N$), a parallel version of the FFT is applied, see [16, 17]. The parallelized version of the Fourier–finite–element algorithm can be characterized by the following steps.

1. Generation of the triangulation of the meridian domain Ω_a. Here, different algorithms of mesh generation, different meshes on distinct processors and a further inner parallelization of mesh generation via subdomains of Ω_a can be employed, cf. [6, 17]. Mesh refinement near corners of the boundary or of the interface can be used for approximating singularities of the 2D–solution in the meridian plane Ω_a and, consequently, for approximating edge singularities or interface singularities in 3D. For simplicity, we consider subsequently only the case that on all processors the same mesh is employed.

2. Computation of the Fourier coefficients $f_0^c(r,z)$, $f_k^c(r,z)$ and $f_k^s(r,z)$ $(k=1,2,\ldots,N)$ of $f(r,\varphi,z)$ at all cubature points $(r,z)\in\omega_f$ of the linear forms $f_k(w_h)$. Additionally, for nonhomogeneous Dirichlet data $g\not\equiv 0$, the Fourier coefficients $g_0^c(r,z)$, $g_k^c(r,z)$ and $g_k^s(r,z)$ $(k=1,2,\ldots,N)$ are to be computed at the nodes $(r,z)\in\gamma_g$, where mostly $\gamma_g=\gamma_{ah}$ is taken.
Concerning f, the parallelization of this step is organized as follows: Choose $n=2^m$ (m: positive integer), taking into account the condition that $N<n/2$, and apply FFT (see e.g. [22]) for calculating sums of the type (4.2). The input data of the FFT are N_f vectors of the type $\{f(r_i,\varphi_j,z_i)\}_{j=0}^{n-1}$, $i=1,2,\ldots,N_f$, where $N_f=|\omega_f|$ denotes the number of cubature points of f. These vectors are decomposed with respect to j into P parts (P: number of processors) and each part is assigned to one processor, e.g. the k–th $(k=0,1,\ldots,P-1)$ processor holds the section $\{f(r_i,\varphi_j,z_i)\}_{j=kn/P}^{(k+1)n/P-1}$, $i=1,2,\ldots,N_f$. The parallel FFT is now applied to N_f vectors of length n, which is described in [16, 17]. It yields the Fourier coefficients $f_{k'}^c(r,z)$ and $f_{k'}^s(r,z)$, $(r,z)\in\omega_f$, on the processor k, where k' arises from k by bit-swapping ($k'=P-1-k$, cf. [17]).

3. Simultaneous generation of the finite-element schemes approximating the 2D-BVPs (2.8) on the processors. Here, the k–th processor $(k=0,1,\ldots,P-1)$ holds the systems corresponding to k', with k' from step 2. These systems are given by

$$\begin{aligned} B_{0h}\breve{u}_{0h}^c &= F_{0h}^c, \\ \left.\begin{aligned}(\widetilde{B}_{0h}+k'^2D_h)\,\breve{u}_{k'h}^c &= F_{k'h}^c \\ (\widetilde{B}_{0h}+k'^2D_h)\,\breve{u}_{k'h}^s &= F_{k'h}^s\end{aligned}\right\} &\quad \text{for } k'=1,2,\ldots,N, \end{aligned} \tag{4.3}$$

with square matrices B_{0h} $(n_0\times n_0)$ and $\widetilde{B}_{0h}$, D_h $(n_1\times n_1)$. The matrix B_{0h} is the stiffness matrix assigned to the differential operator $-\Delta_{r,z}=-\{\frac{1}{r}\frac{\partial}{\partial r}(r\frac{\partial\cdot}{\partial r})+\frac{\partial^2\cdot}{\partial z^2}\}$ involved in the 2D problems (2.8), (3.8), and $\widetilde{B}_{0h}$ is obtained from B_{0h} by taking zero boundary conditions on γ_{0h}. The matrix D_h is a mass matrix according to the term containing $1/r^2$ in the equation (3.8) for $k'\neq 0$. Furthermore, F_{0h}^c, $F_{k'h}^c$ and $F_{k'h}^s$ $(k'=1,2,\ldots,N)$ are the right-hand sides of the systems and depend on $f_{k'}^c$, $g_{k'}^c$ and $f_{k'}^s$, $g_{k'}^s$, respectively. The components of the vectors $\breve{u}_{0h}^c$ (n_0 components) and $\breve{u}_{k'h}^c$, $\breve{u}_{k'h}^s$ (n_1 components, $k'=1,2,\ldots,N$) represent the nodal values of the finite-element approximations u_{0h}^c, $u_{k'h}^c$, $u_{k'h}^s$ $(k'=1,2,\ldots,N)$ of the unknown Fourier coefficients u_0, $u_{k'}^c$, $u_{k'}^s$, cf. (3.8) and (4.1).

4. The $2N+1$ systems of equations (4.3) are solved simultaneously on the processors by some iterative solver. Here, the conjugate gradient method with preconditioning (incomplete Cholesky factorization, see [18]) is applied. This yields the vectors of nodal values $\breve{u}_{0h}^c$ (on the processor $k=0$) and $\breve{u}_{k'h}^c$, $\breve{u}_{k'h}^s$ (on the processor k, $k=1,2,\ldots,P-1$), where $\breve{u}_{0h}^c$ is given on $\omega_{ah}\cup\gamma_{0h}$ and $\breve{u}_{k'h}^c$, $\breve{u}_{k'h}^s$ $(k'=1,2,\ldots,N)$ are given on ω_{ah}.

5. The nodal values of the solution u_{hN} from (3.7) are computed by Fourier synthesis of the vectors $\breve{u}_{0h}^c$, $\breve{u}_{k'h}^c$, $\breve{u}_{k'h}^s$ $(k'=1,2,\ldots,N)$ obtained in step 4. The number of nodal points for the Fourier synthesis can be chosen different from n (cf. step 2), i.e. we can take $n'=2^{m'}$ with a positive integer m'. For the computation of sums of the type (4.1), but now with $\varphi=\varphi_j$ $(j=0,1,\ldots,n'-1)$, with $\breve{u}_{hN}$ instead of u_{hN} and with $\breve{u}_{0h}^c$, $\breve{u}_{k'h}^c$, $\breve{u}_{k'h}^c$ instead of u_{0h}^c, u_{kh}^c, u_{kh}^s, respectively, the parallel FFT from [16, 17] is

employed. This yields N_u vectors of the type $\{\breve{u}_{hN}(r_i,\varphi_j,z_i)\}_{j=0}^{n'-1}$, $i=1,2,\ldots,N_u$, where the part $\{\breve{u}_{hN}(r_i,\varphi_j,z_i)\}_{j=kn'/P}^{(k+1)n'/P-1}$, $i=1,2,\ldots,N_u$, is assigned to the k-th ($k=0,1,\ldots,P-1$) processor. The vectors $\{\breve{u}_{hN}(r_i,\varphi_j,z_i)\}_{j=0}^{n'-1}$, $i=1,2,\ldots,N_u$, represent the nodal values of the Fourier–finite–element approximation u_{hN} of the solution u from (2.5).

For the choice of N there are several possibilities. Thus, N can be a given fixed value. This enables us to distribute a priori the computation of the $2N+1$ Fourier coefficients e.g. to $P := N+1$ processors (i.e., each of $\breve{u}_{0h}^c$ and $(\breve{u}_{kh}^c, \breve{u}_{kh}^s)$, $k=1,2,\ldots,N$, are assigned to one processor).
Moreover, error estimators with respect to both discretization parameters h and N can be used for defining the mesh, with sufficiently small h, and large N. E.g., an appropriate choice of N can be derived from the "exhaustion" of the data f, g by

$$\frac{\|f-f_N^h\|}{\|f\|} < \varepsilon \quad \text{and} \quad \frac{\|g-g_N^h\|}{\|g\|} < \varepsilon, \tag{4.4}$$

with sufficiently small $\varepsilon > 0$ and suitably chosen norms $\|\cdot\|$. In (4.4), f_N^h and g_N^h denote Fourier sums truncated at the level N, which are calculated approximately (with respect to h) by means of numerical integration.

5 Convergence analysis and error estimates

A rigorous convergence analysis of $\hat{u}_{hN} \to \hat{u}$ as $N \to \infty$ and $h \to 0$ is given in the paper [14], where regular solutions $\hat{u} \in H^2(\hat{\Omega})$ of the Dirichlet problem ($\hat{g}=0$ on $\partial\hat{\Omega}$) for the elliptic equation

$$-\sum_{i=1}^{3}\frac{\partial}{\partial x_i}(\hat{p}\frac{\partial \hat{u}}{\partial x_i}) = \hat{f} \ \text{ in } \hat{\Omega}, \ \ \hat{f} \in L_2(\hat{\Omega}), \tag{5.1}$$

with C^2-smooth boundary, are taken into account. The coefficient p may depend on $(r,\varphi,z) \in \Omega$ and, therefore, the Fourier coefficients $u_k(r,z)$ of $u(r,\varphi,z)$ are the solutions of a coupled system of variational equations in 2D. Nevertheless, the rate of convergence can be determined by

$$\|\hat{u}-\hat{u}_{hN}\|_{H^l(\hat{\Omega})} \le C\|\hat{u}\|_{H^2(\hat{\Omega})}\{N^{-2+l}+h^{2-l}\} \text{ for } l=0,1. \tag{5.2}$$

The well-known technique of using Cea's lemma and estimates of the interpolation type leads to some undesired coupling of the discretization parameters N and h. By using other projections, the coupling can be avoided. This is shown in [14] for $\hat{u} \in H^2(\hat{\Omega})$.
In [3] related Fourier-finite-element approximations of the interpolation type are studied, with special emphasis on anisotropic Sobolev spaces and finite elements on the one-dimensional meridian domain $\Omega_a = (0,1)$.
For the Poisson equation ($\hat{p} \equiv 1$) and axisymmetric domains with re-entrant edges, which cause singularities of the solution, the rate of the convergence is studied in [8]. By means of some explicit non-tensor product singularity function given in [7] and mixed projections of the solution $\hat{u}$ into the Fourier-finite-element space, the error estimate

$$\|\hat{u}-\hat{u}_{hN}\|_{H^l(\hat{\Omega})} \le C\|\hat{f}\|_{L_2(\hat{\Omega})}\{N^{-2+l}+h^{(2-l)\lambda}\} \text{ for } l=0,1 \tag{5.3}$$

is derived. If Θ_0 denotes the interior angle of the corner generating the axisymmetric edge, we have $\lambda = 1$ for $0 < \Theta_0 \leq \pi$ (regular solutions), and $\lambda := \pi/\Theta_0$ for $\pi < \Theta_0 < 2\pi$ (re-entrant edge). Appropriate mesh grading in the neighbourhood of the corner of $\partial\Omega_a$ (the triangle diameter h_T goes with $R^{1-\mu}$, where R is the distance to the corner, $0 < \mu \leq 1$) leads again to the error estimate (5.3), but now with $\lambda = 1$ for $\pi < \Theta_0 < 2\pi$. The proof of (5.3) and numerical experiments for studying the rate of convergence of the Fourier-finite-element approximation, especially under the aspect of mesh grading, are given in [8]. In [23, 11] the analysis of interface problems involving piecewise constant coefficients $\hat{p}$, cf. (2.1), as well as adapted Fourier-finite-element approximations are studied. Special emphasis is given to the numerical treatment of curved interfaces and of singularities due to interface edges.

6 Numerical example

We consider the boundary value problem $-\Delta_3\hat{u} = \hat{f}$ in $\hat{\Omega}$, $\hat{u} = \hat{g}$ on $\partial\hat{\Omega}$. The domain $\hat{\Omega}\backslash\Gamma_0$ is generated by the meridian domain Ω_a which is, together with some initial triangulation, represented in Fig. 2a. Thus, Ω_a is a square where a small sector is removed. This sector has the vertex $E_a = (r_e, z_e)$ at the centre of the square (side length $2r_e$) and the symmetry line at $z = z_e$. The interior angle of the re-entrant edge of $\hat{\Omega}$ is denoted by Θ_0. The data $\hat{f}$ and $\hat{g}$ are defined such that the solution $\hat{u}$ is

$$\begin{aligned} \hat{u} &:= r^\gamma R^\lambda \sin(\lambda\Theta)\Psi(\varphi, R), \quad \gamma > 1, \\ \Psi(\varphi, R) &:= R - \ln\{4\sinh^2(\frac{R}{2}) + 4\sin^2(\frac{\varphi}{2})\} = \sum_{k=1}^{\infty} \frac{2}{k} e^{-kR} \cos k\varphi, \end{aligned} \tag{6.1}$$

where Θ $(0 < \Theta < \Theta_0)$ and R $(0 \leq R)$ are local polar coordinates with respect to E_a. Especially, by means of the parameter ε define

$$\Theta_0 := \frac{2\pi}{1+\varepsilon}, \quad \lambda := \frac{\pi}{\Theta_0} = \frac{1+\varepsilon}{2} \quad \text{for } 0 < \varepsilon < 1. \tag{6.2}$$

Then, choose $\varepsilon := 0.01$, i. e. $\lambda = 0.505$ $(\Theta_0 \approx 1.98\pi)$ and $\gamma := 1.1$, $r_e = z_e := 0.5$. The factor r^γ, $\gamma > 1$, guarantees that $\hat{u}$ has a sufficiently fast decrease near the x_3-axis such that $\hat{f} \in L_2(\hat{\Omega})$, with $\hat{f} := -\Delta_3\hat{u}$. For $\varepsilon = 0.01$ (Θ_0 near 2π) the regularity of $\Psi(\cdot, R)$ is given by $\Psi(\cdot, R) \in H^{1-\lambda+\tau}(-\pi, \pi)$ for $0 < \tau < 0.005$, cf. [8].
We apply now the Fourier-finite-element method to this boundary value problem. Cover the meridian domain $\overline{\Omega}_a$ with triangles and use five consecutive levels of triangulations T_h $(h = h_i,\ i = 1, 2, 3, 4, 5)$, which are, for simplicity, hierarchic. The first triangulation $(h = h_1)$ is shown in Fig. 2a and the others are obtained by successive dividing each triangle into four smaller ones. Concerning the quadrature with respect to φ and (r, z), formulas with sufficiently high accuracy are utilized, e.g. at least 2^{10} quadrature points within $[-\pi, \pi]$. The underlying hypothesis for the approximate measuring of the rate of convergence is

$$||\hat{u} - \hat{u}_{hN}||_{H^l(\hat{\Omega})} \approx CN^{-\alpha} + Mh^\beta, \ l \in \{0, 1\}, \tag{6.3}$$

where the constants C and M are assumed to be approximately the same for three consecutive levels of N $(N_i = N,\ N_{i-1} = N - 1,\ N_{i-2} = N - 2)$ and h $(h_i = h,\ h_{i-1} =$

$2h$, $h_{i-2} = 4h$). For fixed N, the mesh size h is varied and, for fixed h, the truncation parameter N. The exponents α and β are estimated via the quotient $(e_i - e_{i-1})/(e_{i-1} - e_{i-2})$ assigned to level i, where e_i, e_{i-1} and e_{i-2} ($e_i < e_{i-1} < e_{i-2}$) denote the error norms on the levels i, $i-1$ and $i-2$, with respect to the parameters N or h. The norms e are determined according to $e = e(l, N, h, \mu) := |||\hat{u} - \check{u}_{hN}|||_{H^l(\hat{\Omega})}$, with $\check{u}_{hN}$ and e as the calculated (approximate) values of $\hat{u}_{hN}$ and $||\hat{u} - \hat{u}_{hN}||_{H^l(\hat{\Omega})}$, respectively. The estimation of α and β is carried out for the H^1–norm ($l = 1$) and the L_2–norm ($l = 0$) on quasiuniform triangulations ($\mu = 1$) as well as on meshes with sufficiently fine local grading ($0 < \mu < \lambda$) around the corner point E_a. Here, μ denotes the parameter of grading the mesh mentioned in Section 5 (see e.g. [8]). For simplicity, the mesh grading is realized in some subdomain Ω_a^s which is similar to Ω_a. For $h = h_3$ and $\mu = 0.8\lambda$, the triangulation is shown in Fig. 2b.
In Tables 1 and 2, the values α and β estimated according to (6.3) are given. They can be compared one with each other on different levels of N and h as well as with the corresponding expected values α_{exp} and β_{exp}. The approximate Fourier-finite-element error norm e was observed to be declining with increasing values of N and decreasing values of h. In Fig. 3, the error norms e in H^l ($l = 0, 1$) are shown for $2 \leq N \leq 40$, $h = h_4$ and $h = h_5$, combined with $\mu = 1$ and $\mu = 0.8\lambda$. For $l = 0$ the order of convergence with respect to N^{-1} and h is about twice of that for $l = 1$. In comparison with $\mu = 1$, for values $\mu < \lambda$ the increase of the approximation order is visible, and the decrease of the error was observed, too.
Thus, we can conclude that local mesh grading in the meridian domain is suited for getting more accuracy and a higher approximation order of the Fourier-finite-element method applied to problems with edge singularities.

In order to get some insight into the effectiveness of the parallel computation of the Fourier–finite–element approximation, for different numbers of processors the run time (it includes the time for arithmetic operations and communication) of partial steps of the computational process is measured. For the example previously discussed (see (6.1)), which is provided with complicated data, Table 3 shows the reduction of the run time for a constant problem size under the influence of a growing number of processors involved in the parallel computation. The problem size corresponds to more than 4×10^6 unknown nodal values (16705 nodes, 255 Fourier coefficients), but it also depends on the number of integration points (n=1024). The conjugate gradient method with preconditioning (incomplete Cholesky factorization) was applied to solve the finite–element equations. By means of 128 processors, the Fourier–finite–element approximation was calculated within 30 seconds. The parallel generation of 255 systems of equations requires 60 seconds, 8 seconds for the stiffness matrices and 52 seconds for the right-hand sides.
In Table 3, the first column indicates the number of processors employed, the second array shows the time for the computation of the right-hand sides F^c_{kh}, F^s_{kh} (left column), without the FFT, and the speed-up (run time on one processor/run time on P processors, right column). Clearly, the parallelization accelerates essentially the computational process and the optimal speed-up is taken on (in some cases at least, for technical reasons). The small speed-up for the parallel FFT-analysis (here, the time does not include the time of calculating the function values) is due to the considerable amount of communication time. Because of the special implementation of the parallel version of FEFOP, for the evaluation of solving the systems of equations in 2D, some fictitious speed–up is utilized (cf. [9]). The FFT–synthesis was carried out only for a sufficiently large number of pro-

cessors, since otherwise the solution could not be stored (here, the speed-up is related to the level of 8 processors).
The time needed for solving the kth problem in 2D depends essentially on k. The following observation is in correspondence with the theory: if k is growing then the run time for solving the systems of equations is decreasing rapidly.
For more details on the experiments, further examples including the use of 128 processors and scale-up considerations, see [9].

Acknowledgement: The author thanks Dr. B. Weber and Dipl.-Math. B. Lang (TU Chemnitz-Zwickau, Faculty of Mathematics) for the discussion of this paper, especially of the parallel computation of the Fourier-finite-element approximation, and for carrying out the computation on the parallel computer Parsytec GC/PowerPlus-128.

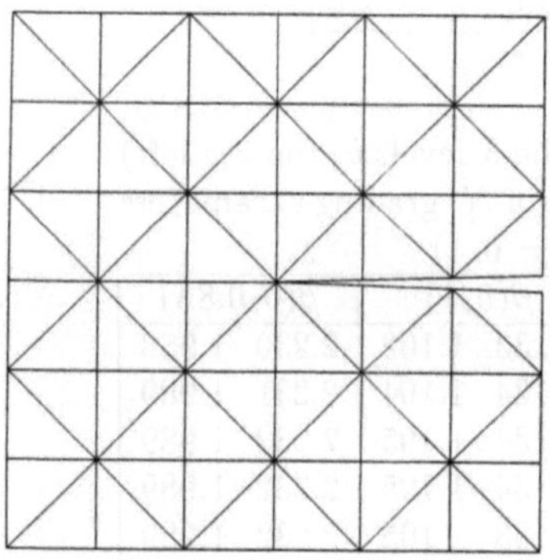

Fig. 2a

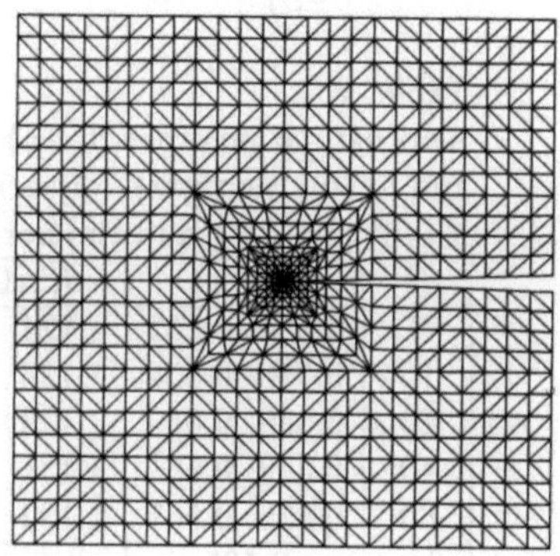

Fig. 2b

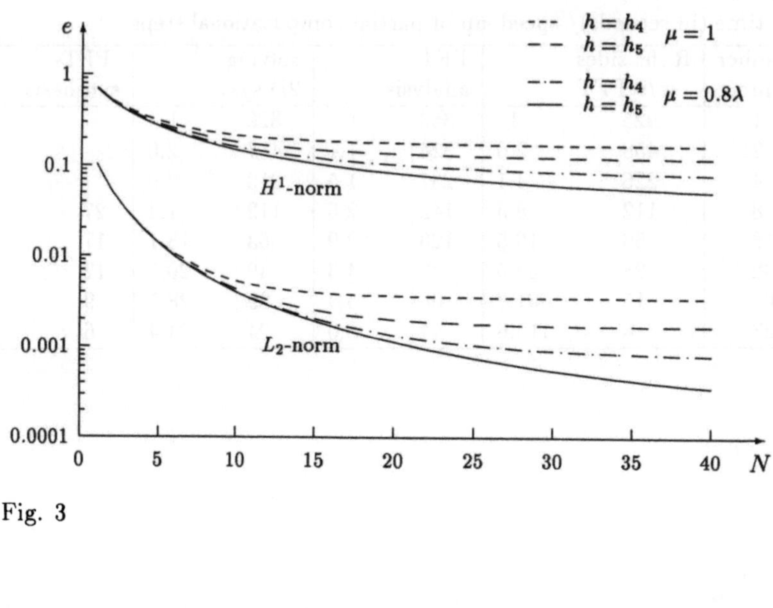

Fig. 3

Table 1
Approximation order $\alpha(l,\mu)$ of $N^{-\alpha}$ on the h-level h_5, for various values of N, grading parameters $\mu = 1$, $\mu = 0.8\lambda$ and the norm in $H^l(l = 0, 1)$

N	$\alpha(1,1)$	$\alpha(1,0.8\lambda)$	$\alpha(0,1)$	$\alpha(0,0.8\lambda)$
5	0.583	0.581	1.460	1.460
10	0.845	0.837	1.755	1.755
15	0.910	0.896	1.811	1.810
20	0.944	0.923	1.850	1.847
30	0.990	0.951	1.900	1.893
40	1.018	0.965	1.932	1.919
50	1.047	0.974	1.960	1.935
60	1.077	0.980	1.985	1.946
α_{exp}	1.0	1.0	2.0	2.0

Table 2
Approximation order $\beta(l,\mu)$ of h^β on the h-levels h_4 (on the left) and h_5 (on the right), for various values of N, grading parameters $\mu = 1$, $\mu = 0.8\lambda$ and the norm in $H^l(l = 0, 1)$

N	$\beta(1,1)$		$\beta(1,0.8\lambda)$		$\beta(0,1)$		$\beta(0,0.8\lambda)$	
5	0.553	0.553	1.124	1.007	1.133	1.102	2.230	1.988
10	0.558	0.556	1.128	1.010	1.134	1.104	2.231	1.989
20	0.560	0.559	1.129	1.013	1.134	1.105	2.232	1.989
40	0.559	0.561	1.131	1.015	1.134	1.105	2.232	1.989
60	0.555	0.561	1.132	1.016	1.133	1.105	2.232	1.989
β_{exp}	0.505		1		1.01		2	

Table 3
Run time (in seconds)/ speed–up of partial computational steps

Number of proc.	R.-h. sides w/o FFT		FFT-analysis		solving 2D-syst.		FFT-synthesis	
1	925	1	353	1	826	1	-	-
2	458	2.0	324	1.1	417	2.0	-	-
4	226	4.1	237	1.5	213	3.9	-	-
8	112	8.3	142	2.5	112	7.4	27	1
16	56	16.5	120	2.9	63	13.1	17	1.6
32	28	33.5	82	4.3	40	20.1	12	2.3
64	15	61.7	66	5.3	29	28.5	9	3.0
128	8	115.6	44	8.0	24	34.4	6	4.5

References

[1] K. E. Buck. *Zur Berechnung der Verschiebungen und Spannungen in rotationssymmetrischen Körpern unter beliebiger Belastung.* PhD thesis, Universität Stuttgart, 1970. Dissertation.

[2] P. Butzer and R. Nessel. *Fourier Analysis and Approximation, Vol. I.* Birkhäuser Verlag, Basel, 1971.

[3] C. Canuto, Y. Maday, and A. Quarteroni. Analysis of the Combined Finite Element and Fourier Interpolation. *Num. Math.*, 39:205–220, 1982.

[4] P. Ciarlet. *The finite element method for elliptic problems.* North–Holland, Amsterdam, 1978.

[5] St. Fischer. Numerische Lösung spezieller 3D–Randwertaufgaben mittels einer gekoppelten FEM–Fourier–Strategie. Diplomarbeit, Technische Universität Karl–Marx–Stadt (Chemnitz), Sektion Mathematik, 1988.

[6] G. Globisch. PARMESH - a parallel mesh generator. *Parallel Computing*, 21(3):509–524, 1995.

[7] B. Heinrich. Singularity Functions at Axisymmetric Edges and their Representation by Fourier Series. *Math. Meth. Appl. Sci.*, 16:837–854, 1993.

[8] B. Heinrich. The Fourier–finite–element method for Poisson's equation in axisymmetric domains with edges. Preprint SPC 94_17, Technische Universität Chemnitz–Zwickau, Fakultät für Mathematik, August 1994. To appear in *SIAM J. Num. Anal.*, 1996.

[9] B. Heinrich, B. Lang, and B. Weber. Parallel computation of Fourier–finite–element approximations and some experiments. Preprint SPC 95_15, Technische Universität Chemnitz–Zwickau, Fakultät für Mathematik, Mai 1995.

[10] B. Heinrich and St. Fischer. Lösung spezieller 3D–Randwertaufgaben mittels gekoppelter Fourier–Finite–Elemente–Methode. Preprint 100, Technische Universität Karl–Marx–Stadt (Chemnitz) , 1989.

[11] B. Heinrich and B. Weber. The Fourier–finite–element method for three–dimensional elliptic problems with axisymmetric interfaces. Preprint SPC 94_22, Technische Universität Chemnitz–Zwickau, Fakultät für Mathematik, November 1994.

[12] J. Y. Lai and J. R. Booker. Application of Discrete Fourier Series to the Finite Element Stress Analysis of Axi–Symmetric Solids. *Int. J. Num. Meth. Eng.*, 31:619–647, 1991.

[13] B. Mercier. *An Introduction to the Numerical Analysis of Spectral Methods.* Springer–Verlag, Berlin Heidelberg, 1989.

[14] B. Mercier and G. Raugel. Résolution d'un problème aux limites dans un ouvert axisymétrique par éléments finis en r, z et séries de Fourier en θ. *R.A.I.R.O. Analyse numérique*, 16(4):405–461, 1982.

[15] B. Nkemzi. Partielle Fourierdekomposition für das lineare Elastizitätsproblem in rotationssymmetrischen Gebieten. Preprint SPC 95_12, Technische Universität Chemnitz–Zwickau, Fakultät für Mathematik, März 1995.

[16] M. Pester. Zur Implementation paralleler Algorithmen - Elementare Operationen mit Matrizen und Vektoren im Hypercube. Preprint 166, Technische Universität Karl–Marx–Stadt (Chemnitz), Sektion Mathematik, 1990.

[17] M. Pester and T. Steidten. Parallel Implementation of the Fourier Finite Element Method. Preprint SPC 94_10, Technische Universität Chemnitz–Zwickau, Fakultät für Mathematik, 1994.

[18] S. Rjasanow. Dokumentation und theoretische Grundlagen zum Programm SOLKLZ. Preprint 15, Technische Universität Karl-Marx-Stadt (Chemnitz), 1986.

[19] K. Runesson and J. R. Booker. Finite element analysis of elastic–plastic layered soil using discrete Fourier series expansion. *Int. J. Num. Meth. Eng.*, 19:473–478, 1983.

[20] E. Schultchen, H. Ulonska, and W. Wurmnest. Statische Berechnung von Rotationskörpern unter beliebiger nicht–rotationssymmetrischer Belastung mit dem Programmsystem ANTRAS–ROT. Forschungsbericht 35–2, Techn. Mitt. Krupp, 1977.

[21] C. Schwab. *Numerische Verfahren zur Untersuchung der magnetohydrodynamischen Stabilität dreidimensionaler Plasmagleichgewichte.* Dissertation, TH Darmstadt, 1991.

[22] H. R. Schwarz. The Fast Fourier Transform for General Order. *Computing*, 19:341–350, 1978.

[23] B. Weber. *Die Fourier–Finite–Elemente–Methode für elliptische Interfaceprobleme in axialsymmetrischen Gebieten.* PhD thesis, Technische Universität Chemnitz–Zwickau, 1994.

[24] W. Weese. *Spannungs– und Verformungsberechnung in Rotationskörpern mit beliebiger Belastung auf der Grundlage der Fourierreihenentwicklung nach der Methode der finiten Elemente.* TU Magdeburg, Sektion Maschinenbau, 1983. Anwenderdokumentation für das Programm ROKO 3/83.

[25] E. L. Wilson. Structural analysis of axisymmetric solids. *AIAA J.*, 3(12):2269–2274, 1965.

[26] O. C. Zienkiewicz. *The Finite Element Method, 3rd ed.* McGraw–Hill, New York, 1977.

Parallel Solvers for coupled FEM-BEM equations with applications to non-linear magnetic field problems

Bodo Heise*

Institute of Mathematics, Johannes Kepler University Linz,
Altenbergerstr. 69, A-4040 Linz, Austria.
e-mail: heise@miraculix.numa.uni-linz.ac.at

Summary

An efficient parallel algorithm for solving nonlinear boundary value problems arising, e.g., in the magnetic field computation, is presented. It is based on both the idea of "nested iteration" (also called Full Multilevel Method) and parallel domain decomposition (DD) solvers for the linear systems suited for computations on MIMD computers with local memory and message-passing principle. It makes use of the parallel data structure of these solvers, the linearization is done by Newton's method, the linear system is solved by CG with DD preconditioning.

The DD approach allows us to couple Finite Element and Galerkin Boundary Element Methods in a unified variational problem. In this way, e.g., magnetic field problems in an infinite domain with Sommerfeld's radiation condition can be modelled correctly. The problem of a nonsymmetric system matrix due to Galerkin-BEM is overcome by transforming it into a symmetric but indefinite matrix and applying Bramble/Pasciak's CG for indefinite systems. For preconditioning, the main ideas of recent DD research are applied.

Test computations on a Power Xplorer parallel system were performed for model problems.

Key words: nonlinear partial differential equations, domain decomposition, finite element methods, boundary element methods, coupling, magnetic field calculations
AMS subject classifications: 65N55, 65N22, 65N30, 65N38, 78A30

1 Introduction

Recently, domain decomposition (DD) methods have found more and more interest, in particular in connection with the rising application of Multiple Instruction Multiple Data (MIMD) parallel computers (cf. the proceedings of the International Symposia on "DD methods for partial differential equations" of the recent years). The DD methods with their inherent parallel data distribution allow parallel computations with a communication cost one order lower than the problem itself. Thus, computing on hundreds of processors can be performed with high efficiency.

Further, DD methods allow the advantages of the Finite Element Method (FEM) to be "married" to those of the Boundary Element Method (BEM) via a unified variational formulation [4, 35, 30, 29]. In particular, in the magnetic field computations for electric machines, we can use BEM for the infinite exterior subdomain of the machine more successfully than the FEM, which is prefered for the subdomains in the interior of the machine due to the complicated structure, currents, and nonlinearities occuring there [30, 29, 27, 28].

*This research has been supported by the German Research Foundation DFG within the Priority Research Programme "Boundary Element Methods" under the grant La 767/1-3.

Nonlinear equations arising from magnetic field simulation and other problems can be successfully linearized by Newton's method [16, 18, 17, 19, 21, 22]. A Newton method with approximate iterative linear solver benefits from the quadratic rate of convergence if the accuracy of the linear solver is controlled well [5, 6, 19, 22]. A possible drawback of Newton's method, the generation of a new Jacobi matrix in each iteration step, nearly vanishes since it runs completely in parallel without any communication between processors [19, 23, 22].

The multilevel idea is applied three times: First, the "nested iteration" [15, 19] allows the nonlinearity to be "captured" on a coarse grid by some Newton iteration steps requiring a short time only. Then on the finer grids only a few Newton iterates are necessary. Second, in the FE subdomains hierarchical extension on a sequence of grids [14] and an inner multigrid preconditioner are applied [11]. Third, it is possible to use multigrid methods for Schur complement preconditioning [22, 24].

The rest of the paper is organized as follows. In Section 2 we describe the nonlinear boundary value problem. In Section 3 we present the parallel nested Newton method. We refer to the preconditioned CG method for solving linear problems in Section 4. The extension of the method to coupled FE/BE problems is discussed in Section 5. Finally, in Section 6 we present field calculations for model problems.

2 Nonlinear electromagnetic field problems

Consider a two-dimensional nonlinear stationary magnetic field problem in a bounded domain $\Omega \subset \mathbf{R}^2$. The variational formulation of the problem can be written as follows:
Find $u \in V = H_0^1(\Omega)$ *such that*

$$a(u,v) = \langle f, v\rangle \quad \forall v \in V, \tag{1}$$

where

$$a(u,v) = \int_\Omega \nu(x,|\nabla u|)\; \nabla^{\mathsf{T}} u \; \nabla v \; dx \ ,$$

and

$$\langle f, v\rangle = \int_\Omega (\mathrm{S}v - \mathrm{H}_{0\mathrm{y}}\frac{\partial v}{\partial \mathrm{x}} + \mathrm{H}_{0\mathrm{x}}\frac{\partial v}{\partial \mathrm{y}})\; dx.$$

Assume that Ω representing the cross–section of some electromagnetic device lies in the x–y–plane of the $\mathbf{R}^3$. Solution u is then the z–component of the vector potential $\vec{\mathrm{A}}$. For details of the physical model developed from the Maxwell equations, cf. [21, 20]. The z–component of the current density is represented by S. Note that in the following we omit the x–dependence of the material function $\nu(x,|\nabla u|)$. We assume that $\bar{\Omega}$ is decomposed into subdomains

$$\bar{\Omega} = \bigcup_{j=1}^{N_M} \bar{\hat{\Omega}}_j, \quad \text{with} \quad \hat{\Omega}_i \cap \hat{\Omega}_j = \emptyset \quad \forall i \neq j.$$

The $\hat{\Omega}_j$'s represent materials with different magnetic properties (iron, copper, air, permanent-magnetic materials) in the cross–section of an electromagnetic device. We assume that the function ν depends on the locus $x \in \Omega$, but ν is always the same function in one of the $\hat{\Omega}_j$, i.e.

$$\nu(x,B) = \nu^{(j)}(B) \quad \text{if} \quad x \in \hat{\Omega}_j.$$

Assuming certain monotonicity and boundedness conditions on functions $\nu^{(j)}(.)$ we can prove the existence and uniqueness of the solution of the variational problem (1) [19, 21]. Standard Finite Element discretization with linear triangular elements has been discussed in [21]. Therein the algorithm for monotonicity preserving spline interpolation of a pointwise given material function $\nu^{(j)}(.)$ is described, error estimates are given, too.

3 The Nested Newton DD Method

We assume that, as in the finite element substructuring technique, Ω is decomposed into p non-overlapping subdomains $\Omega_i, i = 1, \ldots, p$, such that

$$\bar{\Omega} = \bigcup_{i=1}^{p} \bar{\Omega}_i, \quad \text{and} \quad \bar{\hat{\Omega}}_j = \bigcup_{i \in I_j} \bar{\Omega}_i \quad \forall j = 1, \ldots, N_M \tag{2}$$

with index sets fulfilling

$$I_j \subset I^* = \{1, \ldots, p\}, \quad \bigcup_{j=1}^{N_M} I_j = I^*, \quad I_j \cap I_k = \emptyset \, \forall j \neq k,$$

i.e., the subdomains $\hat{\Omega}_j$ determined by the materials may be decomposed further (cf. [10, 9, 11, 13, 12]). The subdomains $\bar{\Omega}_i, i = 1, \ldots, p$, are assigned to p processors of the MIMD computer.

We further assume that in each subdomain $\bar{\Omega}_i$ there is a multilevel sequence of linear finite element discretizations such that this discretization process results in conform triangulations of Ω creating a nested sequence

$$V_1 \subset V_2 \subset \cdots \subset V_l \subset V = H_0^1(\Omega) \tag{3}$$

of spaces of linear finite elements.

We obtain a sequence of variational problems for $q = 1, \ldots, l$:
Find $u_q \in V_q \subset V$ such that

$$a_q(u_q, v_q) = \langle f_q, v_q \rangle \quad \forall v_q \in V_q. \tag{4}$$

and a sequence of equivalent nonlinear FE equations

$$K_q \mathbf{u}_q = \mathbf{f}_q \quad q = 1, \ldots, l, \tag{5}$$

with nonlinear operators $K_q : \mathbf{R}^{N_q} \longrightarrow \mathbf{R}^{N_q}$, solution vectors $\mathbf{u}_q \in \mathbf{R}^{N_q}$ and vectors $\mathbf{f}_q \in \mathbf{R}^{N_q}$.

Any Newton method for solving (5) in $\mathbf{R}^{N_q}$ requires the Fréchet derivative of K_q at a vector $\mathbf{v}_q$,

$$K_q'[\mathbf{v}_q] : \mathbf{R}^{N_q} \rightarrow \mathbf{R}^{N_q}$$

which is a linear operator and can be represented by the Jacobi matrix, cf. [19]. A parallel algorithm connecting the (damped) Newton method with the nested iteration and domain decomposition principles for solving (5) with $q = l$ can be written as follows. The result will be the approximate solution $\mathbf{u}_l^*$ with relative accuracy ε (nested iteration ε).

Algorithm PNN (Parallel Nested Newton)
Step 0
Initialization of the grid number:
(0.)(P) $q := 1$.

Step 1
Set the initial solution for the grid q:

(1.1)(P) IF $q = 1$ THEN $\mathbf{u}_q^0 = \mathbf{0}$;

(1.2)(P) IF $q > 1$ THEN $\mathbf{u}_q^0 = \tilde{I}_{q-1}^q \mathbf{u}_{q-1}^*$;
the initial solution is the interpolation of the best solution on the grid $q - 1$.

(1.3)(P) Initialize the Newton iteration number
$j := 0$.

Step 2
Compute the Jacobi matrix and the defect vector

(2.)(P) $B_{q,0} = K'_q[\mathbf{u}_q^0]$;
$\mathbf{d}_q^1 = \mathbf{f}_q - K_q \mathbf{u}_q^0$

Step 3

(3.)(P) Choose a relaxation parameter τ_q^j with $0 < \tau_q^j \leq 1$ and a relative accuracy parameter ε_{lin} with $0 < \varepsilon_{\text{lin}} < 1$.

Step 4

(4.)(S) Solve the linear defect system

$$B_{q,j}\mathbf{w}_q^{j+1} = \mathbf{d}_q^{j+1}$$

approximately (with relative accuracy ε_{lin}) using a parallelized preconditioned conjugate gradient (PCG) solver. The preconditioning is made by a domain decomposition (DD) method. The result is $\tilde{\mathbf{w}}_q^{j+1}$.

Step 5
Correct the solution:

(5.)(P) $\mathbf{u}_q^{j+1} = \mathbf{u}_q^j + \tau_q^j \tilde{\mathbf{w}}_q^{j+1}$.

Step 6
Control the convergence (parameter c_τ is chosen a priori with $c_\tau < 1$):

(6.1)(P) Compute the new defect vector and the new Jacobi matrix
$\mathbf{d}_q^{j+2} = \mathbf{f}_q - K_q \mathbf{u}_q^{j+1}$;
$B_{q,j+1} = K'_q[\mathbf{u}_q^{j+1}]$.

(6.2)(C) Compute defect norms
$d_q^{j+1} = \|\mathbf{d}_q^{j+1}\|$;
$d_q^{j+2} = \|\mathbf{d}_q^{j+2}\|$;

(6.3)(P) IF $d_q^{j+2} \geq d_q^{j+1}$ THEN $\left(\begin{array}{l} \tau_q^j := \min\left\{ c_\tau \tau_q^j, \dfrac{\tau_q^j d_q^{j+1}}{d_q^{j+1} + d_q^{j+2}} \right\}; \\ \texttt{GOTO } \textit{Step 5} \end{array} \right)$;

(6.4)(P) IF $d_q^{j+2} \leq \varepsilon d_q^1$ THEN $\left(\begin{array}{l} \mathbf{u}_q^\star := \mathbf{u}_q^{j+1}; \\ \texttt{IF } q < l \texttt{ THEN } (q := q+1; \texttt{ GOTO } \textit{Step 1}); \\ \texttt{IF } q = l \texttt{ THEN EXIT}; \end{array} \right)$;

(6.5)(P) Perform a further Newton step:
$j := j + 1$;
GOTO *Step 3*.

In this description, (P) indicates that the step is performed completely in parallel, i.e., independently at the processors. Note that (C) indicates that global communication is necessary. The solver (S) includes parallel independent parts, communication between processors handling neighbouring subdomains, and global communication.

In the parallel PCG solvers, two types of vector distribution are applied [11]. The overlapping type (type 1) stores the values for coupling nodes redundantly on all processors possessing the node. This type is used for approximate solutions $\mathbf{u}_q^j$ and corrections $\tilde{\mathbf{w}}_q^j$. For adding type (type 2) vectors the value for a coupling node is the sum of the entries on all processors possessing the node. Right-hand side $\mathbf{f}_q$ and defects $\mathbf{d}_q^j$ are stored as adding type vectors.

Since functions $\nu(B)$ and $\nu'(B)$ are known explicitly as interpolants of the given data [21], and arguments $B = |\nabla \mathbf{u}_q^j|$ depend locally on the current approximate solution, the Jacobi matrices $K_q'[\mathbf{u}_q^j]$ can be computed by a FE assembling procedure, cf. [19]. This can be done completely in parallel as for linear boundary value problems if the matrices are stored in the standard way [11, 8]. Therefore, the defect vector $\mathbf{d}_q^j$ can be computed completely in parallel, but the computation of its norm requires next-neighbour and global communications.

4 The PCG solver with DD preconditioning

The parallel CG algorithm with DD preconditioning can be implemented in a standard way, cf. [11]. It runs completely in parallel with the exception of the two scalar products, and the preconditioning. For the purpose of describing preconditioning, the linear equation

$$K_q'[\mathbf{u}_q^j]\mathbf{w}_q^{j+1} = \mathbf{d}_q^{j+1}$$

can be rewritten in the standard block form [11, 31]

$$\tilde{K}\begin{pmatrix}\mathbf{w}_C\\ \mathbf{w}_I\end{pmatrix} = \begin{pmatrix}K_C & K_{CI}\\ K_{IC} & K_I\end{pmatrix}\begin{pmatrix}\mathbf{w}_C\\ \mathbf{w}_I\end{pmatrix} = \begin{pmatrix}\mathbf{d}_C\\ \mathbf{d}_I\end{pmatrix} \tag{6}$$

where indices "I" and "C" correspond to the nodes belonging to the interior of subdomains Ω_i and to the coupling boundaries, respectively, and $\dim(\mathbf{w}_I) = N_{I,q}$, $\dim(\mathbf{w}_C) = N_{C,q}$. Note that $K_I = \operatorname{diag}(K_{I,i})_{i=1,2,\ldots,p}$ is a block-diagonal matrix. The DD preconditioner C with

$$C = \begin{pmatrix}I_C & K_{CI}B_I^{-T}\\ O & I_I\end{pmatrix}\begin{pmatrix}C_C & O\\ O & C_I\end{pmatrix}\begin{pmatrix}I_C & O\\ B_I^{-1}K_{IC} & I_I\end{pmatrix} \tag{7}$$

contains the three components C_C, $C_I = \operatorname{diag}(C_{I,i})_{i=1,2,\ldots,p}$, and the regular basis transformation matrix $B_I = \operatorname{diag}(B_{I,i})_{i=1,2,\ldots,p}$, which can be adapted to the problem as well as possible [11].

The following result has been proved in [10, 11].

THEOREM 1 *Let the symmetric and positive definite block preconditioners C_C and C_I satisfy the spectral equivalence inequalities*

$$\underline{\gamma}_C C_C \le S_C + T_C \le \overline{\gamma}_C C_C \quad \text{and} \quad \underline{\gamma}_I C_I \le K_I \le \overline{\gamma}_I C_I \tag{8}$$

with some positive constants $\underline{\gamma}_C$, $\overline{\gamma}_C$, $\underline{\gamma}_I$ and $\overline{\gamma}_I$. Then the ASM-DD preconditioner (7) satisfies the spectral equivalence inequalities

$$\underline{\gamma} C \le \tilde{K} \le \overline{\gamma} C \tag{9}$$

with the spectral equivalence constants

$$\underline{\gamma} = \min\{\underline{\gamma}_C, \underline{\gamma}_I\}\left(1 - \sqrt{\tfrac{\mu}{1+\mu}}\right) \quad \text{and} \quad \overline{\gamma} = \max\{\overline{\gamma}_C, \overline{\gamma}_I\}\left(1 + \sqrt{\tfrac{\mu}{1+\mu}}\right) \tag{10}$$

where $\mu = \rho(S_C^{-1}T_C)$ *denotes the spectral radius of* $S_C^{-1}T_C$ *, with* $S_C = K_C - K_{CI}K_I^{-1}K_{IC}$ *and* $T_C = K_{CI}(K_I^{-1} - B_I^{-T})K_I(K_I^{-1} - B_I^{-1})K_{IC}$ *.*
For the spectral condition number $\kappa(C^{-1}\tilde{K})$ *of* $C^{-1}\tilde{K}$*, the two-side estimate*

$$\left(\underline{\gamma}_1 / \overline{\gamma}_1\right)(\sqrt{\mu} + \sqrt{1+\mu})^2 \leq \kappa(C^{-1}\tilde{K}) \leq \left(\overline{\gamma}_1 / \underline{\gamma}_1\right)(\sqrt{\mu} + \sqrt{1+\mu})^2 \quad (11)$$

holds with $\underline{\gamma}_1 = \min\{\underline{\gamma}_C, \underline{\gamma}_I\}$ *and* $\overline{\gamma}_1 = \max\{\overline{\gamma}_C, \overline{\gamma}_I\}$ *.*

5 The FEM – BEM coupling

The coupling of FEM and BEM via non-overlapping DD is described, e.g., in [4, 35]. The coupled Galerkin finite element / boundary element discretization of linear and nonlinear elliptic problems is discussed, e.g., in [30, 29, 27, 28, 25], see also [26].

We assume that in the domain decomposition (2) the subdomain Ω_p is the unbounded exterior subdomain. As a consequence, Ω is unbounded. We assume that the complement of $\bar{\Omega}$,

$$\Omega_0 = \mathbf{R}^2 \setminus \bar{\Omega},$$

is a simply connected bounded domain, and that we have homogeneous Dirichlet boundary conditions on $\Gamma_D = \partial\Omega_0$. Formally the problem can be written as

$$\begin{aligned} -\text{div}\,(\nu(x, |\nabla u(x)|)\nabla u(x)) &= \mathrm{S}(x) + \frac{\partial \mathrm{H}_{0\mathrm{y}}(x)}{\partial \mathrm{x}} - \frac{\partial \mathrm{H}_{0\mathrm{x}}(x)}{\partial \mathrm{y}}, \quad x \in \Omega \\ u(x) &= 0, \quad x \in \Gamma_D \\ |u(x)| &\to 0 \quad \text{for } |x| \to \infty. \end{aligned} \quad (12)$$

Let us assume that the coefficient $\nu = \nu_p$ is a constant in Ω_p, and that the right-hand side vanishes there, i.e., that the relations

$$\text{supp S} \subset \bar{\Omega}_-, \quad \text{supp H}_0 \subset \bar{\Omega}_- \quad \text{with} \quad \bar{\Omega}_- = \bigcup_{i=1}^{p-1} \bar{\Omega}_i$$

hold. This is satisfied, e.g., if Ω_p represents an exterior air (or vacuum) domain without currents or permanent magnets. For simplicity, we assume that the problem is properly scaled such that the inequality

$$\text{diam}\,(\mathbf{R}^2 \setminus \Omega_p) < 1$$

holds. Then the BEM can be applied for Ω_p.

Further, we may assume that there is an index r with $0 \leq r < p$ such that $\nu = \nu_i, i = r+1, \ldots, p-1$ are constants in $\Omega_i, i = r+1, \ldots, p-1$, and the right-hand side vanishes in $\Omega_i, i = r+1, \ldots, p-1$. Then the BEM can be applied for the "interior" subdomains $\Omega_i, i = r+1, \ldots, p-1$, too.

Then we get the following variational formulation [27, 28]:

Find $(\lambda, u) \in V := \Lambda \times U_0$:

$$a(\lambda, u; \eta, v) = \langle F, v\rangle \quad \forall(\eta, v) \in V, \quad (13)$$

where

$$\begin{aligned} a(\lambda, u; \eta, v) &:= a_B(\lambda, u; \eta, v) + a_F(u, v) \\ a_B(\lambda, u; \eta, v) &:= \sum_{i=r+1}^{p-1} \nu_i\{\langle D_i u_i, v_i\rangle_{\Gamma_i} + \frac{1}{2}\langle \lambda_i, v_i\rangle_{\Gamma_i} + \langle \lambda_i, K_i v_i\rangle_{\Gamma_i} + \end{aligned}$$

$$\langle\eta_i, V_i\lambda_i\rangle_{\Gamma_i} - \langle\eta_i, K_i u_i\rangle_{\Gamma_i} - \frac{1}{2}\langle\eta_i, u_i\rangle_{\Gamma_i}\} +$$

$$\nu_p\{\langle D_p u_+, v_+\rangle_\Gamma - \frac{1}{2}\langle\lambda_p, v_+\rangle_\Gamma + \langle\lambda_p, K_p v_+\rangle_\Gamma +$$

$$\langle\eta_p, V_p\lambda_p\rangle_\Gamma - \langle\eta_p, K_p u_+\rangle_\Gamma + \frac{1}{2}\langle\eta_p, u_+\rangle_\Gamma\}$$

$$\begin{aligned} a_F(u,v) &:= \sum_{i=1}^{r}\int_{\Omega_i}\nu(x,|\nabla u|)\nabla^\top u(x)\nabla v(x)\,dx \\ \langle F, v\rangle &:= \sum_{i=1}^{r}\int_{\Omega_i}(\mathrm{S}v - \mathrm{H}_{0\mathrm{y}}\frac{\partial v}{\partial \mathrm{x}} + \mathrm{H}_{0\mathrm{x}}\frac{\partial v}{\partial \mathrm{y}})\,dx \\ \langle\lambda_i, v_i\rangle_{\Gamma_i} &:= \int_{\Gamma_i}\lambda_i v_i\,ds \end{aligned}$$

and

$$v_+ = v|_\Gamma, \quad u_+ = u|_\Gamma, \quad v_i = v|_{\partial\Omega_i}, \quad u_i = u|_{\partial\Omega_i}.$$

Let the spaces U_0 and Λ be defined as follows:

$$\begin{aligned} U_0 &:= \{u \in H^1(\Omega_F): \ u|_{\Gamma_{BE}} \in H^{1/2}(\Gamma_{BE}), u|_{\partial\Omega_0} = 0\} \\ \Lambda &:= H^{-1/2}(\Gamma_{BE}), \end{aligned}$$

with the notations

$$\Gamma_i = \partial\Omega_i, \ \Gamma = \partial\Omega_p,$$

$$\bar{\Omega}_F = \bigcup_{i=1}^{r}\bar{\Omega}_i, \ \Gamma_{BE} := \bigcup_{i=r+1}^{p}\partial\Omega_i \setminus \Gamma_D, \ \Gamma_{FE} := \bigcup_{i=1}^{r}\partial\Omega_i \setminus \Gamma_D, \ \Gamma_C := \Gamma_{BE} \cup \Gamma_{FE}.$$

The standard boundary integral operators are defined by the relations

$$\begin{aligned} (V_i v_i)(x) &= \int_{\Gamma_i} s(x,y)v_i(y)\,d\Gamma_y \\ (K_i u_i)(x) &= \int_{\Gamma_i}\frac{\partial}{\partial n_y}s(x,y)u_i(y)\,d\Gamma_y \\ (K_i' v_i)(x) &= \int_{\Gamma_i}\frac{\partial}{\partial n_x}s(x,y)v_i(y)\,d\Gamma_y \\ (D_i u_i)(x) &= -\frac{\partial}{\partial n_x}\int_{\Gamma_i}\frac{\partial}{\partial n_y}s(x,y)u_i(y)\,d\Gamma_y \end{aligned} \tag{14}$$

where n_x and n_y denote the normal outward direction, and the fundamental solution

$$s(x,y) = -\frac{1}{2\pi}\ln|x-y|. \tag{15}$$

of the Laplacian in two-dimensional case.

Then, the problem may be discretized using a standard nodal FE basis, corresponding basis functions for u on Γ_{BE}, and suitable basis functions for approximating λ on Γ_{BE} (cf., e.g., [30, 27]). This discretization results in a nonlinear system (we omit the grid index q in the following, and set $\dim(\mathbf{u}_\Lambda) = N_\Lambda$, $N = N_\Lambda + N_C + N_I$)

$$K\begin{pmatrix}\mathbf{u}_\Lambda\\ \mathbf{u}_C\\ \mathbf{u}_I\end{pmatrix} = K_F\begin{pmatrix}\mathbf{u}_C\\ \mathbf{u}_I\end{pmatrix} + K_B\cdot\begin{pmatrix}\mathbf{u}_\Lambda\\ \mathbf{u}_C\end{pmatrix} = \begin{pmatrix}\mathbf{f}_\Lambda\\ \mathbf{f}_C\\ \mathbf{f}_I\end{pmatrix},$$

where the nonlinear operator $K : \mathbf{R}^N \longrightarrow \mathbf{R}^N$ can be split up into a nonlinear operator $K_F : \mathbf{R}^{N_C+N_I} \longrightarrow \mathbf{R}^N$ originating from a_F and a linear operator $K_B : \mathbf{R}^{N_\Lambda+N_C} \longrightarrow \mathbf{R}^N$ originating from a_B.

Therefore, the linear Newton defect system with the Jacobi matrix $K'[\mathbf{u}]$ is given by

$$K'\left[\begin{pmatrix}\mathbf{u}_\Lambda\\ \mathbf{u}_C\\ \mathbf{u}_I\end{pmatrix}\right]\cdot\begin{pmatrix}\mathbf{w}_\Lambda\\ \mathbf{w}_C\\ \mathbf{w}_I\end{pmatrix} = K_F'\left[\begin{pmatrix}\mathbf{u}_C\\ \mathbf{u}_I\end{pmatrix}\right]\cdot\begin{pmatrix}\mathbf{w}_C\\ \mathbf{w}_I\end{pmatrix} + K_B\cdot\begin{pmatrix}\mathbf{w}_\Lambda\\ \mathbf{w}_C\end{pmatrix} = \begin{pmatrix}\mathbf{d}_\Lambda\\ \mathbf{d}_C\\ \mathbf{d}_I\end{pmatrix},$$

it can be rewritten in a block form similar to (6),

$$\begin{pmatrix}K_\Lambda & -K_{\Lambda C} & 0\\ K_{C\Lambda} & K_C & K_{CI}\\ 0 & K_{IC} & K_I\end{pmatrix}\begin{pmatrix}\mathbf{w}_\Lambda\\ \mathbf{w}_C\\ \mathbf{w}_I\end{pmatrix} = \begin{pmatrix}\mathbf{d}_\Lambda\\ \mathbf{d}_C\\ \mathbf{d}_I\end{pmatrix}. \tag{16}$$

Here the index "Λ" denotes the unknowns associated with λ on Γ_{BE}. An index "C" denotes vector components corresponding to the nodal basis on Γ_C, i.e., the coupling boundaries between the FE/FE, FE/BE, and BE/BE subdomains.

The nonsymmetric, positive definite system (16) can be approximately solved by Bramble/Pasciak's CG method [1]. The method requires a preconditioner C_Λ which can be inverted easily and which fulfills the spectral equivalence inequalities

$$\underline{\gamma}_\Lambda C_\Lambda \le K_\Lambda \le \overline{\gamma}_\Lambda C_\Lambda, \quad \text{with} \quad \underline{\gamma}_\Lambda > 1. \tag{17}$$

The complete method is described in detail in [30, 27, 25]. We will sketch only the main equations here. With the definitions

$$\begin{aligned}K_1 &= K_\Lambda,\ \mathbf{d}_1 = \mathbf{d}_\Lambda\\ K_{12} &= K_{21}^T = (-K_{\Lambda C}\quad 0)\\ K_2 &= \begin{pmatrix}K_C & K_{CI}\\ K_{IC} & K_I\end{pmatrix},\ \mathbf{d}_2 = \begin{pmatrix}-\mathbf{d}_C\\ -\mathbf{d}_I\end{pmatrix}.\end{aligned}$$

we can reformulate (16) as a symmetric but indefinite system:

$$\begin{pmatrix}K_1 & K_{12}\\ K_{21} & -K_2\end{pmatrix}\begin{pmatrix}\mathbf{w}_1\\ \mathbf{w}_2\end{pmatrix} = \begin{pmatrix}\mathbf{d}_1\\ \mathbf{d}_2\end{pmatrix}. \tag{18}$$

Following Bramble and Pasciak [1] this system can be transformed into

$$M\mathbf{w} = \mathbf{p}, \tag{19}$$

where

$$M := \begin{pmatrix}C_\Lambda^{-1}K_1 & C_\Lambda^{-1}K_{12}\\ K_{21}C_\Lambda^{-1}(K_1 - C_\Lambda) & K_2 + K_{21}C_\Lambda^{-1}K_{12}\end{pmatrix}$$

and

$$\mathbf{p} = \begin{pmatrix}C_\Lambda^{-1}\mathbf{d}_1\\ K_{21}C_\Lambda^{-1}\mathbf{d}_1 - \mathbf{d}_2\end{pmatrix}.$$

Then, the matrix M is symmetric and positive definite with respect to the scalar product $[.,.]$ which is defined by

$$[\mathbf{w},\mathbf{v}] := ((K_1 - C_\Lambda)\mathbf{w}_1, \mathbf{v}_1) + (\mathbf{w}_2, \mathbf{v}_2). \tag{20}$$

Moreover, M is spectrally equivalent to the regularisator R, where

$$R := \begin{pmatrix} I & 0 \\ 0 & K_2 + K_{21}K_1^{-1}K_{12} \end{pmatrix}.$$

Bramble and Pasciak [1] proved the spectral equivalence inequalities

$$\underline{\lambda}[R\mathbf{v},\mathbf{v}] \le [M\mathbf{v},\mathbf{v}] \le \overline{\lambda}[R\mathbf{v},\mathbf{v}] \quad \forall \mathbf{v} \in \mathbf{R}^N, \tag{21}$$

where

$$\underline{\lambda} = \left(1 + \frac{\alpha}{2} + \sqrt{\alpha + \frac{\alpha^2}{4}}\right)^{-1} \quad \text{and} \quad \overline{\lambda} = \frac{1+\sqrt{\alpha}}{1-\alpha} \tag{22}$$

with $\alpha = 1 - (1/\overline{\gamma}_\Lambda)$. Thus, we have to find a preconditioner C_2 for the matrix

$$K_2 + K_{21}K_1^{-1}K_{12} = \begin{pmatrix} K_C + K_{C\Lambda}K_\Lambda^{-1}K_{\Lambda C} & K_{CI} \\ K_{IC} & K_I \end{pmatrix} \tag{23}$$

which is in some sense similar to the matrix in (6).

The preconditioner defined by

$$C_2 = \begin{pmatrix} I_C & K_{CI}B_I^{-T} \\ 0 & I_I \end{pmatrix} \begin{pmatrix} C_C & 0 \\ 0 & C_I \end{pmatrix} \begin{pmatrix} I_C & 0 \\ B_I^{-1}K_{IC} & I_I \end{pmatrix} \tag{24}$$

is spectrally equivalent to $K_2 + K_{21}K_1^{-1}K_{12}$ if

$$\underline{\gamma}_C C_C \le \tilde{S}_C + K_{C\Lambda}K_\Lambda^{-1}K_{\Lambda C} \le \overline{\gamma}_C C_C \tag{25}$$

$$\underline{\gamma}_I C_I \le K_I \le \overline{\gamma}_I C_I, \tag{26}$$

where

$$\tilde{S}_C = K_C - K_{CI}K_I^{-1}K_{IC} + K_{CI}(K_I^{-1} - B_I^{-T})K_I(K_I^{-1} - B_I^{-1})K_{IC}.$$

The following lemma holds [30, 29].

Lemma 1 *If the symmetric and positive definite block preconditioners $C_I = diag(C_{I,i})_{i=1,\dots,r}$ and C_C satisfy the spectral equivalence inequalities (25) and (26) with positive constants $\underline{\gamma}_C$, $\overline{\gamma}_C$, $\underline{\gamma}_I$, $\overline{\gamma}_I$, then the spectral equivalence inequalities*

$$\underline{\gamma}_2 C_2 \le K_2 + K_{21}K_1^{-1}K_{12} \le \overline{\gamma}_2 C_2 \tag{27}$$

hold for the preconditioner C_C defined in (24) with the constants

$$\underline{\gamma}_2 = \min\{\underline{\gamma}_C, \underline{\gamma}_I\}\left(1 - \sqrt{\tfrac{\mu}{1+\mu}}\right) \quad \textit{and} \quad \overline{\gamma}_2 = \max\{\overline{\gamma}_C, \overline{\gamma}_I\}\left(1 + \sqrt{\tfrac{\mu}{1+\mu}}\right), \tag{28}$$

and the number μ which was defined in Theorem 1.

The proof is given in [29]. With (22) we conclude the following theorem.

THEOREM 2 *If the conditions imposed on C_Λ, C_C, C_I, and B_I, especially (17), (25), and (26) are satisfied, then the FE/BE DD preconditioner*

$$C = diag(I_1, C_2) \tag{29}$$

is self-adjoint and positive definite with respect to the inner product $[.,.]$ and satisfies the spectral equivalence inequalities

$$\underline{\gamma}[C\mathbf{v},\mathbf{v}] \le [M\mathbf{v},\mathbf{v}] \le \overline{\gamma}[C\mathbf{v},\mathbf{v}] \quad \forall \mathbf{v} \in \mathbf{R}^N, \tag{30}$$

with the constants

$$\underline{\gamma} = \underline{\lambda}\min\left\{1, \underline{\gamma}_2\right\} \quad \textit{and} \quad \overline{\gamma} = \overline{\lambda}\max\{1, \overline{\gamma}_2\},$$

where $\underline{\lambda}$, $\overline{\lambda}$, $\underline{\gamma}_2$, $\overline{\gamma}_2$ are given in (22) and (28), respectively.

6 Numerical results and conclusions

The algorithm is implemented in the parallel code FEM⊕BEM [8]. The components of the algorithm are chosen in the following way:
C_I: multigrid V-cycle [11];
B_I: implicitly defined by hierarchical extension (formally $E_{IC} = -B_I^{-1} K_{IC}$) [14];
C_C: applying a Bramble/Pasciak/Xu [34, 3] type Schur complement preconditioner (BPX);
C_Λ:

- Scaled single layer potential BE matrix for circular domain which is a circulant [32] (Circ), or

- properly scaled preconditioner as described in [33] by Steinbach based on the BEM mass matrix and the hypersingular operator (Hyper).

Further possibilities to choose the Schur complement preconditioner C_C, e.g. following Bramble/Pasciak/Schatz [2], using the idea of Dryja [7] on the coupling boundaries, and a global crosspoint system (BPS), or by solving a 2D Neumann problem for coupling boundaries via a multigrid V-cycle, and a global crosspoint system (MG), have been discussed in [22, 24].

As an example for a coupled FE/BE method, the three FE subdomains forming a magnet are surrounded by four air subdomains. The latter are surrounded by the exterior BE subdomain which has a circular boundary Γ, as shown in Figure 1. The copper domains, where we assume to be a current density of the strength S, and the iron domain, are squares with the edges being 16cm long. Alternatively, the three FE subdomains of the magnet can be surrounded by the exterior BE subdomain then having a rectangular boundary Γ. In the latter case, the BE matrix storage requires more memory, and the matrix generation is more expensive than in the first case. Each column of Table 1 shows the performance of the nested Newton iteration where the numbers of CG iterations are given for $\varepsilon_{\text{lin}} = 10^{-2}$ (resp. $\varepsilon_{\text{lin}} = 10^{-4}$ after the slash "/"). For this example, the uniqueness of the solution is guaranteed by the radiation condition, even if no Dirichlet boundary Γ_D is present. The radiation condition is implicitely contained in our BE discretization. Figure 2 shows level lines for two different current densities in the FE subdomains. Computations on Power Xplorer and GC Power Plus parallel systems for further examples, including an electric motor and a far field problem, have been documented in [22, 24, 25].

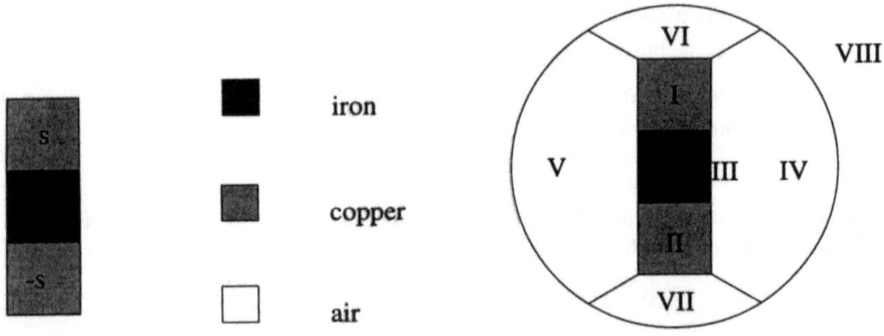

Figure 1: The magnet and the subdomains being used.

Table 1: Calculations for the electric magnet (whole $\mathbf{R}^2$)

Current density (A/mm^2)	1	100	1	100	1	100
Shape of Γ	circle				rectangle	
subdomains (processors)	FEM: I-VII BEM: VIII		FEM: I-III BEM: IV-VIII		FEM: I-III BEM: exterior	
Number of unknowns	68353		18429		16129	
Choice for C_Λ	Circ		Circ + Hyper		Hyper	
Newton iterations 1st grid	3	9	3	9	3	9
Newton iterations 2nd grid	2	2	2	2	2	2
CG iterations 2nd grid	6,7	5,7	7,8	7,8	6,7	6,7
Newton iterations 3rd grid	2	2	2	2	2	2
CG iterations 3rd grid	6,7	6,7	8,8	8,8	6,8	6,8
Newton iterations 4th grid	2	2	2	2	2	3
CG iterations 4th grid	6,8	6,8	8,8	8,8	7,8	7,8
Newton iterations 5th grid	3	3	3	3	3	3
CG iterations 5th grid	6,9/16	7,9/16	7,8/16	7,9/17	7,8/14	7,9,18
Time (system generation)	12.3	13.2	15.6	16.7	18.7	19.1
Time (linear solver)	35.4	39.2	19.5	22.0	18.0	19.2
Total time	48.5	53.2	35.9	39.5	37.5	39.1

Time in seconds, Power Xplorer; $\varepsilon = 10^{-6}$.

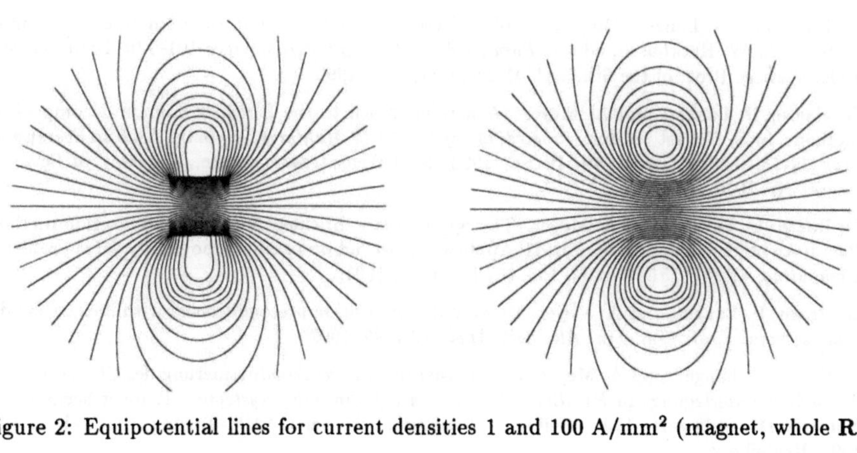

Figure 2: Equipotential lines for current densities 1 and 100 A/mm^2 (magnet, whole $\mathbf{R}^2$)

We may conclude that the proposed Parallel Nested Newton algorithm is able to solve nonlinear field problems very efficiently. The time for repeated parallel system generation in the Newton algorithm is nearly hidden behind the time for the linear solver. The DD-method turned out to be a good basis for the coupling of Finite Element and Boundary Element methods.

References

[1] J. H. Bramble and J. E. Pasciak. A preconditioning technique for indefinite systems resulting from mixed approximation of elliptic problems. *Math. Comput.*, 50(181):1–17, 1988.

[2] J. H. Bramble, J. E. Pasciak, and A. H. Schatz. The construction of preconditioners for elliptic problems by substructuring I – IV. *Mathematics of Computation*, 1986, 1987, 1988, 1989. 47, 103–134, 49, 1–16, 51, 415–430, 53, 1–24.

[3] J. H. Bramble, J. E. Pasciak, and J. Xu. Parallel multilevel preconditioners. *Mathematics of Computation*, 55(191):1 – 22, 1990.

[4] M. Costabel. Symmetric methods for the coupling of finite elements and boundary elements. In C. A. Brebbia, W. L. Wendland, and G. Kuhn, editors, *Boundary Elements IX*, pages 411–420. Springer–Verlag, 1987.

[5] R. S. Dembo, S. C. Eisenstat, and T. Steihaug. Inexact Newton methods. *SIAM J. Numer. Anal.*, 19:400–408, 1982.

[6] P. Deuflhard. Global inexact Newton methods for very large scale nonlinear problems. *IMPACT of Computing in Science and Engineering*, 3:366–393, 1991.

[7] M. Dryja. A capacitance matrix method for Dirichlet problems on polygonal regions. *Numerische Mathematik*, 39(1):51–64, 1982.

[8] G. Haase, B. Heise, M. Jung, and M. Kuhn. FEM⊕BEM - a parallel solver for linear and nonlinear coupled FE/BE-equations. DFG-Schwerpunkt "Randelementmethoden" 94-16, University Stuttgart, 1994.

[9] G. Haase and U. Langer. On the use of multigrid preconditioners in the domain decomposition method. In W. Hackbusch, editor, *Parallel Algorithms for PDEs*, pages 101–110, Braunschweig, 1990. Vieweg. Proc. of the 6th GAMM–Seminar, Kiel, 1990.

[10] G. Haase, U. Langer, and A. Meyer. A new approach to the Dirichlet domain decomposition method. In S. Hengst, editor, *Proceedings of the "5-th Multigrid Seminar" held at Eberswalde, May 14-18, 1990*, pages 1–59, Berlin, 1990. Karl-Weierstrass-Institute, Academy of Sciences. Report-Nr. R-MATH-09/90.

[11] G. Haase, U. Langer, and A. Meyer. The approximate dirichlet domain decomposition method. Part I: An algebraic approach. Part II: Applications to 2nd-order elliptic boundary value problems. *Computing*, 47:137–151 (Part I), 153–167 (Part II), 1991.

[12] G. Haase, U. Langer, and A. Meyer. Domain decomposition preconditioners with inexact subdomain solvers. *J. of Num. Lin. Alg. with Appl.*, 1:27–42, 1992.

[13] G. Haase, U. Langer, and A. Meyer. Parallelisierung und Vorkonditionierung des CG-Verfahrens durch Gebietszerlegung. In *Parallele Algorithmen auf Transputersystemen*, Teubner-Scripten zur Numerik III, Stuttgart, 1992. Teubner. Tagungsbericht der GAMM-Tagung, 31. Mai- 1. Juni 1991, Heidelberg.

[14] G. Haase, U. Langer, A. Meyer, and S. V. Nepomnyaschikh. Hierarchical extension operators and local multigrid methods in domain decomposition preconditioners. *East-West J. Numer. Math.*, 2(3):173–193, 1994.

[15] W. Hackbusch. *Multi-Grid methods and applications*, volume 4 of *Springer Series in Computational Mathematics*. Springer–Verlag, Berlin, 1985.

[16] B. Heise. Multigrid–Newton methods for the calculation or electromagnetic fields. In G. Telschow, editor, *Third Multigrid Seminar, Biesenthal 1988*, pages 53–73, Berlin, 1989. Karl–Weierstrass–Institute, Academy of Sciences. Report R–MATH–03/89.

[17] B. Heise. Mehrgitter–Newton–Verfahren zur Berechnung nichtlinearer magnetischer Felder. Wissenschaftliche Schriftenreihe 4/1991, Technische Universität Chemnitz, 1991.

[18] B. Heise. *Mehrgitter-Newton-Verfahren zur Lösung elektromagnetischer Feldgleichungen.* PhD thesis, Technische Universität Chemnitz, 1991.

[19] B. Heise. Nonlinear field calculations with multigrid-newton methods. *IMPACT of Computing in Science and Engineering*, 5:75–110, 1993.

[20] B. Heise. Sensitivity analysis for nonlinear magnetic field simulation. In H. Bandemer, editor, *Modelling uncertain data*, volume 68 of *Mathematical Research*. Akademie-Verlag, 1993. Proceedings of the GAMM-Workshop in Freiberg, March 21–24, 1992.

[21] B. Heise. Analysis of a fully discrete finite element method for a nonlinear magnetic field problem. *SIAM J. Numer. Anal.*, 31(3):745–759, 1994.

[22] B. Heise. Nonlinear magnetic field simulation with FE/BE domain decomposition methods on MIMD parallel computers. DFG-Schwerpunkt "Randelementmethoden" 94-19, University Stuttgart, 1994.

[23] B. Heise. Nonlinear simulation of electro-magnetic fields with domain decomposition methods on MIMD parallel computers. 1994. Proceedings of MODELLING 94, Prague, August 29 - September 2, 1994; to appear.

[24] B. Heise. Nonlinear field simulation with FE domain decomposition methods on massively parallel computers. 1995. In preparation.

[25] B. Heise and M. Kuhn. Parallel solvers for linear and nonlinear exterior magnetic field problems based upon coupled FE/BE formulations. Inst. of Mathematics, Report 486, Johannes Kepler University Linz, 1995. Submitted to: Proceedings of the GAMM-Workshop on Multilevel Methods, Schloss Meisdorf, September 28 - October 1, 1994.

[26] G. C. Hsiao and W. Wendland. Domain decomposition in boundary element methods. In *Proc. of IV Int. Symposium on Domain Decomposition Methods, (R. Glowinski, Y. A. Kuznetsov, G. Meurant, J. Pérraux, O. B. Widlund eds.), Moscow, May 1990*, pages 41–49, Philadelphia, 1991. SIAM Publ.

[27] M. Kuhn. A parallel solver for exterior problems based on a coupled FE/BE-model. DFG-Schwerpunkt "Randelementmethoden" 94-15, University Stuttgart, 1994.

[28] M. Kuhn. Parallele Lösungsstrategien zur Lösung gekoppelter FE/BE Gleichungen für nichtlineare Magnetfeldprobleme. Diplomarbeit, Technische Universität Chemnitz-Zwickau, 1994.

[29] U. Langer. Parallel iterative solution of symmetric coupled fe/be- equations via domain decomposition. Preprint 217, TU Chemnitz, 1992. Report No. 92–2, DFG-Schwerpunkt "Randelementmethoden".

[30] U. Langer. Parallel iterative solution of symmetric coupled fe/be- equations via domain decomposition. *Contemporary mathematics*, 157:335–344, 1994.

[31] A. Meyer. A parallel preconditioned conjugate gradient method using domain decomposition and inexact solvers on each subdomain. *Computing*, 45:217–234, 1990.

[32] S. Rjasanow. Vorkonditionierte iterative Auflösung von Randelementgleichungen für die Dirichlet–Aufgabe. Wissenschaftliche Schriftenreihe 7, Technische Universität Chemnitz, 1990.

[33] O. Steinbach. Preconditioned iterative solvers for boundary element and domain decomposition methods. Talk given at Oberwolfach conference, October 2 – 8,, 1994.

[34] C. H. Tong, T. F. Chan, and C. J. Kuo. A domain decomposition preconditioner based on a change to a multilevel nodal basis. *SIAM J. Sci. Stat. Comput.*, 12(6):1486–1495, 1991.

[35] W. Wendland. On the coupling of finite elements and boundary elements. In G. Kuhn and H. Mang, editors, *Discretization methods in structural mechanics*, pages 405–414, Berlin, 1990. Springer–Verlag. Proc. IUTUM/IACM Symposium, Vienna, 1989.

COUPLING PROBLEMS IN MICROELECTRONIC DEVICE SIMULATION

R. HIPTMAIR, R.H.W. HOPPE and B. WOHLMUTH [1]

Abstract

The cost-effective design of electronic microstructures requires an advanced modeling and coupled simulation of various physical effects. The classical isothermal approach leads to the basic drift-diffusion model for semiconductor device simulation. In the stationary case, it represents a coupled nonlinear system consisting of a Poisson equation for the electric potential and two continuity equations for the electron and hole flow. We discuss various discretization schemes with special emphasis on mixed finite element methods and we further address efficient numerical solution techniques including adaptive multilevel methods. Finally, to allow for ambient conditions such as external magnetic fields we consider consistent extensions of the classical model and discuss perspectives for their numerical treatment.

1. Introduction

The numerical simulation of microelectronic devices is an integral part of the Computer Aided Design of integrated microsystems. It consists of three main stages: The first stage is concerned with the simulation of the fabrication processes and comprises structure defining processing steps such as lithography, etching, decomposition and thermal oxidation as well as the processing steps of ion implantation and dopand diffusion (cf. e.g. [23]).
The second stage deals with the simulation of the physical behavior of the devices within a unified framework which takes into account not only the electrical behavior but also thermal, magnetic and optical effects depending on what the device is aimed for (cf. e.g. [30], [31]).
The third stage represents the simulation of the global behavior of networks of microstructures under operating conditions of interest (cf. e.g. [19]).

In this paper, the emphasis is on concepts for microelectronic device simulation allowing an efficient and reliable numerical solution of the underlying physical equations. A common feature of the physical modeling is the coupled simulation of carrier and energy transport and the proper incorporation of ambient conditions such as electromagnetic radiation fields (power laser diodes) or galvanomagnetic efforts in the presence of an external quasi-static magnetic field (magnetotransistors).

It has to be noted that in the modeling community there is a controversial debate what such advanced device models should be based on. One common approach uses the so-called hydrodynamic model resulting from a momentum expansion of the Boltzmann equation truncated after the second-order moments (cf. e.g. [20]). Another approach, advocated for example by Wachutka [30], uses the principles of irreversible phenomenological thermodynamics to provide an extended thermodynamic model as a comprehensive framework for a unified treatment which allows a tailored modeling of specific microelectronic devices. The model is such that in the absence of mechanical, thermal, magnetic and optical effects it reduces to the description of the interaction between the electric field

[1] e-mail: hoppe@math.uni-augsburg.de, address: Math.-Nat. Fakultät der Universität Augsburg, Universitätsstr. 14, D–86 159 Augsburg. The first and third author were supported by FORTWIHR, Bavarian Consortium for High Performance Scientific Computing.

and the particle transport as given by the classical drift-diffusion model in semiconductor device simulation. In this sense the drift-diffusion model represents, so-to-say, the kernel of the extended model and any numerical approach for the extended model has to rely on an efficient solution of the drift-diffusion equations.

In this paper, we will advocate the use of mixed finite elements with special emphasis on a technique called mixed hybridization allowing the simultaneous computation of the current densities and the carrier concentrations. We will also indicate its relationship to other widely used discretization techniques with the celebrated Scharfetter-Gummel scheme at prominent place. Further, we will present tools for adaptive grid generation within a mixed finite element approach and give some numerical results for a MOSFET. Finally, we will shortly address the numerical treatment of an extended model incorporating external magnetic fields.

2. Mixed finite element discretization of the drift-diffusion model

In the stationary case, the classical drift-diffusion model describing the distribution of the electron and hole densities n and p under the influence of an electric field consists of two continuity equations for the current densities $\mathbf{j}_n$ and $\mathbf{j}_p$ coupled with a Poisson equation for the electric potential Ψ.

$$\begin{aligned} -\epsilon\Delta\Psi &= q(D+n-p), \\ \operatorname{div}\mathbf{j}_n &= qR(n,p)+f_1, \\ \operatorname{div}\mathbf{j}_p &= -qR(n,p)+f_2 \end{aligned} \tag{2.1}$$

where

$$\begin{aligned} \mathbf{j}_n &= q\mu_n(U_T\operatorname{grad} n - n\operatorname{grad}\Psi), \\ \mathbf{j}_p &= -q\mu_p(U_T\operatorname{grad} p + p\operatorname{grad}\Psi). \end{aligned}$$

Here, D stands for the dopand profile, $R(n,p)$ for the recombination/generation rate and f_1, f_2 are the source terms. Further, ϵ, q, U_T and μ_n, μ_p denote the dielectric permittivity, the elementary charge, the thermal voltage and the mobilities, respectively. We consider (2.1) in a bounded polyhedral domain $\Omega \subset \mathbb{R}^3$ with boundary conditions of Dirichlet type at the Ohmic contacts Γ_D and of homogeneous Neumann type at the isolating part Γ_N of the device.

$$\begin{aligned} &\Psi = \Psi_D, \quad n = n_D, \quad p = p_D \quad \text{on } \Gamma_D, \\ &\tfrac{\partial\Psi}{\partial\nu} = 0, \quad \nu\cdot\mathbf{j}_n = 0, \quad \nu\cdot\mathbf{j}_p = 0 \quad \text{on } \Gamma_N. \end{aligned} \tag{2.2}$$

Typically, the dopand profile D is strongly varying causing a pronounced layer behavior of the electric potential Ψ. Consequently, we may expect local areas with significant change in magnitude of the gradient of Ψ. In view of the constitutive relations for the current densities, the convective transport is dominant in such areas so that the continuity equations represent convection-dominated diffusion-convection problems. It is well known that the numerical solution of such problems requires utmost care in the discretization process. Apart from upwinding techniques for methods based on the primal variational approach (cf. e.g. [4]), finite volume techniques are widely used, because they are based on the principle of current conservation. Related techniques are provided by mixed finite element discretizations allowing the simultaneous computation of the current densities and the carrier concentrations. For simplicity, we consider the continuity equation for only one minority carrier in the following scaled and linearized form

$$\operatorname{div}(\exp(\Psi)\operatorname{grad} u) - \exp(\Psi)Ru = f$$

where we have introduced the Slotboom variable $u = \exp(-\Psi)n$ and R stands for the differential net recombination/generation rate. The mixed approach is based on the dual formulation of the problem which can be obtained by introducing the current density $\mathbf{j} = \exp(\Psi)\operatorname{grad} u$ as an additional unknown. In this way, the second order elliptic partial differential equation can be formally written as a first order system. In contrast to the primal approach we rely on the weak formulation of the relationship between he current density $\mathbf{j}$ and the Slotboom variable u whereas the continuity equation is treated in the strong L^2-sense. This gives rise to a system of two variational equations:

Find $(\mathbf{j}, u) \in H_{\Gamma_N}(\operatorname{div};\Omega) \times L^2(\Omega)$ such that

$$\begin{aligned} \int_\Omega \exp(-\Psi)\mathbf{j}\cdot\mathbf{q}\,dx + \int_\Omega \operatorname{div}\mathbf{q}\,u\,dx &= \int_{\Gamma_D} \exp(-\Psi)n_D\nu\cdot\mathbf{q}\,d\sigma, \ \mathbf{q} \in H_{\Gamma_N}(\operatorname{div};\Omega) \\ \int_\Omega \operatorname{div}\mathbf{j}\,v\,dx - \int_\Omega \exp(\Psi)Ruv\,dx &= \int_\Omega fv\,dx, \ v \in L^2(\Omega) \end{aligned} \tag{2.3}$$

where $H_{\Gamma_N}(\operatorname{div};\Omega) = \{\mathbf{q} \in (L^2(\Omega))^3 \mid \operatorname{div}\mathbf{q} \in L^2(\Omega),\ \nu\cdot\mathbf{q}|_{\Gamma_N} = 0\}$.

In the following we will focus on an approach called dual or mixed hybridization which takes advantage of ideas from domain decomposition and can be carried out both in the continuous and discrete setting of the problem. We will concentrate on the discrete setting and refer to [16] for a treatment of the continuous case. For that purpose we consider the standard mixed discretization by means of the lowest order Raviart-Thomas approximation. Given a simplicial triangulation $\mathcal{T}_h$ of Ω, the Slotboom variable u will be approximated by piecewise constants resulting in the ansatz space

$$W_0(\Omega;\mathcal{T}_h) = \{v_h \in L^2(\Omega) \mid v_h|_K \in P_0(K),\ K \in \mathcal{T}_h\}$$

where $P_k(K)$, $k \geq 0$, denotes the polynomials of degree $\leq k$ on K. An associated approximation of the dual variable $\mathbf{j}$ can be obtained by requiring the divergence of the discrete flux to be piecewise constant, too. This requirement leads in a natural way to the lowest order Raviart-Thomas element $RT_0(K) = P_0(K)^3 + \mathbf{x}P_0(K)$, $\mathbf{x} = (x_1, x_2, x_3)^T$. We note that any vector field from $RT_0(K)$ is uniquely determined by 4 degrees of freedom which can be chosen as the normal component of the flux in the center of gravity of the faces of K. Setting

$$RT_{0;\Gamma_N}(\Omega;\mathcal{T}_h) = \{\mathbf{q}_h \in H_{\Gamma_N}(\operatorname{div};\Omega) \mid \mathbf{q}_h|_K \in RT_0(K),\ K \in \mathcal{T}_h\},$$

the mixed discretization requires the computation of a pair $(\mathbf{j}_h, u_h) \in RT_{0;\Gamma_N}(\Omega;\mathcal{T}_h) \times W_0(\Omega;\mathcal{T}_h)$ as the solution of (2.3) when restricted to $RT_{0;\Gamma_N}(\Omega;\mathcal{T}_h)$ and $W_0(\Omega;\mathcal{T}_h)$, respectively. The concept of mixed hybridization originally due to Fraeijs de Veubeke [11] (cf. also [7]) is to decompose the saddle point problem arising from the standard mixed approach into less bulky pieces by considering the "broken" Raviart-Thomas ansatz space

$$RT_{0;\Gamma_N}^{-1}(\Omega;\mathcal{T}_h) = \{\mathbf{q}_h \in \prod_{K\in\mathcal{T}_h} RT_0(K) \mid \nu\cdot\mathbf{q}_h|_{\Gamma_N} = 0\}.$$

We note that $RT_{0;\Gamma_N}^{-1}(\Omega;\mathcal{T}_h)$ is not a subspace of $H_{\Gamma_N}(\operatorname{div};\Omega)$, since the continuity of the normal components $\nu\cdot\mathbf{q}_h$ of vector fields $\mathbf{q}_h \in RT_{0;\Gamma_N}^{-1}(\Omega;\mathcal{T}_h)$ across the interelement boundaries of the triangulation is no longer guaranteed. Instead, the continuity constraints are taken care of by appropriate Lagrangian multipliers. Denoting by $\mathcal{F}_h$ the set of faces of $\mathcal{T}_h$ an appropriate multiplier space is given by

$$M_{0;\Gamma_D}(\Omega;\mathcal{T}_h) = \{\mu_h \in L^2(\mathcal{F}_h) \mid \mu_h|_F \in P_0(F),\ F \in \mathcal{F}_h,\ \mu_h|_F = 0,\ F \in \mathcal{F}_h \cap \Gamma_D\}.$$

Then, in terms of the original primal variable n_h the mixed hybridization requires the computation of a triple $(\mathbf{j}_h, n_h, \lambda_h) \in RT^{-1}_{0;\Gamma_N}(\Omega; \mathcal{T}_h) \times W_0(\Omega; \mathcal{T}_h) \times M_{0;\Gamma_D}(\Omega; \mathcal{T}_h)$ such that

$$\begin{aligned}
&\sum_{K\in\mathcal{T}_h} \int_K \exp(-\Psi)\mathbf{j}_h \cdot \mathbf{q}_h \, dx + \sum_{K\in\mathcal{T}_h} \int_K \operatorname{div} \mathbf{q}_h \exp(-\Psi) n_h \, dx \\
&- \sum_{K\in\mathcal{T}_h} \int_{\partial K} \exp(-\Psi)\lambda_h \nu \cdot \mathbf{q}_h \, d\sigma = \sum_{K\in\mathcal{T}_h} \int_{\partial K\cap\Gamma_D} \exp(-\Psi) n_D \nu \cdot \mathbf{q}_h \, d\sigma, \ \mathbf{q}_h \in RT^{-1}_{0;\Gamma_N}(\Omega; \mathcal{T}_h), \\
&\sum_{K\in\mathcal{T}_h} \int_K \operatorname{div} \mathbf{j}_h v_h \, dx - \sum_{K\in\mathcal{T}_h} \int_K R n_h v_h \, dx = \int_\Omega f v_h \, dx, \ v_h \in W_0(\Omega; \mathcal{T}_h), \\
&\sum_{K\in\mathcal{T}_h} \int_{\partial K} \mu_h \nu \cdot \mathbf{j}_h \, d\sigma = 0, \ \mu_h \in M_{0;\Gamma_D}(\Omega; \mathcal{T}_h).
\end{aligned} \tag{2.4}$$

The purely local character of the bases of $RT^{-1}_{0;\Gamma_N}(\Omega; \mathcal{T}_h)$ and $W_0(\Omega; \mathcal{T}_h)$ strongly suggests static condensation of both the discrete flux $\mathbf{j}_h$ and the discrete carrier concentration n_h resulting in the Schur complement system

$$S_h \lambda_h = t_h. \tag{2.5}$$

Theorem 2.1 *The Schur complement S_h can be assembled from its local contributions S_h^K, $K \in \mathcal{T}_h$, resulting from static condensation on the elements' level. For interior elements $K \in \mathcal{T}_h \cap \Omega$ we obtain*

$$S_h^K = -D_K^1(\alpha_K^{-1} U_K + \beta_K T_K) D_K^2 \tag{2.6}$$

where D_K^1, D_K^2, U_K and T_K are the 4×4 matrices

$$\begin{aligned}
&D_K^1 = (|F_i|\delta_{ij})_{i,j=1}^4, \ D_K^2 = \left(\left(\int_{F_i} \exp(-\Psi) \, d\sigma\right) \delta_{ij} \right)_{i,j=1}^4, \ U_K = (\nu_i \cdot \nu_j)_{i,j=1}^4 \\
&T_K = \eta_K \eta_K^T, \ \eta_{K,i} = \mathbf{p} \cdot \nu_i - \alpha_K^{-1} \int_K \exp(-\Psi) \mathbf{x} \cdot \nu_i \, dx, \ \mathbf{p} \in F_i, \ 1 \le i \le 4
\end{aligned}$$

and the constants α_K, β_K are given by

$$\begin{aligned}
&\alpha_K = \int_K \exp(-\Psi) \, dx, \ \beta_K = (\gamma_K \int_K R \, dx + 9\alpha_K |K|)^{-1} \int_K R \, dx, \\
&\gamma_K = \int_K \exp(-\Psi) \xi_K \cdot \xi_K \, dx, \ \xi_K(x) = \mathbf{x} - \alpha_K^{-1} \int_K \exp(-\Psi) \mathbf{x}' \, dx'.
\end{aligned}$$

For elements $K \in \mathcal{T}_h$ attached to the boundary Γ_D obvious modifications apply.

Proof: Since the Schur complement does not depend on the choice of the bases in $RT^{-1}_{0;\Gamma_N}(\Omega; \mathcal{T}_h)$ and $W_0(\Omega; \mathcal{T}_h)$, we pick a local basis $\mathbf{q}_K^{(i)}$, $1 \le i \le 4$, of the discrete flux space according to $\mathbf{q}_K^{(i)} = \mathbf{e}_i$, $1 \le i \le 3$, $\mathbf{q}_K^{(4)} = \xi_K$ where $\mathbf{e}_i$ stands for the i-th canonical basis vector of $\mathbb{R}^3$. Observing that the $\mathbf{q}_K^{(i)}$ are orthogonal with respect to the inner product $(\mathbf{p}_K, \mathbf{q}_K) = \int_K \exp(-\Psi) \mathbf{p}_K \cdot \mathbf{q}_K \, dx$ and that $\operatorname{div} \mathbf{q}_K^{(i)} = 0$, $1 \le i \le 3$, $\operatorname{div} \mathbf{q}_K^{(4)} = 3$, the representation (2.6) follows from straightforward, but tedious computations. □

It can be immediately seen from (2.6) that the matrix $-S_h^K$ is an M-matrix provided the element K has no obtuse angle. Obviously, this property carries over to the global Schur complement $-S_h$.

Following the same line of arguments as in the previous theorem, the right hand side of the Schur complement system can be similarly assembled from local load vectors t_h^K, $K \in \mathcal{T}_h$.

Moreover, the inhomogeneous Dirichlet data can be incorporated into the multiplier by setting $\lambda_h|_F = (\int_F \exp(-\Psi)\, d\sigma)^{-1} \int_F \exp(-\Psi) n_D \, d\sigma$, $F \in \mathcal{F}_h \cap \Gamma_D$.

Though in the process of mixed hybridization we have discarded the discrete current densities $\mathbf{j}_h$ and the discrete carrier concentrations n_D, they can be easily retrieved from the values of the Lagrangian multiplier λ_h. In particular, denoting by $\mathbf{j}_h^K$ the vector of fluxes through the faces of $K \in \mathcal{T}_h$ we find that $\mathbf{j}_h^K$ is readily available as the residual of the local Schur complement equation:

$$\mathbf{j}_h^K = t_h^K - S_h^K \lambda_h^K, \quad K \in \mathcal{T}_h.$$

Note that the residual in general does not vanish, since the actual value of λ_h results from the solution of the global Schur complement system and thus inherits contributions from the other elements.

An important implementational aspect of mixed hybridization - in particular with respect to the construction of iterative solvers - is that the Schur complement system can be shown to be equivalent to a nonconforming Petrov-Galerkin scheme of inverse average type. That equivalence has been observed, among others, by Brezzi, Marini and Pietra [8], [9] and has been exploited, for example, by Bachmann [2] and Reusken [24] in a multigrid approach to the continuity equation. To be more precise, let us denote by Ψ_K^i, $1 \le i \le 4$, the local canonical basis of the lowest order nonconforming Crouzeix-Raviart element $CR_1(K)$ and by $\tilde{\Psi}_K^i$ the scaled basis functions

$$\tilde{\Psi}_K^i = s_i \Psi_K^i, \quad s_i = |F_i|^{-1} \int_{F_i} \exp(-\Psi)\, d\sigma, \; 1 \le i \le 4$$

spanning the scaled element $\widetilde{CR}_1(K)$. We choose

$$\begin{aligned} \widetilde{CR}_{1;\Gamma_D}(\Omega; \mathcal{T}_h) \;&=\; \{v_h \in L^2(\Omega) \mid v_h|_K \in \widetilde{CR}_1(K),\, K \in \mathcal{T}_h, \\ &\qquad v_h|_{F_i} = s_i^{-1} \exp(-\Psi) n_D,\, F_i \in \mathcal{F}_h \cap \Gamma_D\} \end{aligned}$$

as the ansatz space and

$$CR_{1;\Gamma_D}^0(\Omega; \mathcal{T}_h) = \{v_h \in L^2(\Omega) \mid v_h|_K \in CR_1(K),\, K \in \mathcal{T}_h,\, v_h|_{\Gamma_D} = 0\}$$

as the test space in a modified nonconforming Petrov-Galerkin approach to the primal variational formulation of the continuity equation where the modification consists in replacing the coefficient function $\exp(\Psi)|_K$ by its harmonic average over an individual element $K \in \mathcal{T}_h$. Then, the following result can be established where for notational convenience we only consider the case $R = 0$.

Theorem 2.2 *Let $u_h \in \widetilde{CR}_{1;\Gamma_D}(\Omega; \mathcal{T}_h)$ be the solution of the nonconforming Petrov-Galerkin scheme*

$$a_h(u_h, v_h) = l_h(v_h), \quad v_h \in CR_{1;\Gamma_D}^0(\Omega; \mathcal{T}_h) \tag{2.7}$$

where

$$\begin{aligned} a_h(u_h, v_h) \;&=\; \sum_{K \in \mathcal{T}_h} \int_K \left(|K|^{-1} \int_K \exp(-\Psi)\, dx \right)^{-1} \operatorname{grad} u_h \cdot \operatorname{grad} v_h \, dx, \\ l_h(v_h) \;&=\; \sum_{K \in \mathcal{T}_h} (3|K|)^{-1} \int_K f\, dx \int_K \left(\exp(-\Psi) \left(|K|^{-1} \int_K \exp(-\Psi)\, dx' \right)^{-1} - 4 \right) v_h \, dx \end{aligned}$$

Then, if λ_h is the solution of the Schur complement system (2.5) in case $R = 0$ and Π_h denotes the L^2-projection onto the multiplier space, we have $\lambda_h = \Pi_h u_h$.

Proof: The proof is the 3D analogue of that one given in [16; Thm. 4.2] in the 2D case (cf. also [15; Satz 3.2.2]). □

As said before, the equivalence with the nonconforming Petrov-Galerkin scheme (2.7) can be used by taking advantage of efficient multilevel iterative solvers designed for nonconforming finite element discretizations (cf. e.g. [6]). In the sequel, we will not follow this aspect but instead elaborate on the relationship between mixed methods and the Scharfetter-Gummel box method widely used in device simulation.

3. Mixed discretization and the Scharfetter-Gummel scheme

It is well known that mixed finite element methods are closely related to finite volume techniques. Such box methods are also appropriate candidates for the discretization of the drift-diffusion model, since they are based on the principle of current conservation. The most prominent box method in device simulation is the Scharfetter-Gummel scheme which is used in many existent program packages (cf. e.g. [10], [27], [28]). In order to establish the relationship between mixed methods and the Scharfetter-Gummel scheme, let us consider the mixed discretization of the continuity equation by means of the lowest order Raviart-Thomas approximation with respect to a hexahedral triangulation. In this case, the Raviart-Thomas element $RT_{[0]}(K)$, $K \in \mathcal{T}_h$, is given by

$$RT_{[0]}(K) = Q_{1,0,0}(K) \times Q_{0,1,0}(K) \times Q_{0,0,1}(K)$$

where $Q_{\nu_1,\nu_2,\nu_3}(K) = \{p \mid p(x) = \sum_{|\alpha_i| \le \nu_i} \gamma_\alpha x_1^{\alpha_1} x_2^{\alpha_2} x_3^{\alpha_3},\ \mathbf{x} \in K\}$. Any vector field from $RT_{[0]}(K)$ is uniquely determined by six degrees of freedom which are again chosen as the normal components in the centers of gravity of the faces of K. As for simplicial triangulations the primal variable is approximated by piecewise constants. Denoting by $RT_{[0];\Gamma_N}(\Omega; \mathcal{T}_h)$ and $W_0(\Omega; \mathcal{T}_h)$ the associated global ansatz spaces, we have to compute $(\mathbf{j}_h, n_h) \in RT_{[0];\Gamma_N}(\Omega; \mathcal{T}_h) \times W_0(\Omega; \mathcal{T}_h)$ as the solution of

$$\begin{aligned} &\int_\Omega \exp(-\Psi)\mathbf{j}_h \cdot \mathbf{q}_h\, dx + \int_\Omega \exp(-\Psi)\operatorname{div} \mathbf{q}_h\, n_h\, dx = \int_{\Gamma_D} \exp(-\Psi) n_D \nu \cdot \mathbf{q}_h\, d\sigma, \\ &\int_\Omega \operatorname{div} \mathbf{j}_h\, v_h\, dx - \int_\Omega R n_h v_h\, dx = \int_\Omega f v_h\, dx, \quad v_h \in W_0(\Omega; \mathcal{T}_h),\ \mathbf{q}_h \in RT_{[0];\Gamma_N}(\Omega; \mathcal{T}_h). \end{aligned} \tag{3.8}$$

By regular refinement we subdivide each element $K \in \mathcal{T}_h$ with vertices P_ν, $1 \le \nu \le 8$, into eight subelements K_ν, $1 \le \nu \le 8$, and evaluate the integrals $\int_K \exp(-\Psi)\mathbf{j}_h \cdot \mathbf{q}_h\, dx$ in (3.8) by means of the quadrature formula

$$\sum_{\nu=1}^{8} (\mathbf{p}_h \cdot \mathbf{q}_h)(P_\nu) \int_K \exp(-\Psi)\, dx. \tag{3.9}$$

Theorem 3.3 *The mixed method (3.8) in combination with the quadrature formula (3.9) is equivalent to the Scharfetter-Gummel discretization of the continuity equation.*

Proof: Considering the 2×2 block coefficient matrix which represents the algebraic formulation of (3.8), the first diagonal block obviously reduces to a diagonal matrix when using (3.9). Then, the equivalence can be easily established by static condensation of the discrete fluxes. □

Another finite volume technique that has been proposed for device simulation is the Baliga-Patankar scheme (cf. e.g. [3], [29]). However, as has been shown in [21], small carrier concentrations cannot be computed in a numerically stable way due to the influence of rounding-off errors.

4. Adaptive grid refinement

As we have already pointed out, for most devices the dopand profile is such that we are faced with a pronounced layer behavior of both the electric potential and the carrier concentrations. An efficient resolution of such layers requires the use of highly nonuniform grids which can be generated by means of adaptive grid refinement. For mixed finite element discretizations of the drift-diffusion model, more or less heuristic criteria for grid refinement based on the gradient of the solutions have been used by Hemker and Molenaar [14] and by van Nooyen [22]. However, for mixed hybridization an efficient and reliable a posteriori error estimator is at hand which can be implemented very cheaply (cf. [17], [18]). This estimator is motivated by a superconvergence result for mixed hybridization due to Arnold and Brezzi [1] which states that under the same regularity assumptions the nonconforming extension $\hat{n}_h$ of the Lagrangian multiplier λ_h from the mixed hybrid approach approximates the primal variable n of a higher order than does n_h. This superconvergence result supports the following saturation assumption

$$\|n - \hat{n}_h\|_0 \leq \beta \|n - n_h\|_0, \quad 0 \leq \beta < 1 \tag{4.10}$$

where $\|\cdot\|_0$ stands for the standard L^2-norm. In practice, instead of λ_h and n_h we only have some iterative approximations $\tilde{\lambda}_h$ and $\tilde{n}_h$ at hand. If we denote by $\hat{\tilde{n}}_h$ the nonconforming extension of $\tilde{\lambda}_h$ and further assume

$$\|n - n_h\|_0 \leq \sigma \|n - \tilde{n}_h\|_0, \quad 0 \leq \sigma, \ \beta\sigma < 1, \tag{4.11}$$

then from (4.10) and (4.11) we can easily deduce the following two-sided estimate of the total error $\|n - \tilde{n}_h\|_0$:

$$\begin{aligned} \|n - \tilde{n}_h\|_0 &\geq (1 + \beta\sigma)^{-1} \left(\|\hat{\tilde{n}}_h - \tilde{n}_h\|_0 - C\|n_h - \tilde{n}_h\|_0 \right), \\ \|n - \tilde{n}_h\|_0 &\leq (1 - \beta\sigma)^{-1} \left(\|\hat{\tilde{n}}_h - \tilde{n}_h\|_0 + C\|n_h - \tilde{n}_h\|_0 \right). \end{aligned} \tag{4.12}$$

For adaptive grid refinement we use the easily computable elementwise contributions of $\|\hat{\tilde{n}}_h - \tilde{n}_h\|_0$, determine the mean value and mark an element for refinement, if the local contribution exceeds the mean value by a certain problem-dependent margin. The refinement process itself is performed according to the strategy proposed by Bornemann et al. [5].

5. Numerical results for a MOSFET

For an illustration of the application of adaptive grid refinement in device simulation let us consider the results of the mixed hybrid approach to the drift-diffusion model in case of a MOSFET.

As shown in Figure 1, the MOSFET is a three layer device consisting of two highly doped n regions equipped with the source and drain contacts and a lower doped p region called bulk. Between source and drain the semiconductor material is coated with a thin oxide layer equipped with the gate contact. If a sufficiently high voltage is applied to the gate, an inversion layer is created in the bulk region close to the interface between the silicon and the oxide. In case of a potential difference between source and drain there is a current from source to drain in the n-channel whose magnitude can be controlled by the voltage applied at the gate.

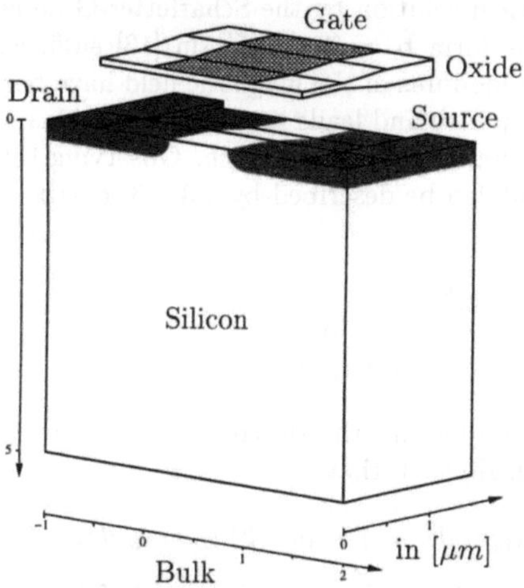

Fig. 1: Typical geometry of a MOSFET

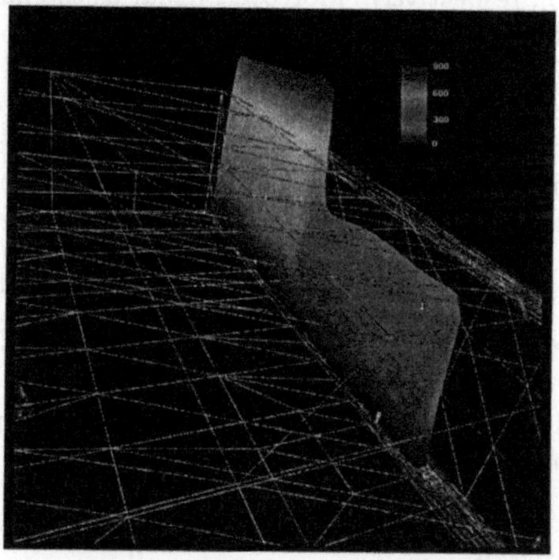

Fig. 2: Scaled electron density in a vicinity of an n-channel MOSFET

The drift-diffusion equations have been solved by an approximate Newton method, namely the well-known Gummel iteration. The resulting linearized problems have been treated by mixed hybridization with respect to an adaptively generated hierarchy of simplicial triangulations along the lines described in the previous sections.
Figure 2 represents a visualization of the electron density and the underlying triangulation in a vicinity of the n-channel. The layer behavior is clearly visible as it is the pinch-off point close to the drain region. Moreover, we observe that the layer is neatly resolved by the adaptively generated triangulation.

6. Extended model for galvanomagnetic carrier transport

Integrated bipolar magnetotransistors are semiconductor devices that are used as magnetosensors converting a magnetic field into an electronic signal. The magnetic field sensitivity is caused by galvanomagnetic phenomena which have to be modeled on the basis of the galvanomagnetic carrier transport in semiconductors. The physical modeling of the transport equations describing the flow of the electrons and holes has to incorporate the deflection of the carriers by the Lorentz force in the interior of the device (cf. e.g. [25], [26]). In particular, the constitutive equations for the electron and hole current densities must be augmented by appropriate magnetic field dependent terms. Denoting by $\mathbf{j}_\alpha^0$, $\alpha = n$ or $\alpha = p$, the current densities without magnetic field, by $\mathbf{b}$ the magnetic field vector and by μ_α^* the Hall mobilities, we use

$$\mathbf{j}_\alpha = \mathbf{j}_\alpha^0 + \left(1 + (\mu_\alpha^* |\mathbf{b}|)^2\right)^{-1} \left(\mu_\alpha^* \mathbf{b} \times \mathbf{j}_\alpha^0 + \mu_\alpha^* \mathbf{b} \times (\mu_\alpha^* \mathbf{b} \times \mathbf{j}_\alpha^0)\right). \tag{6.13}$$

The extended model then consists of the system (2.1) with $\mathbf{j}_n$, $\mathbf{j}_p$ given by (6.13). Further, the boundary conditions (2.2) have to be modified slightly for taking into account the action of the magnetic field on the device (see e.g. [25] for details).

The existence and uniqueness of weak solutions and the stability of the thermal equilibrium have been established by Gajewski and Gröger [12] and by Gajewski and Gärtner [13], respectively. Moreover, as far as the numerical solution by the Scharfetter-Gummel scheme is concerned, for a magnetic field of the form $\mathbf{b} = (0,0,b_z)^T$, in [13] sufficient conditions in terms of the grid geometry and the modulus of the magnetic field have been derived guaranteeing that the scheme is still dissipative and leads to positive solutions. A similar analysis can be carried out for the mixed finite element approach. Observing that for $\mathbf{b} = (0,0,b_z)^T$ the action of the magnetic field can be described by a 3×3 matrix B_α according to $\mathbf{j}_\alpha = B_\alpha \mathbf{j}_\alpha^0$,

$$B_\alpha = \left(1 + (\mu_\alpha^* b_z)^2\right)^{-1} \begin{pmatrix} 1 & -\mu_\alpha^* b_z & 0 \\ \mu_\alpha^* b_z & 1 & 0 \\ 0 & 0 & 1 + (\mu_\alpha^* b_z)^2 \end{pmatrix},$$

the mixed discretization of the continuity equation for e.g. the electron flux requires the computation of $(\mathbf{j}_h, n_h) \in RT_{0;\Gamma_N}(\Omega; \mathcal{T}_h) \times W_0(\Omega; \mathcal{T}_h)$ such that

$$\int_\Omega \exp(-\Psi) B_n^{-1} \mathbf{j}_h \cdot \mathbf{q}_h \, dx + \int_\Omega \exp(-\Psi) \operatorname{div} \mathbf{q}_h n_h \, dx = \int_{\Gamma_D} \exp(-\Psi) n_D \nu \cdot \mathbf{q}_h \, d\sigma,$$
$$\int_\Omega \operatorname{div} \mathbf{j}_h v_h \, dx - \int_\Omega R n_h v_h \, dx = \int_\Omega f v_h \, dx, \quad v_h \in W_0(\Omega; \mathcal{T}_h), \ \mathbf{q}_h \in RT_{0;\Gamma_N}(\Omega; \mathcal{T}_h).$$

Then, if we perform mixed hybridization, in the process of static condensation of the discrete current densities and the discrete carrier concentrations the local Schur complement can be explicitly computed in terms of the local grid geometry and the modulus of the magnetic field. This allows to state conditions under which the M-matrix property of $-S_h$ is retained. Details will be given in a forthcoming paper.

Acknowledgments: The authors wish to thank G. Wachutka for fruitful discussions on the subject.

REFERENCES

[1] D.N. Arnold and F. Brezzi, *Mixed and nonconforming finite element methods: implementation, post-processing and error estimates.* M^2AN Math. Modelling Numer. Anal. 19, 7-35 (1985).

[2] B. Bachmann, *Adaptive Mehrgitterverfahren zur Lösung der stationären Halbleitergleichungen.* Dissertation, Universität Zürich (1993).

[3] B.R. Baliga and S.V. Patankar, *A new finite-element formulation for convection-diffusion problems.* Numer. Heat Transfer 3, 393-409 (1980).

[4] R. Bank, J.F. Bürgler, W. Fichtner and R.K. Smith, *Some upwinding techniques for finite element approximations of convection-diffusion equations.* Numer. Math. 58, 185-202 (1990).

[5] F. Bornemann, B. Erdmann and Kornhuber, *Adaptive multilevel methods in three space dimensions.* Int. J. Numer. Methods Eng. 36, 3187-3203 (1993).

[6] D. Braess and R. Verfürth, *Multigrid methods for nonconforming finite element methods.* SIAM J. Numer. Anal. 27, No. 4, 979-986 (1990).

[7] F. Brezzi and M. Fortin, *Mixed and Hybrid Finite Element Methods.* Springer, Berlin–Heidelberg–New York, 1991.

[8] F. Brezzi, L.D. Marini and P. Pietra, *Two dimensional exponential fitting and application to drift-diffusion models.* SIAM J. Numer. Anal. 26, 1347-1355 (1989).

[9] F. Brezzi, L.D. Marini and P. Pietra, *Numerical simulation of semiconductor devices.* Comp. Math. Appl. Mech. Eng. 75, 493-514 (1989).

[10] J.F. Bürgler, *Discretization and Grid Adaption in Semiconductor Device Modeling.* Hartung-Gorre, Konstanz, 1990.

[11] B. Fraeijs de Veubeke, *Displacement and equilibrium models in the finite element method.* In: Stress Analysis, C. Zienkiewicz and G. Holister (eds.), John Wiley and Sons, New York, 1965.

[12] H. Gajewski and K. Gröger, *Semiconductor equations for mobilities based on Boltzmann statistics or Fermi-Dirac statistics.* Math. Nachr. 140, 7-36 (1989).

[13] H. Gajewski and K. Gärtner, *On the discretization of van Roosbroeck's equations with magnetic field.* ETH Zürich, Integrated Systems Laboratory, Techn. Rep. No. 94/14 (1994).

[14] P.W.H. Hemker and H. Molenaar, *A multigrid approach for the solution of the 2D semiconductor equations.* IMPACT Comput. Sci. Engrg. 2, 219-243 (1990).

[15] R. Hiptmair, *Gemischte Finite-Elemente-Diskretisierung der Kontinuitätsgleichungen aus der Bauelementsimulation.* Master Thesis, Technische Universität München, 1992.

[16] R. Hiptmair and R. H. W. Hoppe, *Mixed finite element discretization of continuity equations arising in semiconductor device simulation.* In: Proc. Conf. Math. Modelling and Simulation of Electrical Circuits and Devices, R.E. Bank et al. (eds.), p. 197-217 Birkhäuser, Basel, 1994.

[17] R.H.W. Hoppe and B. Wohlmuth, *Adaptive multilevel techniques for mixed finite element discretizations of elliptic boundary value problems.* Submitted to SIAM J. Numer. Anal..

[18] R.H.W. Hoppe and B. Wohlmuth, *Efficient Numerical Solution of Mixed Finite Element Discretizations by Adaptive Multilevel Methods.* to appear in Apl. Mat..

[19] W. Kampowsky, P. Rentrop and W. Schmidt, *Classification and numerical simulation of electric circuits.* Surv. Math. Ind. 2, 23-65 (1992).

[20] P.A. Markowich, Chr.A. Ringhofer and Chr. Schmeiser, *Semiconductor Equations.* Springer, Berlin-Heidelberg-New York, 1990.

[21] F. Montrone, *The method of Baliga-Patankar in 3D device simulation.* In: Proc. Conf. Math. Modelling and Simulation of Electrical Circuits and Devices, R.E. Bank et al. (eds.), p. 251-265 Birkhäuser, Basel, 1994.

[22] R.R.P. van Nooyen, *Some aspects of mixed finite element methods for semiconductor simulation.* Ph.D. Thesis, University of Amsterdam, 1992.

[23] M. Paffrath, W. Jacobs, W. Klein, E. Rank, K. Steger, U. Weinert and U. Wever, *Concepts and algorithms in process simulation.* Surv. Math. Ind. 3, 149-183 (1993).

[24] A. Reusken, *Multigrid applied to mixed finite elements schemes for current continuity equations.* Preprint, Technical University Eindhoven (1990).

[25] C. Riccobene, G. Wachutka, J.F. Bürgler and H. Baltes, *Two-dimensional numerical modeling of dual-collector magnetotransistors: evidence for emitter efficiency modulation.* Sensors and Actuators A, 31, 210-214 (1992).

[26] C. Riccobene, G. Wachutka, J.F. Bürgler and H. Baltes, *Operating principle of dual collector magnetotransistors studied by two-dimensional simulation.* IEEE Trans. Electron Devices ED-41, 32-43 (1994).

[27] D.L. Scharfetter and H.K. Gummel, *Large-signal analysis of a silicon read diode oscillator.* IEEE Trans Electron Devices ED-16, 64-77 (1969).

[28] S. Selberherr, *Analysis and Simulation of Semiconductor Devices.* Springer, Berlin-Heidelberg-New York, 1984.

[29] N. Shigyo, T. Wada and S. Yasuda, *Discretization problem for multidimensional current flow.* IEEE Trans. CAD, CAD-8, 1046-1050 (1989).

[30] G. Wachutka, *Unified framework fir thermal, electrical, magnetic and optical semiconductor device modeling.* COMPEL 10, 311-321 (1991).

[31] G. Wachutka, *Problem-oriented modeling of microtransducers: state of the art and future challenges.* Sensors and Actuators A, 41-42, 279-283 (1994).

Error Estimates of Galerkin FEM for a System of Coupled Helmholtz Equations in One Dimension

F. Ihlenburg[1] and Ch. Makridakis[2]

Summary

The quality of discrete solutions of the reduced wave equation depends on the relation between the physical parameters (frequency, material constants) and the numerical parameters (stepsize, order of polynomial approximation). For a one-dimensional model of fluid-solid interaction, we analyze continuous and discrete stability (inf-sup-condition) state an asymptotic error estimate for a Galerkin FEM (h-p-method). We then proceed to the discussion of preasymptotic error behavior. We show that the finite element error is generally polluted and investigate the dependence of the pollution term on the wavenumber k and on the material properties of different acoustic media in the coupled problem. In the numerical evaluation, we focus on the sensitivity of the FEM-solution to the material constants in the coupled problem. We conclude that, while there are no strictly 'forbidden frequencies' for the FE-FE-coupling approach, the numerical solution is generally ill-conditioned if the wavenumber is close to eigenvalues of the 'interior' problem.

1 Introduction

The classical problems of time-harmonic elastic wave scattering or vibrations of an elastic body embedded in an acoustic medium, resp., lead to coupled sets of Helmholtz equations. Usually, a Helmholtz elasticity equation describes the elastic body (*solid*) in the finite interior domain whereas the wave pattern in the acoustic medium (*fluid*) is determinded by a Helmholtz equation in an infinite exterior domain. At infinity, an absorbing boundary condition is given and transmission conditions are prescribed at the fluid-solid interface (*wet surface*).

A number of different methodologies have been proposed for the numerical solution of coupled Helmholtz problems, including application of BEM, FEM and IFEM (Infinite FEM) – see [14] for a recent overview. Several numerical approaches involve the introduction of an artificial boundary Γ_a at finite range from the solid, replacement of the radiation condition by an absorbing condition at Γ_a, and subsequent approximation of the coupled problem inside Γ_a by a finite element method. Thus, Abboud and Pinsky [15] formulate variational probems for fluid-solid interaction using a Robin boundary condition to damp out (partially, i.e. in first approximation) outgoing waves at Γ_a. Generally, a sufficiently accurate numerical approximation of an exact Dirichlet-to-Neumann condition must be used for wave absorption – see Keller and Givoli [7] for a most recent reference.

Besides the methodology discussed above, the solution of coupled Helmholtz equations by FEM is also part of other numerical approaches to the original finite-infinite problem, like: - FE-BE-coupling (see, e.g., Hsiao [9]); - FE-FE-IFE-coupling (see Burnett [4], Demkowicz/Gerdes [6]).

Due to the oscillatory character of the solutions, acoustic computations tend to be very costly. Hence there is strong interest in reliable error estimation which helps to prevent

[1]Institute for Physical Science and Technology, University of Maryland at College Park, College Park MD 20742, USA

[2]Department of Mathematics, University of Crete, 714 09 Heraklion, Greece

insufficient resolution of the solutions without leading to expensive solutions that overkill the reasonable tolerance. From the mathematical point of view, the problem is challenging since the sesquilinear form arsing from the Helmholtz equation is, generally, indefinite. Asesment of stability is not trivial, and careful analysis of the stability conditions, both for the continuous and for the discrete problem, is highly pertinent to the proof of precise a priori error estimates.

In this paper, we present the main results of a recent study [12] on a one-dimensional model problem for acoustic fluid-solid interaction. We then discuss the asymptotic error estimates on the background of recent work [9]–[11] on the error analysis of Galerkin FEM for the (uncoupled) Helmholtz equation. In the numerical evaluation we focus on new results on the frequency-dependent sensivity of the coupled solutions on the material properties of the fluid and solid, resp.

2 A Model Problem for Acoustic Fluid-Structure Interaction

Let on the real axis a domain $\Omega = \Omega_1 \cup \Omega_2 \cup \Omega_3$, where $\Omega_1 = [0, x_1]$, $\Omega_2 = [x_1, x_2]$ and $\Omega_3 = [x_2, L]$, $0 < x_1 < x_2 < L$, be given. Consider on Ω the coupled system of Helmholtz equations

$$p_{xx} + k^2 p = -g_1, \text{ in } \Omega_1 \tag{2.1}$$
$$u_{xx} + k^2 n u = -f, \text{ in } \Omega_2 \tag{2.2}$$
$$\tilde{p}_{xx} + k^2 \tilde{p} = -g_2 \text{ in } \Omega_3, \tag{2.3}$$

with boundary conditions

$$p_x(0) + ikp(0) = 0, \tag{2.4}$$
$$\tilde{p}_x(L) - ik\tilde{p}(L) = 0, \tag{2.5}$$

and transmission conditions

$$p_x(x_1) - k^2 u(x_1) = 0 \tag{2.6}$$
$$p(x_1) + a u_x(x_1) = 0 \tag{2.7}$$
$$\tilde{p}_x(x_2) - k^2 u(x_2) = 0 \tag{2.8}$$
$$\tilde{p}(x_2) + a u_x(x_2) = 0. \tag{2.9}$$

This is a model for a simple case of time-harmonic one-dimensional acoustic fluid-solid interaction. The "fluid" regions, Ω_1, Ω_3 have material properties B (bulk modulus) and c_f (speed of sound in fluid); the "solid" region Ω_2 has properties E (Hook's modulus) and c_s. In the non-dimensional model (2.1 - 2.9), we have $n = (c_f/c_s)^2$ and $a = E/B$. The real parameter $k = \omega/c_f$, where ω is a given frequency, is the wavenumber in the fluid.

Define the space $\mathcal{H} = H^1(\Omega_1) \times H^1(\Omega_2) \times H^1(\Omega_3)$, where $H^1(\Omega_i)$ are the usual Sobolev spaces, and consider the variational problem: Find $\mathcal{U} = (p, u, \tilde{p}) \in \mathcal{H}$ such that for all $\mathcal{V} \in \mathcal{H}$

$$\mathcal{B}(\mathcal{U}, \mathcal{V}) = (\mathcal{F}, \mathcal{V}), \tag{2.10}$$

where

$$\mathcal{B}(\mathcal{U}, \mathcal{V}) = \int_{\Omega_1} p_x \bar{q}_x dx - k^2 \int_{\Omega_1} p\bar{q} dx - k^2 u(x_1)\bar{q}(x_1) - ikp(0)\bar{q}(0)$$

$$+ \quad k^2 \int_{\Omega_2} a u_x \bar{v}_x dx - k^4 \int_{\Omega_2} g u \bar{v} dx - k^2 p(x_1)\bar{v}(x_1) + k^2 \tilde{p}(x_2)\bar{v}(x_2) \quad (2.11)$$

$$+ \quad \int_{\Omega_3} \tilde{p}_x \bar{\tilde{q}}_x dx - k^2 \int_{\Omega_3} \tilde{p}\bar{\tilde{q}} dx + k^2 u(x_2)\bar{\tilde{q}}(x_2) - ik\tilde{p}(L)\bar{\tilde{q}}(L)$$

and

$$(\mathcal{F}, \mathcal{V})_{o,\mathcal{H}} = (g_1, q) + k^2(f, v) + (g_2, \tilde{q}) . \quad (2.12)$$

This problem is a weak formulation of the coupled boundary value problem (2.1-2.9). By eq (2.12), an inner product is defined on $\mathcal{H}$; we denote the induced L^2-norm by $\|\cdot\|_{o,\mathcal{H}}$. A weighted norm is then defined on $\mathcal{H}$ by

$$\|\mathcal{U}\|_{1,\mathcal{H}} = \left(\|\mathcal{U}_x\|^2_{o,\mathcal{H}} + k^2 \|\mathcal{U}\|^2_{o,\mathcal{H}} \right)^{1/2} .$$

With these definitions, we immedeately have continuity as

$$|B(\mathcal{U}, \mathcal{V})| \leq \beta \, \|\mathcal{U}\|_{1,\mathcal{H}} \, \|\mathcal{V}\|_{1,\mathcal{H}} \quad \forall \mathcal{U}, \mathcal{V} \in \mathcal{H} \quad (2.13)$$

with constant β not depending on k. Existence and uniqueness of the variational problem (2.10) is assured by

Theorem 2.1 [12] *Let $\mathcal{F} \in L^2$. Then there exists a unique solution of (2.10). Furthermore, $\mathcal{U} \in \mathcal{H}^2$, and*

$$\|\mathcal{U}_{xx}\|_{o,\mathcal{H}} \leq C_s (1 + k) \|\mathcal{F}\|_{o,\mathcal{H}} . \quad (2.14)$$

We remark that it is assumed (cf.

the physical meaning of a) that $a > 0$. It follows from the proof that the stability constant C_s is $O(a^{-1})$.

For data that is measurable in the dual norm, existence and uniqueness follows from the inf-sup-condition

Theorem 2.2 [12] *The bilinear form $B(\cdot,\cdot)$ defined on $\mathcal{H} \times \mathcal{H}$ satisfies*

$$\sup_{\mathcal{V} \in \mathcal{H}} \frac{\text{Re } B(\mathcal{U}, \mathcal{V})}{\|\mathcal{V}\|_{1,\mathcal{H}}} \geq \gamma \frac{1}{k} \|\mathcal{U}\|_{1,\mathcal{H}} , \quad \forall \mathcal{U} \in \mathcal{H} \quad (2.15)$$

where γ is a positive constant, independent of k.

From Theorem 2.2 we conclude by well-known theory that, if $\mathcal{U}$ is the solution of (1.6),

$$\|\mathcal{U}\|_{1,\mathcal{H}} \leq k \frac{1}{\gamma} \|\mathcal{F}\|'_{1,\mathcal{H}}$$

holds, where $\|\cdot\|'_{1,H}$ is the norm in the dual space of $\mathcal{H}$.

3 Finite Element Solution and Asymptotic Error Estimates

We define the subspaces $S_h^{d,i} \subset H^1(\Omega_i)$ as the usual finite element approximation spaces with stepwidth h and approximation order d. Assume for simplicity that the degree of approximation is constant, $d = s_o$, over the elements in the regions Ω_i. As usual, the finite element solution $\mathcal{U}_{fe} \in \mathcal{S}_h$ solves the variational equality

$$B(\mathcal{U}_{fe}, \Phi) = (\mathcal{F}, \Phi)_{0,\mathcal{H}} \quad (3.1)$$

for all $\Phi \in \mathcal{S}_h$. Here, existence and uniqueness is concluded from the discrete inf-sup-condition which holds asymptotically as

Theorem 3.1 [12] *Assume that h is small enough such that* $k^2 \frac{h}{s_0} \leq c$, *where c is an appropriately chosen constant. Then there exists a constant* $\tilde{\gamma} > 0$, *independent of h and k, such that*

$$\sup_{\mathcal{V}_h \in \mathcal{S}_h} \frac{\text{Re } B(\mathcal{U}_h, \mathcal{V}_h)}{\|\mathcal{V}_h\|_{1,\mathcal{H}}} \geq \tilde{\gamma} \frac{1}{k} \|\mathcal{U}_h\|_{1,\mathcal{H}} \tag{3.2}$$

holds for all $\mathcal{U}_h \in \mathcal{S}_h$.

The error estimate is given asymptotically as a quasioptimal relation in the $\mathcal{H}$-norm:

Theorem 3.2 [12] *With the assumptions of the previous theorem,*

$$\|\mathcal{U} - \mathcal{U}_{fe}\|_{1,\mathcal{H}} \leq C_{opt} \inf_{\Phi \in \mathcal{S}_h} \|\mathcal{U} - \Phi\|_{1,\mathcal{H}} \tag{3.3}$$

holds with $C_{opt} > 0$, *independent of* k, h.

The proof of this theorem essentially combines the L^2-data stability theorem of the previous section with a standard approximation result. From the proof it follows that the constant c bounding $k^2 h$ in the assumption of Theorem 3.1 is generally of order $O(a^{-2})$.

Theorems 3.1 and 3.2 affirmatively answer the basic question if the finite element solution is well defined. However, due to the assumption on $k^2 h$, the theorems do not cover, generally, the meshsizes used in practical computations. Furthermore, numerical experiments show that the quasioptimal estimate (3.3) does not hold if k is large and h is not small in the sense of the assumption. To clarify this point, we next discuss results for a related uncoupled problem, returning to the coupled problem in the numerical evaluation.

4 Preasymptotic Estimates. The Effect of Numerical Pollution.

The assumption that $k^2 h$ is suficiently small would require very fine meshes in applied computations unless the wavenumber k is small itself. A rule in numerical acoustics is to resolve the wavelength λ by a certain number of elements. Thus, $\lambda/h = 10$ (equivalently, $kh = \pi/5$) is quoted as "acceptable resolution" in [7]. It is easy to see that this rule is sufficient for approximability. Indeed, consider the most simple case of a sinusoidal wave $u(x) = \sin kx$. Then, in usual L^2-norm on $(0, 1)$, $\|u''\|/\|u\| = k^2$ and the relative error of the best approximation is, by standard theory,

$$\inf_{v \in S_h} \frac{\|u - v\|}{\|u\|} \leq Ch^2 \frac{\|u''\|}{\|u\|} = C(hk)^2 .$$

From the quasioptimal estimate (3.3) it then follows that the finite element error is also of order $O((kh)^2)$.

On the other hand, it is well-known [3] that the relative error of the finite element solution is of order $O(k^3 h^2)$ on meshes with constant resolution λ/h. The same error is observed in the H^1-norm if h is not small [9]. Hence the quasioptimal result of Theorem 3.2. generally (i.e. for all k) does not hold if just hk is constrained.

A preasymptotic analysis of the Galerkin FEM for the Helmholtz equation is given in [9, 10]. All statements are proven with assumptions on the magnitude of hk only. It is shown that for sufficiently regular solutions oscillating with frequency k, the estimate [10]

$$\tilde{e}_1 = \frac{|u - u_{fe}|_1}{|u|_1} \leq C_1(p) \left(\frac{hk}{2p}\right)^p + C_2(p) k \left(\frac{hk}{2p}\right)^{2p} , \tag{4.1}$$

holds, where $|\cdot|_1$ is the H^1-seminorm and p is the order of polynomial approximation. For $p = 1$ we have (cf. also [9])

$$\tilde{e}_1 \leq C_1 hk + C_2 k^3 h^2 \tag{4.2}$$

if $hk < 1$. Comparing (4.2) to the quasiasymptotic estimate we see that a pollution term of order $O(k^3h^2)$ is contained in the preasymptotic estimate. We also see that (4.2) generalizes (3.3). Indeed, taking out hk in we see that the estimate (4.2) reduces to (3.3) if k^2h is small.

For illustration of the error behavior in H^1-norm, see Fig. 1. The finite element error is plotted for $k = 100$, piecewise linear Galerkin FE-solution of the Helmholtz equation on $\Omega = (0, 1)$, with Dirichlet and Robin boundary conditions (see [9] for details).

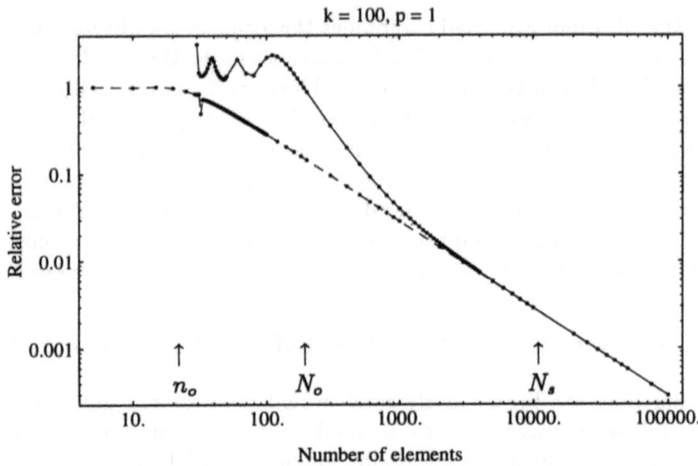

Figure 1: **Convergence pattern of uncoupled case:** Relative errors of the best approximation and the finite element solution in H^1-seminorm for $p = 1$ and $k = 100$

We distinguish three intervals on the n-axis (number of elements):

(1) $\underline{n \leq n_o}$: Point $n = n_o$ is the 'limit of resolution' [7] (two elements per wavelength, $\underline{hk = \pi}$). To the left of n_o, the error of the best approximation is 100%. Both the finite element solution and the best approximation do not approximate the exact solution.

(2) $\underline{n_o < n \leq N_o}$: Point[3] $n = N_o$ is the minimal meshsize for which the error is (and stays, if the mesh is refined) below 100%. In this interval, the error of the best approximation goes down with rate -1 as h decreases, whereas the error of the finite element solution oscillates with amplitudes of more than 100%. The finite element solution is well-defined but it does not approximate the exact solution. The pollution term in the error estimate (4.2) is dominant in this region. For the quoted

[3]Of course the notation of "points", as used here for illustration of the error behavior, has to be understood in a diffuse sense.

'acceptable resolution' of $\lambda/n = 10$ (read at $n = 5k/\pi$ in Fig. 1), the minimal error is $\approx 15\%$ whereas the finite element error still amounts to $\approx 100\%$.

(3) $N_o < n \leq N_s$: To the right of $n = N_s$, the influence of the pollution term in the error is negligible. As a consequence of Theorem 3.2., the formula to compute N_s is

$$N_s = \frac{1}{c^*}k^2$$

where, theoretically, c^* is a small constant. Numerical experience [9, 11, 12] shows that practically c^* needs not to be small, but a constraint on k^2h *is necessary for quasioptimality* in the H^1-norm. To the left of N_s, the convergence behavior of the finite element solution is still governed by the pollution term. Note that the prevalence of this term leads to a *superoptimal* rate of convergence: one observes algebraic decay n^{-2} compared to n^{-1} for the best approximation.

This third region is generally the region of practical interest. The finite element error is sufficently small while the mesh is still reasonably coarse. In particular, the error is within this region if one constrains the magnitude of k^3h^2 – cf. estimate (4.2). With a constraint of this form, reliability of the finite element solution is assured for all k.

(4) $n > N_s$: The condition for quasioptimal convergence of the finite element solution is satisfied. The solution has reached a rate of convergence which is close to the optimal rate n^{-1}.

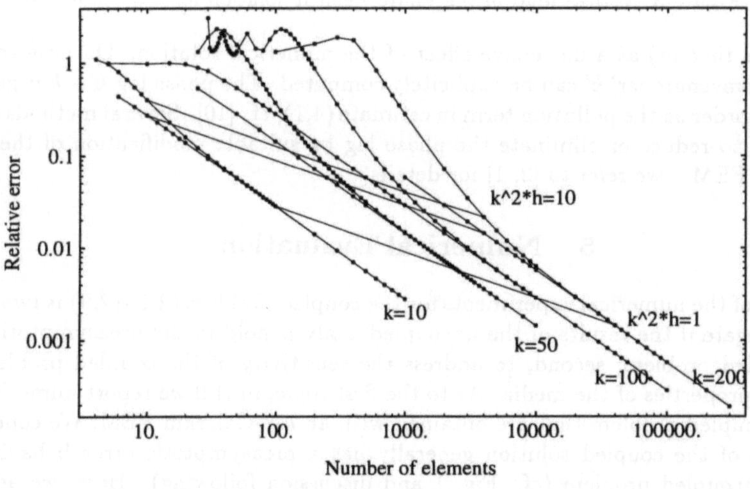

Figure 2: **Quasioptimality, uncoupled case:** Relative error of the finite element solution and the best approximation in H^1-seminorm for $k = 10, 50, 100$ and $k = 200$. Meshes with $k^2h = 10$ or $k^2h = 1$, resp., are connected on both curves. The plot shows that restriction of k^2h is necessary for quasioptimality.

Now turn to Fig. 2. For the same model problem, the error behavior for low wavenumber is compared to the high wavenumber case. We observe optimal convergence of the fe-solution for low wavenumber; pollution is practically negligible for low k. Also in Fig. 2, the lines $k^2h = \alpha$ are plotted for different α. We see that constraining of k^2h is indeed necessary for quasioptimality – the ratio e_{fe}/e_{ba} is constant along these lines. The optimality constant depends, in general, on the constraint α and also on the boundary conditions, cf. [11].

Comparing plots of the finite element solution to the exact solution, one observes a difference in phase, see Fig. 3. This phenomenon has been well-studied (cf. [11] and

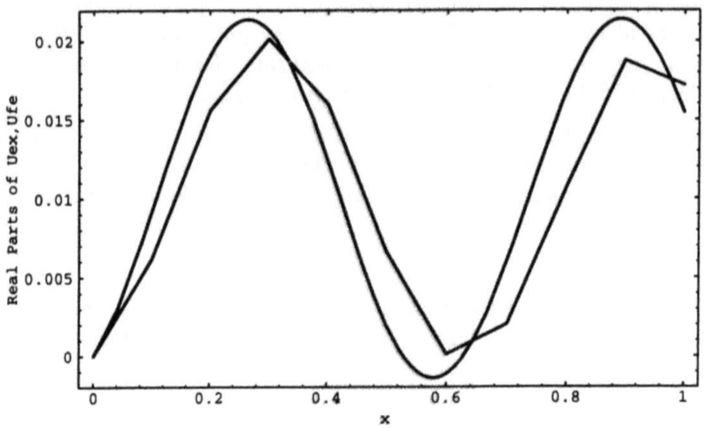

Figure 3: **Phase lead** of the Finite element solution for $k = 10$, $n = 10$.

references therein) as a dispersive effect of the numerical solution. On uniform mesh, a 'discrete wavenumber' k' can be explicitely computed. The phase lag $k' - k$ is generally of the same order as the pollution term in estimate (4.1), cf. [10]. Several methods have been proposed to reduce or eliminate the phase lag by suitable modification of the standard Galerkin FEM – we refer to [2, 1] for details.

5 Numerical Evaluation

The aim of the numerical experiments for the coupled problem (2.1)-(2.9) is twofold: first, to investigate if the results of the uncoupled analysis hold in the preasymptotic range of the coupled problem; second, to address the sensitivity of the coupled problem on the material properties of the media. As to the first topic, in [12] we report numerical results for the coupled problem that are obtained with an h-p-Galerkin FEM. We conclude that the error of the coupled solution generally has a preasymptotic error behavior similar to the uncoupled problem (cf. Fig. 1 and discussion following). Here, we address the sensivity of the numerical pollution w.r. to the material properties. We recall from the theoretical analysis that generally the stability constant is $O(a^{-1})$, where a is the relation of the moduli of elasticity in both media. The constant c^* bounding k^2h in the condition for quasioptimal behavior was shown to be $O(a^{-2})$. For the following numerical investigation, let $\Omega = \Omega_1 \times \Omega_2$, where the solid is given on $\Omega_1 = (0, 1)$ and the fluid on $\Omega_2 = (0, \infty)$. Assume a Dirichlet boundary condition $u(0) = 0$ and let no load be given in the fluid, $g_2(x) = 0$. This is the problem of forced vibrations of a Dirichlet-fixed rod interacting with a fluid at $x = 1$. Via a DtN-condition, this problem can be reduced to the boundary-value problem, cf. Demkowicz [5],

$$\begin{aligned} -u'' - nk^2u &= f \text{ in } (0,1) \\ u &= 0 \quad \text{at } x = 0. \\ u'(1) - ikau(1) &= 0 \quad \text{at } x = 1\,. \end{aligned} \tag{5.1}$$

For $a = 0$, the problem is not well-defined at the eigenvalues

$$\sqrt{n}\,k_j = \left(\frac{\pi}{2} + m\pi\right), \quad m = 1, 2, \ldots$$

In [5], the (continuous) inf-sup constant of (5.1) is computed. For weak damping (i.e. when a is small), it is shown that the coupled problem is ill-conditioned near the eigenvalues of the uncoupled problem. The following results for the preasymptotic error behavior complement this observation on the discrete level. Consider the finite element solution (with data $f = 1$, piecewise linear approximation) near the critical values for $\sqrt{n}k$. In

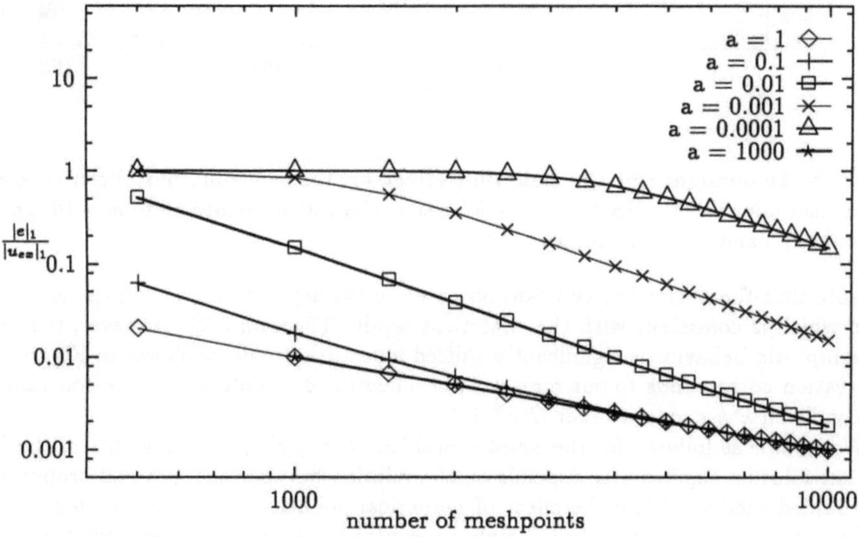

Figure 4: **Dependence of the pollution effect on the damping coefficient:** Relative error in H^1-seminorm for $kn^{1/2} = 10.5\pi$ and different values of material relation a.

Fig. 4, the relative error in H^1-seminorm is plotted for different values of a. The solution converges almost optimally for $a = 1$. In this case, the Helmholtz equation with Dirichlet and Robin condition is given, as discussed in the previous section. No significant pollution is observed since the wavenumber is small. However, the pollution effect grows considerably as $a \to 0$. The results correspond well to our remark that the stability constant is, generally, of order $O(a^{-1})$. The stability constant is contained as a factor in the pollution term of the preasymptotic estimates [9, 10]. In the figure we see that, close to the eigenvalues, the error of the finite element solution has indeed a pollution of this order in the preasymptotic range. Away from the eigenvalues, the solution converges with optimal rate, as the line for $a = 1000$ shows[4].

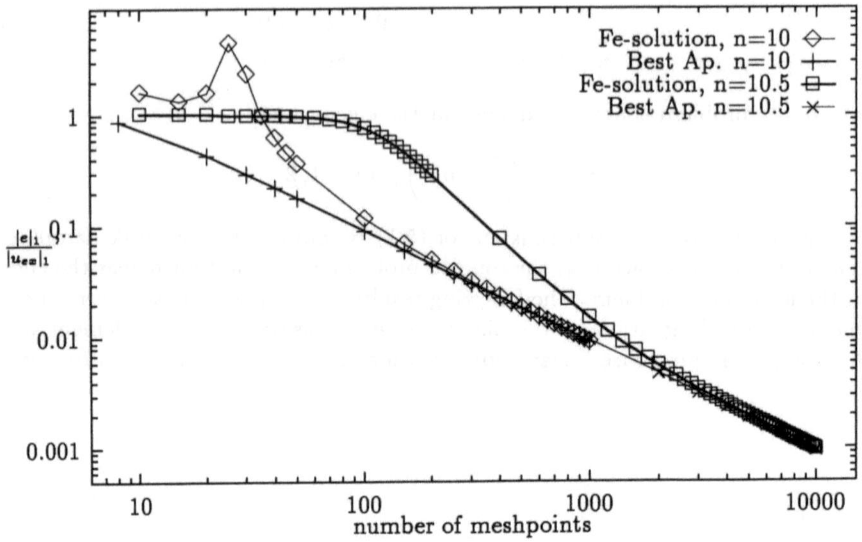

Figure 5: **Dependence of the pollution effect on the wavenumber:** Relative error in H^1-seminorm for $kn^{1/2} = m\pi$ for $m = 10.5$ (close to eigenvalue), compared to $m = 10$ (between eigenvalues); damping coefficient $a = 0.13$.

Note that the optimality constant observed in the asymptotic range does not depend on a, which is consistent with the analytical result (Theorem 3.2). However, the region of asymptotic behavior is significantly shifted when a is small, as shown in Fig. 5. This observation corresponds to our remark after Theorem 3.1. noting that the constant c in the condition $k^2h < c$ is of order $O(a^{-2})$.

We resume as follows: for the coupled problem, the quality and reliability of the finite element solution significantly depends on the relation between the material properties of the coupled media. While the effect of numerical pollution for the uncoupled problem is of practical importance only for high wavenumber, the finite element solution of the coupled, fluid-solid-interaction problem can be considerably polluted also for small k. The pollution occurs close to eigenvalues of the interior problem. In the general case, these eigenvalues are not known a priori. Hence, though there are no 'forbidden frequencies' for FE-FE-discretisation, careful a posteriori analysis is necessary to assure reliability of the computational results.

Acknowledgement: The first author was supported by ONR Grant N00014-93-I-0131. The authors are most grateful to Prof. Ivo Babuška for numerous discussions which have provided valuable insight into the problem considered.

[4]for large a, the Robin boundary condition tends to the Dirichlet condition; the correspondig eigenvalues are mp, $m = 1, 2, \ldots$

References

[1] Babuška, I., Ihlenburg, F., Paik, E.T., Sauter, S.A.: *A generalized finite element method for solving the Helmholtz equation in two dimensions with minimal pollution*, Comput. Meths. Appl. Mech. Eng., to appear

[2] Babuška, I.; Sauter, S.: *Is the pollution effect of the FEM avoidable for the Helmholtz equation considering high wave numbers*, Technical Note BN-1172 (1994), IPST, U of Maryland at College Park

[3] Bayliss, A.; Goldstein, C.I.; Turkel, E.: *On accuracy conditions for the numerical computation of waves*, J. Comp. Phys. 59 (1985), 396-404

[4] Burnett, D.S., *A three-dimensional acoustic infinite element based on a prolate spheroidal multipole expansion*, J. Acoust. Soc. Am. 96(1994), 5, 2798-2816

[5] L. Demkowicz, *Asymptotic convergence in finite and boundary element methods: part I: theoretical results*, Computers Math. Applic. 27, 12, 69-84 (1994)

[6] L. Demkowicz, K. Gerdes, *Solution of 3D Laplace and Helmholtz equations using hp infinite element methods*, TICAM report 95-04, U of Texas at Austin

[7] Givoli, D.; Keller, J.B. *Special finite elements for use with high-order boundary conditions*, Comp. Methods Appl. mech. Engrg. 119 (1994), 199-213

[8] I. Harari and T.J.R. Hughes, *Galerkin/least squares finite element methods for the reduced wave equation with non-reflecting boundary conditions in unbounded domains*, Comp. Meth. Appl. Mech. Eng. 98 (1992) 411-454

[9] Hsiao, G.C., *The coupling of boundary element and finite element methods* ZAMM 70 (1990) 6, T493-T503

[10] Ihlenburg, F., Babuška, I., *Finite element solution to the Helmholtz equation with high wave numbers - part I: the h-version of the FEM*, Technical Note BN-1159, IPST, UMCP, 1993

[11] Ihlenburg, F., Babuška, I., *Finite element solution to the Helmholtz equation with high wave numbers - part II: the h-p-version of the FEM*, SIAM J. Numer. Anal., to appear

[12] Ihlenburg, F. Babuška, I., *Dispersion analysis and error estimation of Galerkin finite element methods for the numerical computation of waves*, IJNME, to appear

[13] Makridakis, Ch., Ihlenburg, F., Babuška, I. , *Analysis and finite element methods for a fluid-structure interaction problem in one dimension*, Technical Note BN-1183 (1995) IPST, UMCP

[14] Mackerle, J., *Finite element and boundary element techniques in acoustics - a bibliography (1990-92)*, Finite Elements in Analysis and Design 15 (1994) 263-272

[15] Pinsky P.M.; Abboud, N. N., *Two mixed variational principles for exterior fluid-structure interaction problems*, Comp. Struct. 33 (1989), 621-635

[16] Wendland, W.L., *On asymptotic error estimates for the combined BEM and FEM*. In: Stein, E., Wendland, W., (eds), Finite lement and boundary element techniques from mathematical and engineering point of view, CISM Lecture Notes 301, Udine, Springer-Verlag Wien-New Yor 1988, 273-333

Coupled Geometric Modelling
for the
Analysis of Groundwater Flow and Transport in Fractured Rock

H. Kasper[1], G. Kosakowski[2], T. Taniguchi[3], W. Zielke[1]

Summary

A strategy is proposed for preparing quadrilateral and hexahedral finite elements for the numerical analysis of groundwater flow and transport in three-dimensional fractured rock. We present three methods for the local discretization of fractured media (hexahedral elementsfor volumes and quadrilateral elements for fracture planes). These methods are based on the Delaunay triangulation. Therefore they can be applied to arbitrary convex domains consisting of a set of convex subdomains surrounded by fracture planes.

Introduction

Need for numerical analysis of groundwater flow and transport problems

The necessity of the disposal of nuclear and chemical waste impelled the developement of methods for the analysis of flow and transport in fractured rock. The security of such deposits requires accurate analysis of flow and transport. For this purpose the numerical model must consider geological and artificial features on different scales, like micro- and macro- fractures, geological layers, caverns, channels and other deposite systems, which are located in the analysis domain. Due to the complexity of the analysis domain, the finite element method (FEM) is generally prefered.

Groundwater flows in a three-dimensionally (3D) fractured rock mainly through the fracture planes, however, there is also substanal flow through the rock matrix. Hence both phenomena must be taken into account. It is well known that the quality of numerical approximations is greatly influenced by the geometry of finite element meshes, and a good method is required for modelling of a 3D fractured rock. The simulation is done using the flow and transport code *Rockflow*[6][7]. The solver uses different element types: hexahedron, quadrilateral, and line. In 3D, the hexahedral elements are used to model the rock matrix, quadrilaterals for fracture planes and line elements for boreholes and linear features. These element types can be arbitrarily coupled. The discretization error and the time to assemble the element matrices are minimized by using hexaedron (qadrilateral) elements instead of tetrahedron (triangular) elements.

Aims of this investigation

To a certain extent the geometric modelling method in two-dimensional (2D) problems is already developed for the FEM. This is not the case for 3D-problems. The main reasons are 1) the complexity of the geometry and 2) the requirement of hexahedral elements for the rock matrices and rectangular elements for planes. In addition to these requirements, one-dimensional (1D) elements are also needed to simulate boreholes and other linear features. Therefore a FE-model for the analysis of groundwater flow and transport in a 3D fracture

---

[1] Institut für Strömungsmechanik, Appelstr. 9A, D-30167 Hannover
[2]Geowissenschaftliche Gemeinschaftsaufgaben - Niedersächsisches Landesamt für Bodenforschung, Stilleweg 2, D-30655 Hannover
[3] Faculty of Env. Science & Technology, Okayama University

system includes 1D-, 2D- and 3D-elements. Any fracture plane divides the whole domain of analysis in two sub-domains. The meshes on the fracture plane must coincide with the meshes on two surfaces of these two subdomains. On the other hand, the meshes on these two surfaces are determined by the meshing of the two subdomains. This suggests that the meshing of subdomains which form the whole domain cannot be treated independently. The same argument holds for the generation of 1D elements. The aim of this investigation is to propose a mesh generation method for 3D fractured systems, which includes an arbitrary number of fracture planes as well as structures like caverns and channels. The fundamental tool used for meshing is the Delaunay triangulation.

Geometric complexity of fracture systems

Natural fracture system

Natural rock includes discontinuities and inhomogeneities. These structural heterogeneities can be classified according to their length scales as faults, fractures, fissures and pores. Fig. 1 shows the distribution of heterogeneities at different locations and for various sizes of the investigation area. Between single fractures in region A the rock matrix including pores can be treated as an equivalent continuum. The material of this area is assumed to be homogeneous. A similar model is valid for region B, a sufficiently connected fracture network. Region C is characterized by single fractures embedded in the rock matrix. This area cannot be replaced by an equivalent continuum. Region D can be described as a fracture network, that includes significant discontinuities. The importance of these faults will decrease, if the scale of the investigation area increases. The area between large fractures can be replaced by an equivalent continuum.

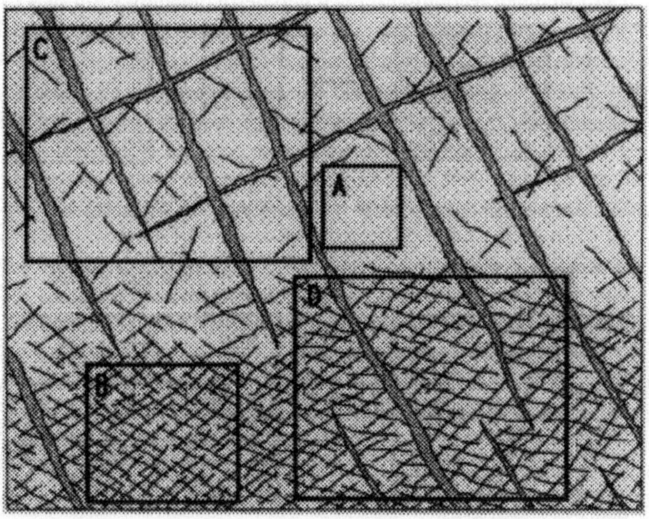

Fig. 1 Simplified natural fracture system

Data base and modelling concepts of natural fracture systems

The data base of natural fracture systems consists mainly of data from boreholes and mining exploration. Therefore, it is 1D and 2D. Kolditz (1994) [5] gives an overview on some databases presently used for the simulation of flow and heat transport in fractured rock. The lack of the third dimension requires some assumptions for modelling of the geometry. Different modelling approaches are used (Berkowitz, 1994 [1]). In *continuum models* a spatial averaging approach based on the concept of a representative elementary volume (REV) is used to simplify the calculation. In this case the averaging volume covers a medium with a sufficient great number of connected fractures (Fig. 1, Region A and B). It is possible to replace the fracture system by an equivalent continuum with effective parameters e.g. the coefficient of hydraulic conductivity and an effective porosity for a corresponding porous medium. The *stochastic model* concept uses a statistical interpretation of the data which leads to a fracture distribution. Distribution functions of frequency and orientation of fracture groups are used to generate fracture networks. *Hierarchical models* are used to describe fractal properties of fracture networks. *Deterministic model* concepts depend on the assumption, that fracture groups can be represented by mean values of frequency and distribution. This leads to idealizations such as parallel fracture systems (Fig1., Region C). *Hybrid models* combine continuum representation of matrix and deterministic/stochastic representation of primary fractures in the formation (Fig 1., region D). One typical example for the stochastic modeling approach is illustrated in Fig. 2. Here the fracture planes are convex planar polygons located in 3D-space. This example is a test case to compare the influence of different FE meshes to the numerical results and will be discussed in the last section of this paper.

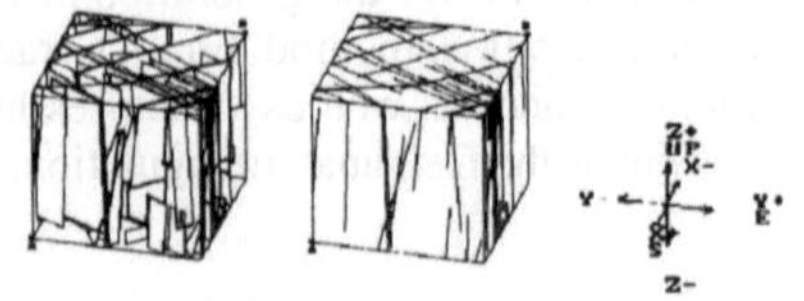

Fig. 2 Stochastic model of fractured claystone used as a test case for meshing

Mesh generation

The Delaunay triangulation is a geometrical subdivision of a convex domain, definded by a set of nodes, into triangles and tetrahedra. The method requires only the coordinates of nodes, and the subdivision is uniquely determined by these coordinates. The generated triangles and tetrahedra can be easily improved by relocating nodes and/or adding new nodes. Further it is possible for the 2D-case to modify the triangulation by the swapping of edges.

2 $^1/_2$-D Mesh generation

We consider the application of the Delaunay triangulation for a mesh generator for an arbitrary set of fracture planes in the 3D space. We call these cases 2½ D problems, because rock matrix is left for discretization. The geometry of each fracture is a planar polygon and all polygons may intersect each other. Taniguchi & Ohta (1994) [3] proposed a method to triangulate arbitary 2D domains into triangular meshes with constraint edges inside the domain. These constraint edges represent the fractures. The method proposed here is extension to 3D space. It consists of the following steps;

(1) Calculate all intersection points between different two polygons and pick up the fracture line between these two polygons.
(2) Calculate all intersection points of all fracture lines.
(3) Apply the Delaunay triangulation for each given polygon independently and introduce also all intersection points of this polygon.
(4) Gather the results of the triangulation
(5) Modify the triangulation by use of the method shown in Fig. 3 a)
(6) Transform the triangular mesh by the method shown in Fig. 3 c)
The result of this procedure is shown in Fig. 8.

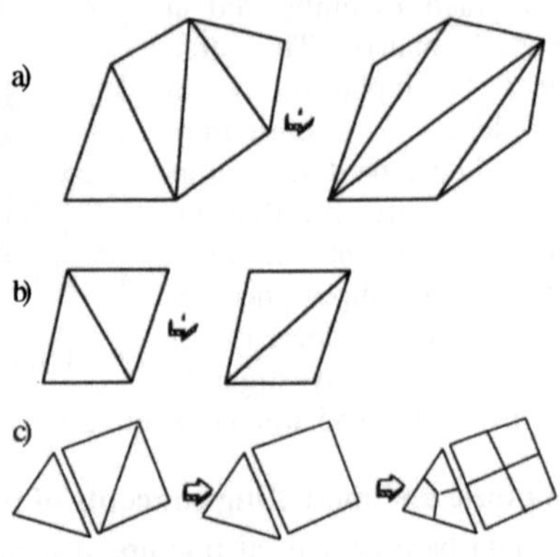

Fig. 3 Modification of triangular meshes

Delaunay triangulation

Now, we consider the application of the Delaunay triangulation as a mesh generator for the domain containing fracture planes. The geometry of the domain is a cube, and a number of planes are located in the domain as to divide it into convex subdomains, i.e. convex polyhedra. Each subdomain is surrounded by convex polygons, and it is defined using the intersection points where three planes cross each other. This suggests that the Delaunay triangulation can be easily used as a mesh generator, and we find two ways of its application; the first way is the direct application to all intersection points in the domain, and the second way is the application of the method for edge nodes in each subdomain. But in both ways there are unresolved problems. (1) There is no way to generate appropriate tetrahedra whose triangular faces can express all fracture planes. (2) A unique triangulation for the polygonal surfaces of each subdomain is not avaible, since the polygon is used as a part of a fracture plane which is common for adjacent two subdomains. Taniguchi (1994) [4] proposed a triangulation method using the second way. On the other hand, if we take the first way, we have to modify the generation method for the tetrahedra in the domain so that triangular surfaces of modified tetrahedra can express all fracture planes. This modification is not avaible. For this reason we take the second way.

Triangulation of the 3D- fracture system

Assume that all nodes are given which are located at the intersection of three planes. The method proposed by Taniguchi et al (1994) [4] consists of following steps for the triangulation of the domain;

(1) Order all nodes using the distance from the origin of the coordinate system.
(2) Apply the Delaunay triangulation for nodes in each subdomain using node ordering as described above.
(3) Gather the results of triangulation of all subdomains.
(4) Divide each tetrahedron into four hexahedra by introducing additional nodes as shown in Fig. 4 a). The geometry of generated hexahedral elements is necessarily distorted. The aim of this investigation is to obtain hexahedral meshes with improved geometry.

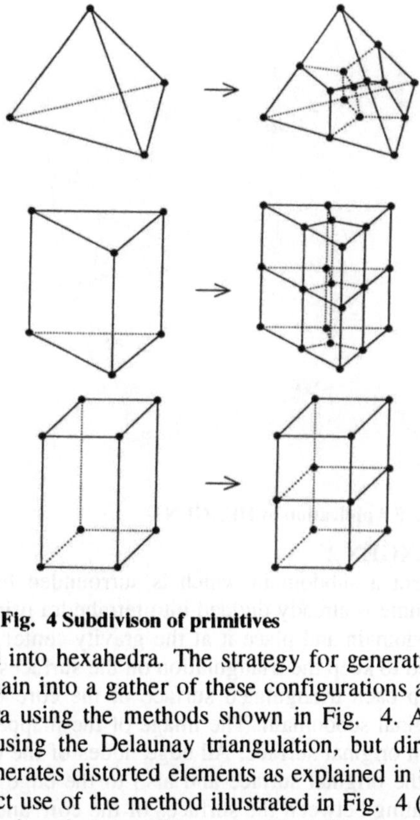

Fig. 4 Subdivison of primitives

Generation of hexahedral meshes

Any tetrahedron, triangular prism and hexahedron can be easily divided into hexahedral elements as shown in Fig. 4 a), b) and c), respectively. It suggests that if a convex domain is divided into a set of these subdomains, the whole domain can be subdivided into hexahedra. The strategy for generating hexahedral meshes is (1) to divide the whole domain into a gather of these configurations and (2) to transform them consecutive into hexahedra using the methods shown in Fig. 4. Any convex domain can be divided into tetrahedra using the Delaunay triangulation, but direct subdivision of these tetrahedra into hexahedra generates distorted elements as explained in the previous section. Fig. 5 is an example of the direct use of the method illustrated in Fig. 4 (a). Then, we aim to place the triangulated and shrunk domain (we call it a „core") at the center of the original domain and fill the space between the core and the surface of the original domain by hexahedral elements with improved geometry. In this section we present three different strategies.

HEXGEN1

We select one subdomain, which is surrounded by a set of fracture planes. Further, we assume that this subdomain is divided into tetrahedra. Then, all polygons forming the surface of the subdomain are triangulated, and the triangulation of each surface coincides with the triangulation of the surface of its adjacent subdomain.

We form a core for the triangulated subdomain and place it at the center of gravity of the original subdomain. Then, the subdomain is divided into the core and the skin part which locates between the core and the surface of the original subdomain. At this stage, we keep the triangulation on the surface of the original subdomain. Connect all edge nodes of the core with corresponding edge nodes of the triangles on the original surface, and divide the skin part into a set of triangular prisms. Since the core is already divided into tetrahedra, the volume of the original subdomain at this stage is divided into tetrahedra and triangular prisms. Using the subdivision of tetrahedra and triangular prisms into hexahedra as shown in Fig. 4 a) and b), the volume of the original subdomain is divided into hexahedra. Above procedure is repeated for all subdomains which form the domain. Fig. 5 shows an example of the application of HEXGEN 1.

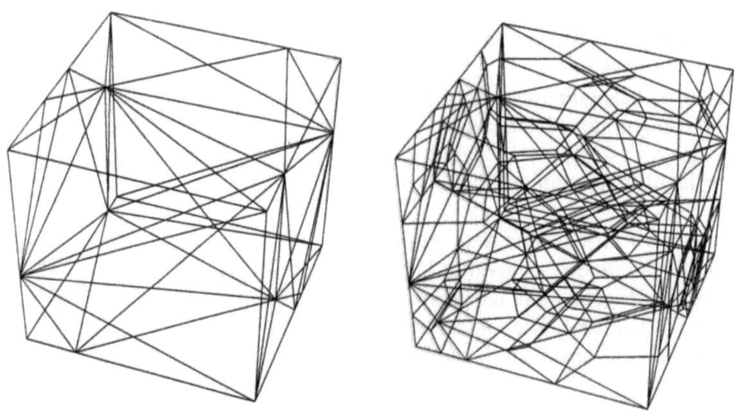

Fig. 5 Application of HEXGEN1

HEXGEN 2

Select a subdomain which is surrounded by a set of fracture planes. We assume that the volume is already divided into tetrahedra using the Delaunay triangulation. Form a core of the subdomain and place it at the gravity center of the original volume. In HEXGEN 2 we don't need to keep the triangulation on the surfaces of the original subdomain.

Map each triangulated surface of the core onto the corresponding polygonal surface of the original subdomain. The image of the mapped configuration is placed at the gravity center of each original surface. All edge nodes of the core are connected to corresponding edge nodes of the original surface and also to the edge nodes of the mapped image. Then, the skin part locating between the surfaces of the core and the original subdomain is divided into polygonal prisms and triangular prisms. Each polygonal prism in the skin part is subdivided into triangular prisms by using the triangulation of the surfaces of the core, and the skin is divided into triangular prism in the final stage. Above procedure indicates that whole volume of the subdomain is divided into tetrahedra and triangular prisms. Then, they can be subdivided once again into hexehedra using the method shown in Fig. 4. The above procedure is repeated for all subdomains which are included in the domain. Fig. 6 shows the subdivision into hexahedra using HEXGEN2.

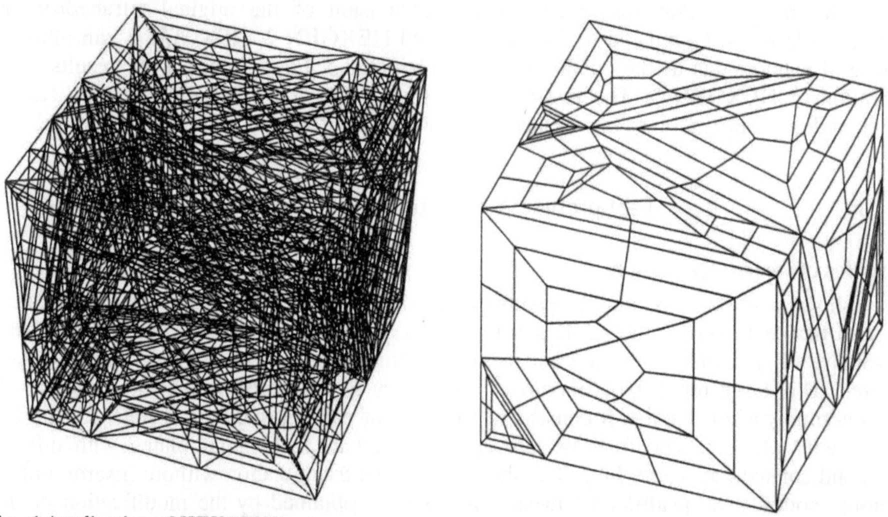

Fig. 6 Application of HEXGEN2

HEXGEN 3

Select one tetrahedron in the subdomain which is generated using the Delaunay triangulation. The following procedure is repeated for all tetrahedra which are included in the domain.
Form a shrinked core of the original tetrahedron, and place it at the gravity center of the original tetrahedron. Map four triangular surfaces of the core onto the corresponding triangular

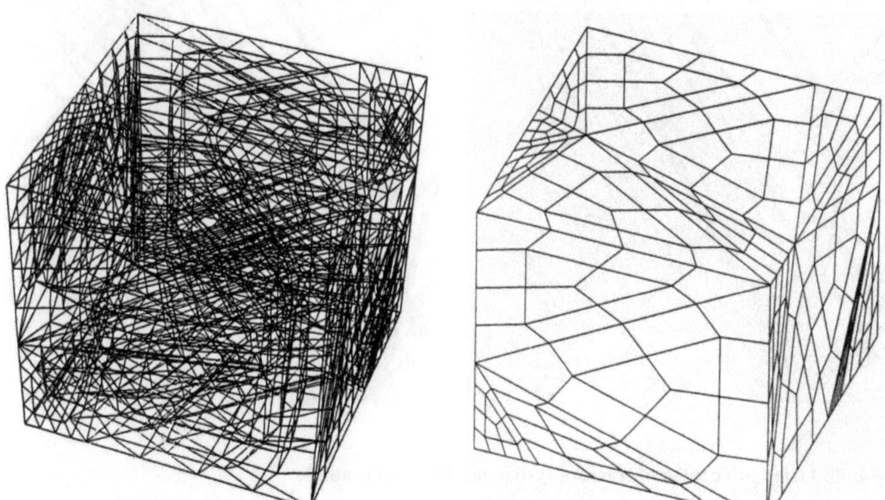

Fig. 7 Application of HEXGEN3

faces of the original tetrahedron. Connect each pair of triangles on the surfaces of the core and the original tetrahedron and form four triangular prisms in the skin part. Consecutively, add two nodes on each edge of the original tetrahedron, and connect them to edge nodes forming four triangles on the triangular faces of the original tetrahedron. Then, the residuum of the skin part between triangular faces of the core and the original tetrahedron is divided into a tetrahedral core, triangular prisms and hexahedra. Then, the use of all subdivisons shown in Fig. 4 to these sub-domains results in the subdivision of the original tetrahedron into hexahedral elements. Compared to HEXGEN 1 and HEXGEN 2, HEXGEN 3 can introduce hexahedral subdomains at the subdivison into subdomains. Fig. 7 shows the results of the application of HEXGEN 3. The test problems shown in Fig´s 6,7,8 are the same fracture system, and the comparison of the meshes emphasizes the difference between these meshes.

Test problems and numerical results

Examples of mesh generation

Fig. 8 shows a fracture system approximating landfill in claystones. This system is presently used as a first test case for the 2.5-D mesh generation method. The aim of this test is to show the validity of this method as a mesh generator for fracture systems with several 10.000 fractures. This large number of fractures is necessary to calculate equivalent continua. The system is simply a small cube. It consists of four sets of parallel fractures (130/0, 35/0, 60/15, 160/0, trend/ plunge). The fractures are approximated as rectangular plates with different extend and constant aperture. Fig. 8 a) shows the initial triangulation without insertion of any additional nodes. The quadrilateral mesh (Fig. 8 b) is obtained by the modification method shown in Fig. 3.

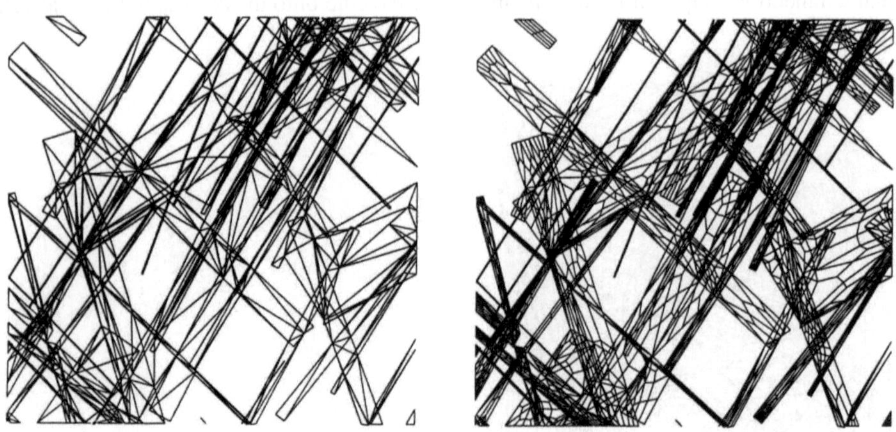

Fig. 8 Initial triangular mesh of fracture system used for test example

Note that all methods proposed in the previous section are based on the Delaunay triangulation, and can therefore applied without any restriction to any convex domain consisting of a set of convex subdomains. Compared to the method proposed by Taniguchi et al (1994) [4], they produce hexahedral meshes with a improved geometry, especially in the portion along fracture planes, since hexahedral meshes in the portion along fracture planes are generated from triangular prisms and/or hexahedra. On the other hand, the method by

Taniguchi et al (1994) [4], generates hexahedral elements directly from tetrahedra. Several examples of the subdivision into hexahedral finite elements are shown in Fig. 9.

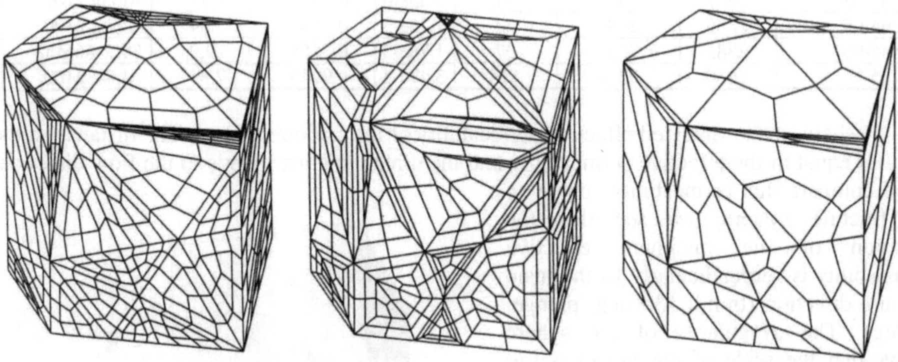

Fig. 9 Examples of hexahedral meshes

Numerical results

The mesh for the landfill (Fig. 8) has been used to analyse natural fracture system in order to obtain equivalent continua. For this purpose several meshes have been calculated, whereby the flow direction had been fixed by assigning constant heads to all boundary nodes on the six faces of the cube. From Tab. 1 we can see the influence of the discretization effect on the flux rate. Muen5-01 is a hybrid mesh consisting of triangles and quadrilaterals. The flux rate of this mesh becomes smaller compared to Muen5, which is a triangular mesh. The meshes Muen5-02 and Muen5-05 differ in the number of combined triangles. This results in different numbers of nodes and elements, but the numerical results are nearly the same. Fig. 10 shows a plane view of the velocity field of directed flux to the positive x-direction. We can see large differences of the velocities in different fractures. There are also some single fractures with dominant flow.

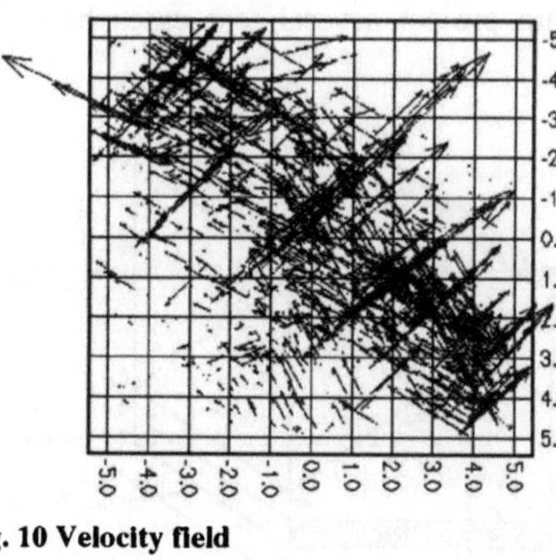

Fig. 10 Velocity field

Table 1: Comparison of different FE-meshes (**Boundary Conditions:** m=0.01, direction of flow (trend, plunge): $\alpha,\beta = 0$ Deg, permeability for single fracture $k_f = 2.5$ m/s, aperture of single fracture b=0.002 m)

mesh	number of nodes	number of elements	number of iterations	Tol	Σ |Q|	time to solve [s]	CPU
Muen5	545	973	11	9.83e-6	1.0000	1.76	Alpha 2100
Muen5-01	545	722	11	0.24e-4	0.9212	11.0	DX2-66Mhz
Muen5-02	2637	2443	37	9.66e-6	0.7552	4.83	Alpha 2100
Muen5-05	2885	2691	37	1.76e-5	0.7558	5.34	Alpha 2100
Muen5-05	2885	2691	37	1.76e-5	0.7558	29.0	DX2-66Mhz
Muen5-N	3680	3310	114	3.47e-3	0.9939	7.34	Alpha 2100

Fig. 11 shows the normalized effective permeability of the fracture system. The distance to the origin is equal to the effective permeability and the direction corresponds to the flow direction. If we compare this permeability sphere to the fracture system, we see that the direction of the minimum effective permeability is perpendicularly to the main fracture direction (trend: 130 deg, plunge: 0 deg). The irregularity of the sphere shows that the scale of the model is too small to define an equivalent continuum. On the other hand we realize the influence of the geometry and boundary conditions. This influence decreases, if we use larger-scale model.

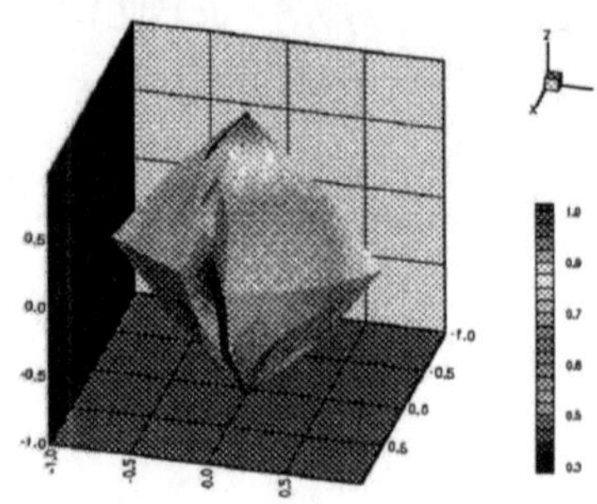

Fig. 11 Effective permeabilities

Within the scope of the international Hydrocoin project several test cases had

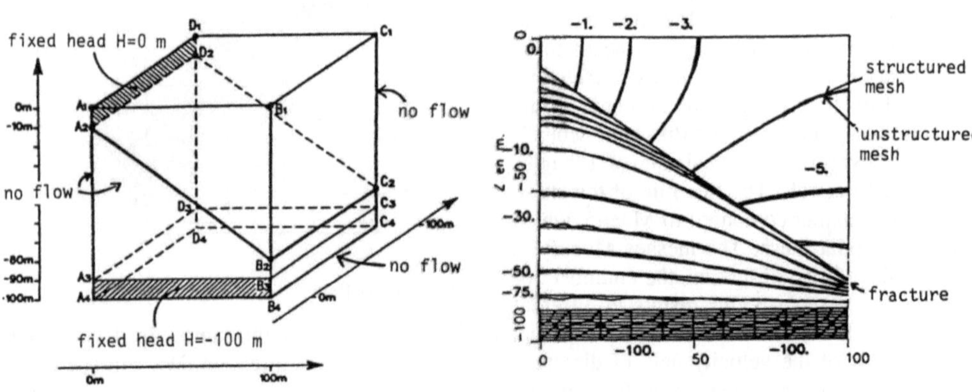

Fig. 12 Sauty cube

been defined for the verification of computer codes for groundwater flow. One of these test cases, the so called Sauty-cube, is shown in Fig. 12. In this system there is only one fracture plane. The boundary condition is illustrated in the same figure. The numerical results are different for unstructured and classical structured meshes, respectively.

For the further improvement of the generated hexahedral elements, we can introduce the Laplacian smoothing technique, which relocates the nodes in a way, that each internal node is repositioned at the center of gravity of the finite elements. Fig. 13 shows the application of this procedure to the Sauty-cube.

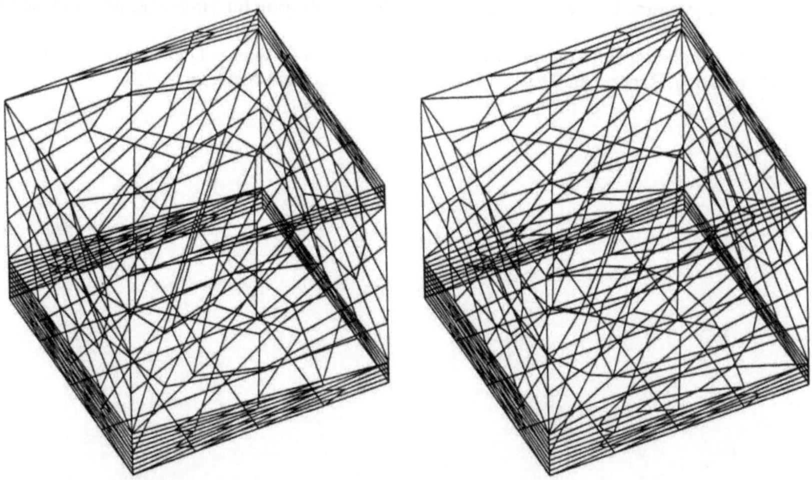

Fig. 13 Improved geometry

Conclusions

Several methods for generating rectangular meshes for fracture planes and hexahedral meshes for the rock matrix are shown in this paper. All of are based on the Delaunay Triangulation. Methods are proposed for preparing finite element models for the groundwater flow and transport problems in three-dimensional fractured rocks. At the same time, some methods are explained for the further improvement of the geometry of generated finite element meshes.

Acknowledgments

This research has been funded in part by the Deutsche Forschungsgemeinschaft grant Cl 121/1-1 to C. Clauser and H. J. Neugebauer, providing partial support for the second author. Tom Doe (Golder Associates Inc.) kindly granted permission to use the FracMan software in this study. The authors appreciate the continuous support and encouragement of Suzanne Hurter and Christoph Clauser (Geowissenschaftliche Gemeinschaftsaufgaben, Niedersächsisches Landesamt für Bodenforschung). A special thank you is extended to Olaf Kolditz (University of Hannover) for inspiring discussion and help.

References

[1] Berkowitz, B. 1994: Modelling flow and contaminant transport in fractured media. in M. Y. Corapeioglu (ed), Advances in porous media. Vol.2, Elsevier, Amsterdam.

[2] T. Taniguchi and Y. Fukuoka, Grid generation of arbitrary 3D surface, Boundary Element Method-Fundamentals and Appl., pp. 353-362, 1992.

[3] T. Taniguchi and C. Ohta, Application of Delaunay-triangulation to arbitrary 2D domain with straight boundaries, Proc. of Japan Society of Civil Engng., 432, 1991, in Japanese.

[4] T. Taniguchi, E. Fillion, J.P. Sauty and W. Zielke, Fast mesh generation for groundwater flow analysis in 3D fracture system, Proc. 4th Conf. on Num. Grid. Gen. in Comp. Fluid Dynamics and rel. Fields, pp. 665-676, 1994.

[5] Kolditz, O., Modellierung von Strömungs- und Transportprozessen im geklüfteten Festgestein, Report No. 112281, Niedersächsisches Landesamt für Bodenforschung - Geowissenschaftliche Gemeinschaftsaufgaben, 1994.

[6] Wollrath, J., Ein Strömungs- und Transportmodell für klüftiges Gestein und Untersuchungen zu homogenen Ersatzsystemen, Dissertation, Report No. 28/1990, Institut für Strömungsmechanik und ERiB.

[7] Kröhn, K.P., Simulation von Transportvorgängen im klüftigen Gestein mit der Methode der Finiten Elemente, Dissertation, Report No. 29/1991, Institut für Strömungsmechanik und ERiB.

Subspace decomposition methods for solving the Euler equations

Edgar Katzer

Graduiertenkolleg: "Modellierung, Berechnung und Identifikation mechanischer Systeme"
Otto-von-Guericke-Universität, Box 4120, D 39016 Magdeburg, Germany

Summary

A new approach for solving the steady Euler equations of gas dynamics is presented. A finite volume discretisation on triangular grids is introduced which avoids the use of reconstruction procedures on dual control grids. Triangles of the original grid are control elements for the flux balance. No time marching method is employed. The flux equations are solved by least square minimization of the residual norm.

The minimization approach enables a consistent formulation of subspace decomposition methods. Subspaces are defined by carefully constructed prolongations on coarser grids. The present work extends results obtained in [1] for linear systems and generalizes the frequency decomposition approach [2], [3]. Convergence rates for sequential and parallel subspace decomposition methods depend only slightly on the grid size. They compare favourably with results for linear systems. Convergence speeds decrease with increasing number of subspaces.

1 Introduction

The numerical solution of flow equations is still a challenging problem. The Navier-Stokes equations are a singularly perturbed system. For high Reynolds numbers they reduce to the Euler equations. It is expected that efficient solution methods for the Euler equations are a prerequisite for efficient solution of the Navier-Stokes equations.

The Euler equations for steady two-dimensional flows yield:

$$\partial_x f(u) + \partial_y g(u) = 0 \ , \tag{1}$$

where $u = (v_x, v_y, \rho)^{\mathrm{T}}$ is the vector of flow velocities and density and f and g are the conservative fluxes in x and y direction:

$$\begin{aligned} f &= (\rho v_x^2 + p, \rho v_x v_y, \rho v_x)^{\mathrm{T}} \\ g &= (\rho v_x v_y, \rho v_y^2 + p, \rho v_y)^{\mathrm{T}} \ . \end{aligned}$$

The total enthalpy is assumed to be constant and for perfect gas with $\gamma = 1.4$ we obtain:

$$H := \frac{\gamma}{\gamma - 1}\frac{p}{\rho} + \frac{1}{2}(v_x^2 + v_y^2) = H_\infty \ .$$

The static pressure is given by:

$$p = \frac{\gamma - 1}{\gamma} \rho \left(H_\infty - \tfrac{1}{2}(v_x^2 + v_y^2) \right) \quad .$$

The quantity

$$s := \frac{p}{p_\infty} / \left(\frac{\rho}{\rho_\infty} \right)^\gamma$$

measures the entropy.

This isoenergetic form of the Euler equations is a nonlinear coupled system. The type of such a system is determined by linear combinations of the Jacobians $\partial_u f$ and $\partial_u g$. The Euler equations are never elliptic, because these 3×3 matrices always have a real eigenvalue. Recall that for elliptic systems the Jacobians must have pure complex eigenvalues (see [4], sect. 1.3). For subsonic flows the Euler equations are of composite elliptic/hyperbolic character. For supersonic flows the Jacobians have three real eigenvalues and the system is hyperbolic. The partly hyperbolic character of subsonic flows is obvious, if we remember that the entropy is advected along streamlines. The streamlines represent charateristic directions associated with the real eigenvalue. For supersonic flows there are three characteristic directions, two of which are associated with Mach waves and one with the streamlines.

The composite character of subsonic flows complicates the efficient solution of the Euler equations considerably. Therefore time marching method are rather popular, even then when the asymptotic steady state solution is of only interest, see [5], [6]. Time step limitations make this approach rather expensive and fast converging algorithms are required.

The Euler equations for isoenergetic flows are even more complex then for homentropic flows where constant entropy is assumed. In that case the Euler equations are equivalent to the gasdynamic equations, see [7]. For subsonic flows, these equations are elliptic and a multigrid approach could be considered for fast solution. In the present approach with composite elliptic/hyperbolic character, the numerical algorithm should efficiently solve elliptic and hyperbolic equations simultaneously. Standard multigrid methods would not be efficient and a robust method is required which is also efficient for the hyperbolic case.

Upwind discretizations which follow the flow direction have been used successfully e. g. by Hemker and Spekreijse [8], Dick [9], Mulder [10] and others. For multigrid methods, a robust smoother is needed which solves the convection dominated phenomena. Flow aligned or alternating line Gauß-Seidel smoothers have been applied in [8] and [9]. A semi-coarsening approach is introduced in [10] for treating characteristic flow directions aligned with the grid. An interesting approach for subsonic flows has been presented by Ta'asan in [11]. The elliptic and hyperbolic parts of the equations are decoupled and treated separately. Efficient solution methods are used for each case.

A different approach is presented here. The author has presented in [1] robust methods based on a subspace decomposition of the solution space for solving linear elliptic and hyperbolic systems. Convergence rates are uniformly bounded for all flow directions and types of the equations. These subspace decomposition methods are generalizations of the frequency decomposition approach of Hackbusch [2], [3]. An overview on subspace decomposition methods is given by Xu [12].

In the present paper, the subspace decomposition approach is extended to the Euler equations. As a first step preliminary results for a simple test case are presented. We

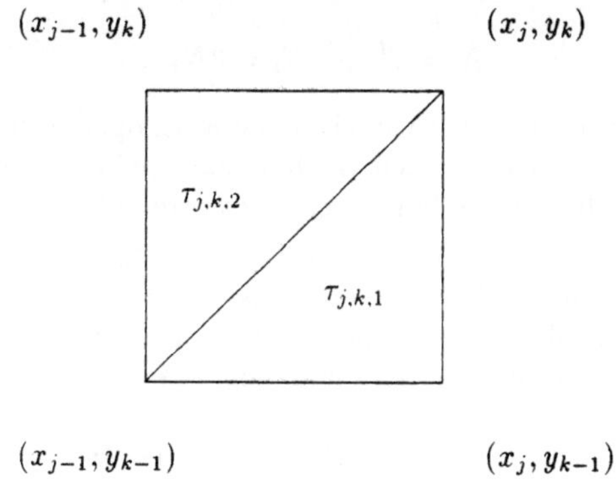

Figure 1: Subdivision of a square into two triangles

consider the subsonic flow through a channel with a circular arc of height 10% at the lower wall. This is a standard test case for the Euler equations.

A finite volume discretization is important for correct resolution of shocks and discontinuities. The integral formulation of the Euler equations yields the flux across sufficiently smooth control elements τ with boundary $\partial\tau$:

$$\Phi(\tau) = \int_{\partial\tau} f dy - g dx = 0 \ . \tag{2}$$

In contrast to standard finite volume methods, we do not introduce a second or dual mesh of control elements across which the flux integrals are approximated. The dual grid requires the reconstruction of the flux values at the boundary. The present approach avoids the use of such a second grid and needs no reconstruction procedure. We define a least square minimization problem by minimising the euclidean norm of the flux residuals. Details are given in section 2. The subspace decomposition approach is presented in section 3 and boundary conditions are given in section 4 together with numerical results.

2 Discrete minimization problem

The conservation equations (2) are discretized by a finite volume method on a triangular grid. Triangles are defined by subdividing an equidistant grid by diagonals running from southwest to northeast as in figure 1. The grid with step size $h_l = 1/N_2$ is defined by a body fitted coordinate transformation of the index set:

$$Z = \{ (j,k) \in \mathbb{Z}^2 \,|\, 0 \le j \le N_1 \ , \ \ 0 \le k \le N_2 \}$$

to the computational domain in $[0,1] \times [-1,2]$:

$$\begin{aligned} x_j &= -1.0 + j h_l \\ y_k &= y_0(x_j) + k h_l (1.0 - y_0(x_j)) \ , \end{aligned}$$

where:

$$N_2 = 2^l \quad , \quad N_1 = 3N_2 \quad .$$

The contour of the lower wall, $y_0(x)$ is given by a straight wall with a circular arc of thickness 10% in $x \in [0,1]$. Notice, that the boundary points are treated as inner points. Even with Dirichlet conditions the boundary points are not eliminated.

The discretization defines a nodal space $U = \{u : Z \to \mathbf{R}^3\}$ of flow variables $(v_x, v_y, \rho)^{\mathrm{T}}$ and a regular triangulation $\mathcal{T}$. The flux integrals (2) of the Euler equations are applied on the triangles τ as control elements. The integrals along the edges are approximated by the trapezoidal rule. We optain a discrete nonlinear operator from the nodal space to the flux space:

$$F : U \to \mathcal{F} \quad , \quad \mathcal{F} := \{\oplus : \mathcal{T} \to \mathbf{R}^3\}$$
$$F(u)\,(\tau) := \Phi(\tau) \quad , \quad \tau \in \mathcal{T} \quad . \tag{3}$$

As there are more triangles involved than nodes, the flux system $F(u) = 0$ cannot be solved exactly. Instead, a least square minimization problem based on the euclidean norm could be defined by $||F(u)||^2 \to \min$.

Boundary conditions are included into the minimization process. The discrete boundary conditions are given in flux type formulation. Vanishing *advective* flux through the wall is approximated by the midpoint rule. For a node $r_j = (x_j, y_k)$ at the wall, the straight line between mid points $r_{j+1/2} = \frac{1}{2}(r_{j+1} + r_j)$ of adjacent triangle edges serves as a control edge. The boundary condition at node r_j is approximated by:

$$B(u)(r_j) := \int_{r_{j-1/2}}^{r_{j+1/2}} \tilde{f}dy - \tilde{g}dx = \tilde{f}_j\,(y_{k+1/2} - y_{k-1/2}) - \tilde{g}_j\,(x_{j+1/2} - x_{j-1/2}) = 0 \quad , \tag{4}$$

$$\tilde{f}_j = \tilde{f}(u(r_j)) \quad , \quad \tilde{g}_j = \tilde{g}(u(r_j)) \quad .$$

Dirichlet type boundary conditions, as e. g. for the entropy, are specified by:

$$B_s(u)(r_j) := (s_\infty - s(r_j))\,\mathrm{length}(r_{j+1/2} - r_{j-1/2}) \quad .$$

Similar Dirichlet boundary conditions are defined for the inflow $v_{x,\infty}$, s_∞ and the static pressure p_{ex} at the outflow. Flux and boundary residuals are simultaneously minimized by:

$$E(u) := \tfrac{1}{2}\,(\,||F(u)||^2 + ||B(u)||^2\,) \to \min$$
$$E(u) \le E(w) \quad , \quad \forall\, w \in U \quad . \tag{5}$$

The solution of this nonlinear minimization problem could be obtained by standard methods as e. g. a conjugate gradient method. In the next section a faster converging method based on a subspace decomposition approach is presented.

3 Subspace decomposition methods

3.1 Minimization problems on subspaces

Domain decomposition methods are well known for parallel computing (see e. g. [13]). The computational region is subdivided into smaller regions based on geometrical considerations. A more general approach is applied here. The nodal space U is subdivided into subspaces:

$$U = U_1 + U_2 + \ldots + U_K \ . \tag{6}$$

These subspaces are not related to a geometrical decomposition of the grid. They are defined by carefully constructed prolongations on coarse grid spaces. Details are given in the next section.

The minimization approach allows a consistent formulation of subspace decomposition methods. On each subspace U_κ, coarse grid corrections are defined by restricting the minimization problem on these subspaces. For given u the correction v_κ is obtained by minimizing

$$E(u + v_\kappa) \leq E(u + v) \ , \quad \forall\, v \in U_\kappa \ .$$

This nonlinear minimization problem is assumed to be simpler and cheaper to solve than the original problem. The solution defines nonlinear correction operators by:

$$\begin{aligned} N_\kappa(u) &:= v_\kappa \\ G_\kappa(u) &:= u + v_\kappa \ . \end{aligned} \tag{7}$$

The subspace corrections may be applied sequentially yielding a sequential subspace decomposition method:

$$u^{n+1} = G_K G_{K-1} \ldots G_1(u^n) \ ,$$

$$\Phi^{SEQ} = \prod_{\kappa=1}^{K} G_\kappa \ . \tag{8}$$

A parallel subspace decomposition method is defined by:

$$\Phi^{PAR} = I + \sum_{\kappa=1}^{K} \alpha_\kappa N_\kappa \ , \tag{9}$$

where:

$$u^{n+1} = u^n + \sum_{\kappa=1}^{K} \alpha_\kappa v_\kappa \ , \quad v_\kappa = N_\kappa(u^n) \ .$$

The coefficients α_κ depend on u^n and are optimized by a small minimization problem operating on the space $U_\alpha = \mathrm{Span}\,\{ v_\kappa \,|\, 1 \leq \kappa \leq K \}$:

$$E(u^{n+1}) \leq E(u^n + \sum_{\kappa=1}^{K} \tilde{\alpha}_\kappa v_\kappa) \ , \quad \forall\, \tilde{\alpha}_\kappa \in \mathbf{R} \ . \tag{10}$$

Here, the subspace corrections v_κ are independent of each other and may be calculated in parallel on different processors. The efficiency depends on the solution strategy of the subspace problems. At the moment, a standard nonlinear conjugate gradient method is applied. As the condition of the subspace problems is different for different subspaces, a varying number of subiterations are needed for calculating v_κ. This leads to some load imbalance. Two remedies overcomes this problem. One approach is the use of a solution method which is independent of the condition of the problem. A direct solution method would be expensive, but a multigrid strategy based on recoursive application of the subspace decomposition approach is expected to be more efficient. A second approach is to restrict the number of subiterations. A fixed number of iterations improves the load balance but decreases the overall efficiency because then the subspace corrections are appoximate solutions with different accuracy. Furthermore the small problem on U_α is not calculated in parallel. An estimate of the coefficients α_κ based on data of the preceding two iterates have been used for further improvement of the load balance. With these improvement a parallel efficiency of 0.77 on nine processors is presented in [1].

3.2 The subspaces

The subspaces are defined recoursively by a sequence of prolongations on coarse grids. The coarse grid spaces use shifted grids with nodes outside the computational domain. A larger index set is defined with additional nodes:

$$\overline{Z} := \{ (j,k) \in \mathbb{Z}^2 \mid -1 \leq j \leq N_1 + 1 \, , \quad -1 \leq k \leq N_2 + 1 \} \ .$$

Four coarse index sets are defined by:

$$\begin{aligned} Z_0 &= \{(j,k) \in \overline{Z} \,|\, j \text{ even} \, , \; k \text{ even} \} \\ Z_1 &= \{(j,k) \in \overline{Z} \,|\, j \text{ odd} \, , \; k \text{ even} \} \\ Z_2 &= \{(j,k) \in \overline{Z} \,|\, j \text{ odd} \, , \; k \text{ odd} \} \\ Z_3 &= \{(j,k) \in \overline{Z} \,|\, j \text{ even} \, , \; k \text{ odd} \} \ . \end{aligned} \tag{11}$$

These coarse index sets cover the fine index set $Z \subset Z_0 \cup Z_1 \cup Z_2 \cup Z_3 \subset \overline{Z}$ and have additional nodes outside of Z. Coarse grid spaces are given as nodal spaces on coarse index sets:

$$V_\kappa = \{ v : Z_\kappa \to \mathbf{R}^3 \} \ , \quad 0 \leq \kappa \leq 3 \ .$$

Subpaces of U are defined by prolongations on the coarse spaces:

$$p_\kappa : V_\kappa \to U \ , \quad 0 \leq \kappa \leq 3$$

$$U_\kappa = \text{Range}\,(p_\kappa) \ ,$$

which are given in stencil notation by:

$$\begin{aligned} p_0 &= \tfrac{1}{2} \begin{bmatrix} 0 & 1 & 1 \\ 1 & 2 & 1 \\ 1 & 1 & 0 \end{bmatrix} \ , \quad p_1 = \tfrac{1}{2} \begin{bmatrix} 0 & 1 & -1 \\ -1 & 2 & -1 \\ -1 & 1 & 0 \end{bmatrix} \\ p_2 &= \tfrac{1}{2} \begin{bmatrix} 0 & -1 & -1 \\ 1 & 2 & 1 \\ -1 & -1 & 0 \end{bmatrix} \ , \quad p_3 = \tfrac{1}{2} \begin{bmatrix} 0 & -1 & 1 \\ -1 & 2 & -1 \\ 1 & -1 & 0 \end{bmatrix} \ . \end{aligned} \tag{12}$$

In the system case, the prolongations are applied on each component of (v_x, v_y, ρ) separately. Note that these prolongations operate on corresponding shifted coarse grids.

Parts of the stencils outside of the fine grid Z are ignored, e. g. at the boundary $x = 2$ we have the prolongation stencils:

$$p_0 = \frac{1}{2}\begin{bmatrix} 0 & 1 & - \\ 1 & 2 & - \\ 1 & 1 & - \end{bmatrix} , \quad p_1 = \frac{1}{2}\begin{bmatrix} 0 & - & - \\ -1 & - & - \\ -1 & - & - \end{bmatrix} .$$

The central node of stencil p_1 in grid Z_1 is prolongated to a node in $\overline{Z} - Z$ outside of the fine grid and is ignored. But this node influences nodes at the boundary with weights $-1/2$ and could not be totally omitted. It was found essential to include the additional degrees of freedom associated with these nodes for obtaining a robust method. Without these additional degrees of freedom the subspace decomposition method was not robust with four subspaces alone but needed four additional subspaces, see [14].

Notice that prolongation p_0 is the seven point prolongation which interpolates linearly in x and y direction and along the southwest/northeast diagonal. These directions coincide with the triangle edges. Three additional prolongations, $p_1, \ldots, p_3$, are defined in analogy to the frequency decomposition approach ([2, 3]). They are based on p_0 and use alternating signs of the coefficients with weight $-1/2$. These prolongations are no more interpolations because they interpolate along only one of the triangle edges and oscillate along the other two edges. This is an essential feature for obtaining a robust iteration for the solution of hyperbolic problems. In that case, errors which propagate along characteristics aligned with triangle edges are hardly elliminated by standard multigrid methods. The prolongations $p_1, \ldots, p_3$, together with p_0 are modelled for treating these characteristic directions. The semi-coarsening approach of Mulder [10] is an alternative concept for treating grid aligned characteristics.

The subspaces V_κ may be further subdivided into four subspaces: $V_{\kappa,0}, V_{\kappa,1}, V_{\kappa,2}, V_{\kappa,3}$ based on coarser grids and prolongations. This process is recoursively repeated on 4^m subspaces. For a recoursive construction we introduce a multiindex:

$$\kappa = (\kappa_1, \kappa_2 \ldots \kappa_m) , \quad 0 \leq \kappa_i \leq 3$$

and truncated multiindexes of length μ:

$$\kappa(\mu) = (\kappa_1, \kappa_2 \ldots \kappa_\mu) , \quad 1 \leq \mu \leq m .$$

We decompose the index sets

$$Z_\kappa \subset Z_{\kappa,0} \cup Z_{\kappa,1} \cup Z_{\kappa,2} \cup Z_{\kappa,3} .$$

By recoursion over the length μ of κ we obtain

$$Z_{\kappa(\mu)} \subset \bigcup_{\kappa_\mu=0}^{3} Z_{\kappa(\mu-1),\kappa_\mu} .$$

This defines coarse grid nodal spaces:

$$V_{\kappa(\mu)} = \{ v : Z_{\kappa(\mu)} \rightarrow \mathbf{R}^3 \} ,$$

and prolongations on corresponding coarse spaces:

$$p_{\kappa(\mu)}^{\kappa(\mu-1)} : V_{\kappa(\mu)} \rightarrow V_{\kappa(\mu-1)}$$

$$p_\kappa = p_{\kappa_1} \circ p^{\kappa_1}_{\kappa_1,\kappa_2} \circ p^{\kappa_1,\kappa_2}_{\kappa_1,\kappa_2,\kappa_3} \circ \ldots \circ p^{\kappa(m-1)}_{\kappa(m)} \quad .$$

Subspaces are defined by:

$$U_\kappa = \text{Range}\,(p_\kappa)$$

with

$$U = \sum_\kappa U_\kappa \quad , \quad \kappa \in [\,0:3\,]^m \quad .$$

The subindex sets $Z_{\kappa(\mu)}$ and prolongations $p^{\kappa(\mu-1)}_{\kappa(\mu)}$ are straightforward generalizations of (11) and (12). The present investigation is restricted to level $m \leq 3$ with up to 64 subspaces.

4 Results

Preliminary numerical results for a simple test case are presented. We consider the subsonic flow through a channel of size 1×3 with a circular arc of height 10% at the lower wall at $x \in [\,0,1\,]$ (see figure 2). Inflow Mach number is $M_\infty = 0.5$. This is a standard test case for the Euler equations.

Boundary conditions at the wall are given by vanishing *advective* flux through the wall. For an element of length Δt tangential to the wall we obtain:

$$\int_t^{t+\Delta t} \tilde{f}dy - \tilde{g}dx = 0 \quad .$$

Here, the advective fluxes contain no pressure term for the momentum equations:

$$\begin{aligned} \tilde{f} &= (0, \rho v_x v_y, \rho v_x)^{\mathrm{T}} \\ \tilde{g} &= (0, \rho v_y^2, \rho v_y)^{\mathrm{T}} \quad . \end{aligned}$$

Dirichlet type boundary conditions are prescribed at the inflow boundary with specified x velocity and entropy:

$$v_x\,(x = -1) = v_{x,\infty}$$

$$s\,(x = -1) = s_\infty \quad .$$

The exit pressure at the outflow boundary is specified according to isentropic flow with exit Mach number of $M_\infty = 0.5$:

$$p\,(x = +2) = p_{ex} \quad , \quad M_\infty = 0.5 \quad .$$

It was advantageous to prescribe the entropy at the wall:

$$s\,(\text{wall}) = s_\infty \quad .$$

The wall contour is a streamline and the entropy is equal to the inflow entropy. First results without prescribed wall entropy led to numerical errors in the entropy at the wall which were advected downstream and accumulated, thereby totally contaminating the solution. Clearly, the entropy boundary condition prevents the accurate calculation of shocks and transonic flows.

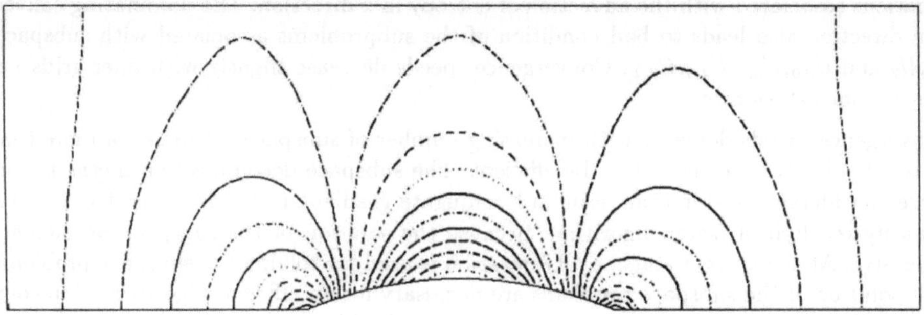

Figure 2: Isobars of Euler solution, $M_\infty = 0.5$

Table1: Convergence of sequential subspace decomposition Φ^{SEQ}

	subspaces		
grid size	4	16	64
6×18	0.74	0.85	0.86
12×36	0.78	0.92	0.95

Table 2: Convergence of parallel subspace decomposition Φ^{PAR}

	subspaces		
grid size	4	16	64
6×18	0.85	0.96	0.96
12×36	0.92	0.97	-

Figure 2 shows isobars for a numerical solution on a 72×24 grid. The wall pressure is in the range $[-0.43, 0.54]$. The solution is symmetric and coincides with a result of Lötstedt [15].

Convergence rates for the sequential and parallel subspace decomposition methods are given in tables 1 and 2. Asymptotic reduction rates of the gradients

$$\rho = \lim_{n \to \infty} ||\text{grad } E(u^n)|| / ||\text{grad } E(u^{n-1})|| \tag{13}$$

are used as a measure of the convergence.

Convergence rates of 0.8 for the sequential and of 0.9 for the parallel approach are obtained with four subspaces. This is in accordance with linear results presented in [1] where corresponding convergence rates of 0.80 and 0.92 were observed for hyperbolic systems. This result could have been expected, because of the hyperbolic character of the Euler equations associated with the advection of entropy in x direction. The dominating convection direction also leads to bad condition of the subproblems associated with subspaces U_0, U_2 and $U_{0,0}, U_{0,2}, U_{2,0}, U_{2,2}$. Convergence speeds decrease slightly with finer grids but they do not deteriorate.

Convergence speeds decrease with increasing number of subspaces. The use of more than 16 to 64 subspaces seems not to be efficient. The subspace decomposition methods converge considerably faster than standard conjugate gradient methods. This does not necessarily result in a faster algorithm, because the solution of the subspace problems is expensive. At the present stage, a CG iteration is used for solving the subspace problems. Fast solution of the subspace problems are necessary for an efficient algorithm. A recoursive application of the subspace decomposition method is proposed for that purpose.

Convergence rates for the parallel approach are worse than for the sequential algorithm, but the method could easily be implemented on parallel computers.

5 Conclusion

A new approach for solving steady inviscid flows is presented. In contrast to time marching methods, we directly solve the steady problem. Former results for linear hyperbolic systems [1] are generalized to the Euler equations.

A finite volume discretization of the flux equations on triangular grids is employed which avoids the use of a dual grid and reconstruction procedures. Flux integrals are approximated on triangles of the primary grid. A least square minimization problem is defined by minimizing the euclidean norm of the flux residuals and the boundary conditions.

The discrete minimization problem is solved by subspace decomposition methods. The nodal space is decomposed into carefully constructed subspaces. The minimization approach allows a consistent formulation of subspace corrections. Sequential and parallel subspace decomposition methods are constructed and compared.

As a simple test case subsonic flow through a channel with 10% circular arc at the lower wall is analysed. The numerical solution compares favourably with published results. First and at the moment preliminary results reveal convergence rates of 0.8 for the sequential approach and 0.9 for the parallel method with four subspaces. Convergence does not deteriorate with decreasing grid step size but decreases with increasing number of subspaces.

The present results demonstrate the ability of the subspace decomposition methods for solving the Euler equations and provide a basis for their efficient solution. Future work is directed towards the improvement of computational speed and verification of the results. The extension to more complicated geometries and flows with shocks is planned. An extension to solving the Navier-Stokes equations and three dimensional flows should be discussed.

Acknowledgement

The present work was supported by the "Deutsche Forschungsgemeinschaft" (German Research Society) and the "Land Sachsen-Anhalt" (Saxony Anhalt) within the programme: "Graduiertenkolleg Modellierung Berechnung und Identifikation mechanischer Systeme" at the Otto-von-Guericke University, Magdeburg. Figure 2 was obtained with help of the graphical program system COMADI of H. Vollmers, DLR Göttingen.

References

[1] Katzer, E.: *A Parallel subspace decomposition method for elliptic and hyperbolic systems.* Tenth GAMM-Seminar Kiel on Fast Solvers for Flow Problems, Kiel Jan. 1994. Proceedings appear in: Hackbusch, W. (ed.): Notes on Numerical Fluid Mechanics. Braunschweig, Wiesbaden: Vieweg, 1995.

[2] Hackbusch, W.: *A new approach to robust multi-grid solvers.* First. Int. Conf. on Industrial and Applied Mathematics. ICIAM '87. Proceedings, Philadelphia, PA: SIAM 1988, pp. 114–126.

[3] Hackbusch, W.: *The frequency decomposition multi-grid method. Part I: Application to anisotropic equations.* Numerische Mathematik 56 (1990) 229-245.

[4] Hackbusch, W.: *Theorie und Numerik elliptischer Differentialgleichungen.* Stuttgart: Teubner, 1986.

[5] Anderson, D. A.; Tanehill, J. C.; Pletcher, R. R.: *Computational fluid mechanics and heat transfer.* Washington etc: Hemisphere Puplishing Corporation, 1984.

[6] Wagner, S. et al. (eds.): *Computational fluid dynamics '94.* Second European Computational Fluid Dynamics Conference, Stuttgart, Sept. 1994. Chichester: J. Wiley & Sons, 1994.

[7] Oswatisch, K.: *Grundlagen der Gasdynamik.* Wien, New York: Springer, 1976.

[8] Hemker, P. W.; Spekreijse, S. P.: *Multiple grid and Osher's scheme for the efficient solution of the Euler equations.* Applied Numer. Math. 2 (1986) 475–493.

[9] Dick, E.: *Multigrid formulation of polynomial flux-difference splitting for steady Euler equations.* J. Computational Physics 91 (1990) 161–173.

[10] Mulder, W. A.: *A high resolution Euler solver based on multigrid, semi-coarsening, and defect correction.* J. Computational Physics 100 (1992) 91–104.

[11] Ta'asan, Sholmo: *Optimal multigrid solvers for steady state inviscid flow problems.* Fourth European Multigrid Conference, EMG '93, Amsterdam, 1993.

[12] Xu, Jinchao: *Iterative methods by space decomposition and subspace correction.* SIAM Review 34 (1992), 581–613.

[13] Keyes, E. D.; Xu, Jinchao (eds.): *Domain Decompositon Methods in Scientific and Engineering Computing.* Proceedings of the Seventh International Conference on Domain Decompositon. Contemporary Mathematics 180, Providence, R. I.: AMS 1994.

[14] Katzer, E.: *A subspace decomposition twogrid method for hyperbolic equations.* Fourth European Conference on Multigrid Methods, EMG '93. In: Hemker, P. W.; Wesseling, P. (esd.): Contributions to Multigrid, pp. 87–98. CWI Tract 103. Amsterdam: Centrum voor Wiskunde en Informatica, 1994.

[15] Lötstedt, Per: *Improved convergence to the steady state of the Euler equations by enhanced wave propagation.* Report No. 144/1992, Uppsala University, Dept. of Scientific Computing.

Domain Decomposition Schemes and Coupling Conditions for Kinetic and Hydrodynamic Equations

Axel Klar
AGTM, FB Mathematik
Universität Kaiserslautern
67663 Kaiserslautern
Germany

Summary

This article is devoted to domain decomposition schemes for kinetic and hydrodynamic equations. A convergence proof for an alternating scheme is given and coupling conditions at the interface between the equations are developed and investigated. In particular for nonequilibrium situations at the interface new coupling conditions are developed by considering interface layers. This leads to kinetic linear half space problems. A fast procedure to solve these problems is given.

1 Introduction

The Boltzmann equation and the more classical gas dynamics equations (such as Euler or Navier-Stokes equations) are used to model hypersonic gas flows. Numerical simulations of such flows are useful for example in the design of space vehicles, in particular in understanding the behavior of the early phases of reentry flights.
Such flows are usually far from any kind of local equilibrium states: real gas effects (and the so many different degrees of freedom involved such as rotational and vibrational energies) as well as the importance of chemical reactions in the energy balance on the vehicle surface demand that variants of the Boltzmann equation be used as first principle equations instead of the Euler or Navier-Stokes equations. However, when the mean free path of molecules becomes small, all numerical methods for the Boltzmann equation become exceedingly expensive in computing time. Therefore, gas dynamics equations should be used whenever possible — in other words, near local equilibrium states in situations where the local mean free path is small and outside of shock and boundary layers. These considerations prompt the use of domain decomposition strategies, where the Boltzmann equation is to be solved only in regions others than those mentioned above. Once the regions described by the gas dynamics equations are determined, the next major problem is the matching of the Boltzmann domain with the Euler or Navier-Stokes domain. This question is far from being an easy one, as the equations to couple and the numerical schemes used to solve them are of very different nature.

The approach usually employed in numerical procedures to solve the coupling problem for kinetic and hydrodynamic equations is the following: The coupling conditions at the interface are determined by equalizing the moments or fluxes at the interface between kinetic and hydrodynamic regions. The solution is then found by an iterative scheme solving in turn the kinetic and hydrodynamic equations using these coupling conditions. This procedure leads to a reasonable solution for near equilibrium situations at the interface. For work on the coupling of Boltzmann- and gas dynamics equations, see (among other references) Bourgat et al. [1] , Lukschin et al. [12]. A more refined approach to find the correct coupling conditions is given for example by Illner and Neunzert [7].
In section 2 we state a convergence proof for this iterative scheme for linearized versions of the above equations.
In section 3 we consider instead of equilibrium situations nonequilibrium states at the interface. In this case the above coupling conditions will not lead to the correct results. Here the matching requires a more exact analysis. As a general principle, the matching can be done by modelling the interface region by a transition layer where some "intermediate equation" (e.g., the linearized Boltzmann equation) is solved, see Golse [3] and Klar [8]. This leads to a kinetic linear half space problem. The asymptotic values and the outgoing flux of the solution of this problem determine the coupling conditions at the interface. Obviously the direct solution of the half space problem would be much too expensive. Moreover from the above we see that only the asymptotic states and the outgoing fluxes are really required. This leads to the development of a fast numerical scheme which computes approximately these two things by a Chapman Enskog type expansion procedure. It makes the approach reasonable from a numerical point of view. We describe it roughly in section 4.
In section 5 a numerical example for the 3-dimensional BGK model is shown comparing different types of coupling conditions.
Concluding the introduction we remark that coupling conditions similiar to the ones described here for gas dynamics can be developed for kinetic semiconductor and drift diffusion equations, see Klar [10].

2 Convergence of the Domain Decomposition Scheme

We investigate in this section the alternating scheme described in the introduction with coupling conditions given by the equality of fluxes at the interface. To simplify, a 1-dimensional geometry, $x \in [-L, L]$, is considered. The physical situation is supposed to be described by the Boltzmann equation linearized in the sense of Grad [6] around a constant Maxwellian state with parameters $\bar{\rho}, \bar{u} = (\bar{u}_1, \bar{u}_2, \bar{u}_3), \bar{T}$. We assume the mean free paths to be ϵ_1 in $[-L, 0]$ and ϵ_2 in $[0, L]$. After shifting velocities $v \to v + \bar{u}$ one obtains the equations

$$\partial_t \varphi_i + (v_1 + \bar{u}_1)\partial_x \varphi_i + \frac{1}{\epsilon_i} Q\varphi_i = 0, \tag{2.1}$$

where the index i=1,2 stands for the the regions $[-L, 0]$ and $[0, L]$ respectively, $\varphi_i = \varphi_i(x, v, t)$, $v = (v_1, v_2, v_3) \in \mathbb{R}^3$, $t \in [0, \infty)$, $\bar{u}_1 \in \mathbb{R}$ and Q is the Boltzmann collision operator linearized around a constant Maxwellian state with parameters $(\bar{\rho}, 0, \bar{T})$, i.e.

$$\bar{M}(v) := \frac{\bar{\rho}}{(2\pi\bar{T})^{3/2}} \, exp\left(-\frac{|v|^2}{2\bar{T}}\right).$$

Q is given by

$$Q\varphi = \nu\varphi - K\varphi,$$

where K is an integral operator and $\nu = \nu(v)$, see, e.g., Grad [6]. The collision invariants are in Grads formulation given by

$$\bar{M}^{1/2}, \bar{M}^{1/2}v_i, \bar{M}^{1/2}|v|^2, \quad i = 1,2,3.$$

The macroscopic moments $\Theta = (\Theta_1, \cdots, \Theta_5) = (\rho, u = (u_1, u_2, u_3), T)$ in D_i are given by

$$\Theta_j(x,t) := \int_{\mathbf{R}^3} \chi_j(v)\varphi_i(x,v,t)dv, \; j = 1, \ldots, 5,$$

where the χ_j are linear independent linear combinations of the collision invariants.
We assume that ϵ_2 in $[0, L]$ is small such that an approximation of the Boltzmann equation by macroscopic equations is valid.
The global kinetic solution in the whole domain $[-L, L]$ is then approximated by the solution of the following coupling problem with possibly overlapping domains:
In the domain $D_1 = [-L, x_0]$, where $x_0 \geq 0$ is a small parameter, we solve the kinetic equation (2.1) with mean free path ϵ_1. In the domain $D_2 = [-x_0, L]$ we solve a macroscopic equation for $\Theta = (\Theta_1, \ldots, \Theta_5)$:

$$\partial_t\Theta = A\partial_x^2\Theta + B\partial_x\Theta \tag{2.2}$$

with $A, B \in \mathbb{R}^{5\times5}$. We suppose the system to be the Euler or Navier Stokes equations linearized around the constant state $(\bar{\rho}, \bar{u}, \bar{T})$. For example the Euler equations are given by $A = 0$ and

$$B = \begin{bmatrix} \bar{u}_1 & \bar{\rho} & 0 & 0 & 0 \\ \frac{\bar{T}}{\bar{\rho}} & \bar{u}_1 & 0 & 0 & 1 \\ 0 & 0 & \bar{u}_1 & 0 & 0 \\ 0 & 0 & 0 & \bar{u}_1 & 0 \\ 0 & \frac{2}{5}c^2 & 0 & 0 & \bar{u}_1 \end{bmatrix},$$

where $c > 0$ defined by $c^2 = \frac{5}{3}\bar{T}$ is the speed of sound. These systems are good approximations of the Boltzmann equation (2.1) in D_2, if the mean free path in D_2 is small. We equip the system with the necessary initial conditions and boundary conditions at $x = -L$ for φ_1 and $x = L$ for Θ. The coupling conditions are for our linearized case given by a linearized Maxwellian as the 'distribution function in the macroscopic region'. We define

$$\varphi_{macro}(x,v,t) \;=\; \bar{M}^{1/2}\left(\frac{\rho(x,t)}{\bar{\rho}} + u(x,t)\frac{v}{\bar{T}} + T(x,t)\frac{|v|^2 - 3\bar{T}}{2\bar{T}^2}\right),$$

where ρ, u, T are the solutions of the macroscopic equations in D_2.
This leads to the following type of coupling conditions at x_0 and $-x_0$: At the point x_0 the distribution function going into the kinetic region is given by $\varphi_{macro}(x_0, v, t)$. The condition is

$$\varphi_1(x_0, v, t) = \varphi_{macro}(x_0, v, t), \quad \forall v_1 + \bar{u}_1 < 0.$$

On the other side at $-x_0$ the boundary values necessary to solve equation (2.2) in D_2 are determined by equalizing the macroscopic fluxes. The conditions are

$$\int_{v_1+\bar{u}_1>0} (v_1+\bar{u}_1)\chi_j(v)\varphi_1(-x_0,v,t)dv$$
$$= \int_{v_1+\bar{u}_1>0} (v_1+\bar{u}_1)\chi_j(v)\varphi_{macro}(-x_0,v,t)dv$$

for $j = 1, \ldots 5$. Here we have to take the correct number of conditons due to the number of characteristics going into D_2.
One could determine Θ as well by the equality of moments at $-x_0$.
The solutions φ_1 and Θ of equations (2.1) in D_1 and (2.2) in D_2 with the above boundary and coupling conditions will be called the solution of the coupling problem. It can be found by the following alternating scheme

Initialization:
$\Theta^{(0)}(-x_0,t) := \chi_+(t)$, where χ_+ is an arbitrary bounded function.

Iterationstep:
Let $\Theta^{(n)}(-x_0,t)$ be known.

1. Solve equation (2.2) in D_2 with the initial and boundary values to get $\Theta^{(n)}(x,t)\ \forall x \in D_2$.
2. Determine $\varphi_1^{(n+1)}(x_0,v,t), v_1+\bar{u}_1 < 0$ according to the coupling condition at x_0.
3. Solve equation (2.1) in D_1 with the given inital and boundary conditions to obtain $\varphi_1^{(n+1)}(x,v,t), \quad \forall x \in D_1$.
4. Determine $\Theta^{(n+1)}(-x_0,t)$ with the coupling condition at $-x_0$.

The following theorem is proven in Klar [9] under slight additional restrictions.

Theorem 2.1 *Let $x_0 \in [0,L)$ and $t \in [0,T]$, T an arbitrary constant, then there is a unique solution (φ_1,Θ) of the coupling problem.*
For the sequences $\varphi_1^{(n)}(x,v,t)$ and $\Theta^{(n)}(x,t)$ of iterative solutions of (2.1) and (2.2) the following is true:

$\varphi_1^{(n)}$ *converges in* $\mathcal{L}^\infty(D_1 \times \mathbb{R}^3, \mathcal{L}^2([0,T]))$ *to* φ_1.
$\Theta^{(n)}$ *converges in* $\mathcal{L}^\infty(D_2, \mathcal{L}^2([0,T],\mathbb{R}^5))$ *to* Θ.

Remark : It is not clear whether the obtained coupled solution is physically correct, i.e. whether the coupled solution really approximates the global kinetic solution. This depends strongly on the situation at the interface and on the type of equations that are used. For the coupling of Boltzmann and Euler equations with near equilibrium conditions at the interface the above conditions are known to give correct results.
However in section 5 it is shown that for the coupling of Boltzmann and Euler equations the above conditions are not sufficient in nonequilibrium situations. Conditions like the one developed in the next section are needed there.
Moreover coupling second order equations like the Naver Stokes system to the Boltzmann equation without overlapping requires the use of first order (in the mean free path) approximations of the distribution function by φ_{macro} to obtain correct results. The above theorem remains true in this case as well. For this type of equations the use of overlapping domains is a possible alternative as shown in Klar [9].

3 Coupling Conditions for Boltzmann and Euler Equations for Nonequilibrium Situations

We describe here exact coupling conditions for nonequilibrium situations around the interface. Again the Boltzmann and Euler equations linearized around a constant state $\bar{\rho}, \bar{u}, \bar{T}$ with $\bar{u} = (\bar{u}_1, \bar{u}_2, \bar{u}_3)$ are considered. We restrict ourselves as in the previous section to a 1-dimensional geometry $x \in [-L, L]$ and refer to the remark at the end of this section for possible extensions to the multidimensional case.
The physical situation is supposed to be described by the same kinetic equations as in section 2 with mean free paths ϵ_1 in $[-L, 0]$ and ϵ_2 in $[0, L]$. We assume here again that ϵ_2 in D_2 is small.
Let $D_1 = [-L, 0]$ and $D_2 = [0, L]$. We solve in D_1 the kinetic equation

$$\partial_t \varphi_1 + (v_1 + \bar{u}_1)\partial_x \varphi_1 + \frac{1}{\epsilon_1} Q\varphi_1 = 0. \tag{3.1}$$

Let $c > 0$ be again the speed of sound. The hydrodynamic equation in D_2 is the compressible linearized Euler equation

$$\partial_t \Theta + B\partial_x \Theta = 0, \tag{3.2}$$

with $x \in [0, L], \Theta = (\rho, u_1, u_2, u_3, T)$. B was defined in the above section.
Now the coupling of (3.1) in D_1 and (3.2) in D_2 is considered and compared to the solution of the Boltzmann equation in the whole domain $[-L, L]$. We concentrate in the following on the coupling conditions at $x = 0$ disregarding the boundary conditions at $x = L$ and $x = -L$.
The system (3.2) is diagonalizable with Eigenvalues $\lambda_i = \bar{u}_1, i = 1, 2, 3, \lambda_4 = \bar{u}_1 + c, \lambda_5 = \bar{u}_1 - c$. One needs at $x = 0$ for the system $0, 1, 4$ or 5 boundary conditions for the characteristic variables according to the value of $\bar{u}_1$, i.e. $\bar{u}_1 < -c, -c < \bar{u}_1 < 0, 0 < \bar{u}_1 < c$ or $\bar{u}_1 > c$. Here and in the following we assume $\bar{u}_1 \neq 0, c, -c$.
Simply using the equality of moments or fluxes at $x = 0$ will not lead to the correct results, as we shall see from the simulations in Section 5. This leads to the consideration of an interface layer between the kinetic and the hydrodynamic region. Asymptotic analysis shows that the following stationary half space problem has to be solved:

$$\begin{aligned} &(v_1 + \bar{u}_1)\partial_x \chi + Q\chi = 0, \quad x \in [0, \infty) \\ &\chi(0, v, t) = \varphi_1(0, v, t), \quad v_1 + \bar{u}_1 > 0 \end{aligned} \tag{3.3}$$

where $\varphi_1(0, v, t)$ is the distribution function in D_1 at $x = 0$. This equation has a unique solution for a suitable collision kernel , see e.g. Coron et al. [2], if - according to the values of $\bar{u}_1$ - a number of constraints is imposed. One needs $5, 4, 1$ or 0 constraints if $\bar{u}_1 < -c$, $-c < \bar{u}_1 < 0, 0 < \bar{u}_1 < c$, $\bar{u}_1 > c$ respectively with $c^2 = \frac{5}{3}\bar{T}$ as before. Solving the half space problem gives the asymptotic value

$$\chi(\infty, v, t) = \left(\frac{a_\infty(t)}{\bar{\rho}} + \frac{b_\infty(t)}{\sqrt{\bar{T}}} \frac{v}{\sqrt{\bar{T}}} + \frac{c_\infty(t)}{\bar{T}} \frac{|v|^2 - 3\bar{T}}{2\bar{T}} \right) \bar{M}^{\frac{1}{2}},$$

with $a_\infty, c_\infty \in \mathbb{R}$ and $b_\infty \in \mathbb{R}^3$. According to the number of constraints one already has $5, 4, 1$ or 0 equations for a_∞, b_∞ and c_∞. . This means that for $\bar{u}_1 < -c$, $-c < \bar{u}_1 < 0$,

$0 < \bar{u}_1 < c$ or $\bar{u}_1 > c$ we obtain $0, 1, 4, 5$ new conditions on the asymptotic values. This fits exactly to what is needed for Eulers equations as already remarked by Golse [3]. We discuss the procedure in more detail for $|\bar{u}_1| > c$.

For $\bar{u}_1 > c$ the half space problem is solved for prescribed incoming fluxes without any constraints. This gives the asymptotic value $\chi(\infty, v, t)$ depending on a_∞, b_∞ and c_∞. The 'macroscopic density function' in D_2 is a linearized Maxwellian with parameters $\rho(x,t), u(x,t), T(x,t)$ given by the solution of Eulers equations:

$$\varphi_{macro}(x, v, t) := \left(\frac{\rho}{\bar{\rho}} + \frac{u}{\sqrt{\bar{T}}} \frac{v}{\sqrt{\bar{T}}} + \frac{T}{\bar{T}} \frac{|v|^2 - 3\bar{T}}{2\bar{T}} \right) \bar{M}^{\frac{1}{2}}.$$

Comparing it at $x = 0$ to $\chi(\infty, v, t)$ one obtains

$$\begin{aligned} \rho(0,t) &= a_\infty(t) \\ u(0,t) &= b_\infty(t) \\ T(0,t) &= c_\infty(t). \end{aligned}$$

Thus the solution of the half space problem gives us the boundary conditions required for Eulers equations with $\bar{u}_1 > c$ at $x = 0$, i.e. $\rho(0,t), u(0,t)$ and $T(0,t)$.

Moreover the outgoing flux $\chi(0, v, t), v_1 + \bar{u}_1 < 0$ gives $\varphi_1(0, v, t), v_1 + \bar{u}_1 < 0$, i.e. the boundary condition at $x = 0$ for the domain D_1.

For $\bar{u}_1 < -c$ the situation is more simple. For $\bar{u}_1 < -c$ we do not need any boundary condition at $x = 0$ for Eulers equation in D_2. To solve the half space problem, however, 5 constraints on the solution are necessary. Comparing the macroscopic density function with parameters given by the solutions of Eulers equations, with the asymptotic value $\chi(\infty, v, t)$, we get the necessary number of constraints. We can then solve the half space problem, which yields $\varphi_1(0, v, t) = \chi(0, v, t)$, $v_1 + \bar{u}_1 < 0$.

For a derivation of these coupling conditions see Klar [8] as well as for a proof of their validity.

Remark: In the multidimensional case one can proceed in the same way. Suppose that the interface Σ divides the computational domain Ω into subdomains Ω_1 and Ω_2. At each point $x \in \Sigma$ one has to solve a one dimensional half space problem with coordinate axis along the unit normal $n(x)$ to Σ at the point x. This will lead for each $x \in \Sigma$ to the correct boundary conditions.

The main task is now to find a computationally reasonable - remark that the half space problem has to be solved at each point of the interface - yet accurate procedure to determine asymptotic states and outgoing distribution of the halfspace problem. This will be done in the next section.

4 Approximate solution of the half space problem

Consider the stationary equation in a halfspace

$$
\begin{aligned}
&(v_1+\bar{u}_1)\partial_x\varphi+\nu\varphi-K\varphi=0\\
&\varphi(0,v)=k(v), v_1+\bar{u}_1>0,
\end{aligned}
\tag{4.1}
$$

where $x\in[0,\infty), v=(v_1,v_2,v_3)\in\mathbb{R}^3, \bar{u}_1\in\mathbb{R}$.
We restrict in the following to the more difficult case $\bar{u}_1>0$ and assume for simplicity the constant state $(\bar{\rho},\bar{u},\bar{T})$ around which we linearize, to be equal to $(1,\bar{u},1)$. As already stated in the last section we need for $\bar{u}_1>c$ no additional condition for the equation to obtain a unique solution. For $0<\bar{u}_1<c$ one condition is needed. Remember that for $-c<\bar{u}_1<0$ four conditions and for $\bar{u}_1<-c$ five conditions are necessary.
The asymptotic state is

$$
\varphi(\infty,v)=\left(a_\infty+b_\infty v+c_\infty\frac{|v|^2-3}{2}\right)\bar{M}^{\frac{1}{2}}.
$$

We will develop now a numerical scheme to compute $a_\infty, b_\infty, c_\infty$.
We consider (4.1) with the condition, that

$$
< v_1\varphi(\infty,v)\bar{M}^{\frac{1}{2}} >
$$

is equal to some constant, if $0<\bar{u}_1<c$ and without any condition if $\bar{u}_1>c$. The adjoint equation is then

$$
\begin{aligned}
-(v_1+\bar{u}_1)\partial_x\psi+\nu\psi-K\psi &= 0, \quad \bar{u}_1>0\\
\psi(0,v) &= 0, \quad v_1+\bar{u}_1<0
\end{aligned}
\tag{4.2}
$$

with 4 conditions if $0<\bar{u}_1<c$ and all 5 conditions if $\bar{u}_1>c$ out of the full set of relations

$$
< (v_1+\bar{u}_1)\psi\bar{M}^{\frac{1}{2}}\begin{pmatrix}1\\ v\\ |v|^2\end{pmatrix} > = \begin{pmatrix}1\\1\\1\end{pmatrix}.
$$

We transform $v\to -v, \bar{u}\to-\bar{u}$ and consider

$$
\begin{aligned}
(v_1+\bar{u}_1)\partial_x\psi+\nu\psi-K\psi &= 0, \quad \bar{u}_1<0\\
\psi(0,v) &= 0, \quad v_1+\bar{u}_1>0
\end{aligned}
\tag{4.3}
$$

with 4, 5 conditions respectively out of the full set of relations

$$
< (v_1+\bar{u}_1)\begin{pmatrix}1\\ v\\ |v|^2\end{pmatrix}\psi\bar{M}^{\frac{1}{2}} > = \begin{pmatrix}-1\\-1\\-1\end{pmatrix}.
$$

By the usual Chapman Enskog procedure one can see that the macroscopic moments

$$
\rho := < \psi\bar{M}^{\frac{1}{2}} >,\ u := < v\psi\bar{M}^{\frac{1}{2}} >, T := < \frac{|v|^2-3}{3}\psi\bar{M}^{\frac{1}{2}} >
$$

approximately solve the stationary linearized Navier-Stokes equations:

$$\begin{aligned}
\partial_x(u_1+\bar{u}_1\rho) &= 0\\
\partial_x(T+\rho+\bar{u}_1u_1-\frac{4}{3}\mu\partial_x u_1) &= 0\\
\partial_x(\bar{u}_1u_i-\mu\partial_x u_i) &= 0, \quad i=2,3\\
\partial_x(3\bar{u}_1T-2\bar{u}_1\rho-2\lambda\partial_x T) &= 0\,.
\end{aligned}$$

μ, λ are the viscosity and conductivity coefficients for the temperature $\bar{T} = 1$. These equations can be solved exactly up to a number of free parameters.
The first approximation ψ_1 for ψ can then be calculated from

$$\begin{aligned}
(v_1+\bar{u}_1)\partial_x\psi_1+\nu\psi_1-K(M_{(\rho,u,T)}) &= 0\\
\psi_1(0,v) &= 0 \quad v_1+\bar{u}_1>0
\end{aligned}$$

where $M_{(\rho,u,T)}(v) := \bar{M}^{\frac{1}{2}}(\rho+u\cdot v+T\frac{|v|^2-3}{2})$, with ρ, u, T the solutions of the above Navier Stokes equations. We determine the free parameters by the following conditions and the requirement that ρ, u, T are finite at infinity:

$$< (v_1+\bar{u}_1)\left\{\begin{array}{c}\psi_1(\infty,v)\\ \psi_1(0,v)\end{array}\right\}\left(\begin{array}{c}1\\ v\\ |v|^2\end{array}\right)\bar{M}^{\frac{1}{2}} >= \left(\begin{array}{c}-1\\ -1\\ -1\end{array}\right),$$

which is the closest analogue to the conditions following equation (4.3).
This procedure can be iterated and we end up with an approximating series $\psi \sim \psi_1 + \ldots + \psi_n$. During the iteration steps inhomogeneous second order equations have to be solved instead of the above NavierStokes system.
By transforming $v \to -v$ and $\bar{u} \to -\bar{u}$ backwards we get an approximation for the solution $\psi(x,v)$of (4.2). The invariance in x of

$$< (v_1+\bar{u}_1)\varphi(x,v)\psi(x,v) >,$$

which can be easily seen using the selfadjointness of the collision operator Q, and the invariance of

$$< (v_1+\bar{u}_1)\bar{M}^{\frac{1}{2}}\left(\begin{array}{c}1\\ v\\ |v|^2\end{array}\right)\varphi(x,v) >$$

gives us the equations we need to determine the asymptotic state

$$\varphi(\infty,v) = (a_\infty+b_\infty v+c_\infty\frac{|v|^2-3}{2})\bar{M}^{\frac{1}{2}}.$$

Remark that the first component of b_∞ is fixed for $0 < \bar{u}_1 < c$ since we imposed a condition on the solution φ of (4.1). The first equation is

$$< (v_1+\bar{u}_1)\varphi(\infty,v)\psi(\infty,v) >=< (v_1+\bar{u}_1)\psi(0,v)\varphi(0,v) >$$

Moreover we use 3 equations for $0 < \bar{u}_1 < c$ (4 equations for $\bar{u}_1 > c$) out of

$$< (v_1+\bar{u}_1)\left(\begin{array}{c}1\\ v\\ |v|^2\end{array}\right)\bar{M}^{\frac{1}{2}}\varphi(\infty,v) > = < (v_1+\bar{u}_1)\left(\begin{array}{c}1\\ v\\ |v|^2\end{array}\right)\bar{M}^{\frac{1}{2}}\varphi(0,v) > .$$

Here we substitute $\varphi(\infty, v)$ as above and $\varphi(0, v) = k(v), \quad v_1 + \bar{u}_1 > 0$.
The function $\psi(0, v)$ is taken from the above approximation. Using the constraints, in particular

$$< (v_1 + \bar{u}_1)\psi(\infty, v)\bar{M}^{\frac{1}{2}} \begin{pmatrix} 1 \\ v \\ |v|^2 \end{pmatrix} > = \begin{pmatrix} 1 \\ 1 \\ 1 \end{pmatrix}$$

provides the desired equations for $a_\infty, b_\infty, c_\infty$.
The outgoing function $\varphi(0, v), v_1 + \bar{u}_1 < 0$ is determined similarly.
A more detailed description of the scheme for a BGK model equation and further references on linear half space problems can be found in Golse/Klar [5].

5 Numerical Results

In this section we investigate the coupling conditions proposed in Section 3 numerically. The coupling conditions described there are compared with the ones obtained by equalizing moments or fluxes at the interface as described in section 2. The coupling conditions obtained by the analysis of the kinetic half space problem are determined by the first step of the numerical scheme mentioned in the introduction and described roughly in section 4. We remark here that for $\bar{u}_1 = 0$ the first step of the scheme reduces to the so called variational method, see e.g. Golse [4] or Loyalka [11].
In the figure the 3-dimensional BGK model is considered, i.e. in D_1 we consider equation (3.1) with $\epsilon_1 = 1$ and substitute the following expression for the collision operator $Q\varphi$ taking $\bar{\rho} = \bar{T} = 1$:

$$\varphi - \left(\int \varphi \bar{M}^{\frac{1}{2}} dv + \sum_{i=1}^{3} v_i \int v_i \varphi \bar{M}^{\frac{1}{2}} dv + \frac{|v|^2 - 3}{2} \int \frac{|v|^2 - 3}{3} \varphi \bar{M}^{\frac{1}{2}} dv \right) \bar{M}^{\frac{1}{2}}$$

with

$$\bar{M} = \frac{1}{(2\pi)^{3/2}} \exp\left(-\frac{|v|^2}{2} \right).$$

In D_2 the Euler equation (3.2) is solved. $\bar{u}_1$ is equal to 1.4, i.e., bigger than $c = \sqrt{\frac{5}{3}}$. The ingoing function at $x = -L$ is $f_+(v, t) = v_1(|v|^2 - 5)\bar{M}^{\frac{1}{2}}$ for $t \in [0, T]$. The figure shows the temperature $\int \frac{|v|^2-3}{3} \varphi_1(x, v, t) \bar{M}^{\frac{1}{2}} dv$ in D_1 and $T(x, t)$ in D_2. The 3 types of coupling conditions are shown together with the kinetic solution in the whole domain with $\epsilon_1 = 1$ and $\epsilon_2 = 0.002$.

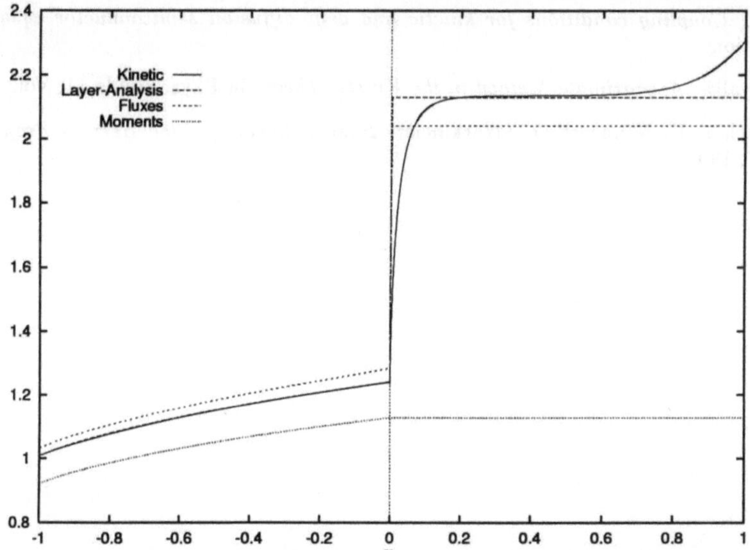

Figure: Temperature distribution for the coupling of BGK and Euler models.

References

[1] J.F. Bourgat, P. Le Tallec, D.Tidriri, Y. Qiu, *Numerical Coupling of Nonconservative or Kinetic Models with the Conservative Compressible Navier Stokes Equations*, Fifth International Symposium on Domain Decomposition methods for P.D.E., SIAM, Philadelphia 1991.

[2] F. Coron, F. Golse, C. Sulem, *A Classification of Well-posed Kinetic Layer Problems*, CPAM, Vol. 41, p. 409, 1988.

[3] F. Golse, *Applications of the Boltzmann Equation Within the Context of Upper Atmosphere Vehicle Aerodynamics*, Computer Meth. in Engineer. and Appl. Mech. Vol. 75, p.299, 1989.

[4] F. Golse, *Knudsen Layers from a Computational Viewpoint*, TTS P 21 (3), 211, 1992.

[5] F. Golse, A. Klar, *A Numerical Method for the Computation of Asymptotic States and Outgoing Distributions for Kinetic Linear Half Space Problems*, to appear in J. Stat. Phys..

[6] H. Grad, *Asymptotic Theory of the Boltzmann Equation II*, Rarefied gas dynamics, Proceedings of the 3. International symposium 1962, ed. J.A. Laurman, Academic Press, p. 26, 1963.

[7] R. Illner, H. Neunzert, *Domain Decomposition: Linking Kinetic and Aerodynamic Descriptions*, AGTM preprint 90, Kaiserslautern, 1993.

[8] A. Klar, *Domain Decomposition for Kinetic Problems with Nonequilibrium States*, to appear in Eur. J. Mech./B Fluids.

[9] A. Klar, *Convergence of Alternating Domain Decomposition Schemes for Kinetic and Aerodynamic Equations* to appear in Math. Meth. Appl. Sci..

[10] A. Klar, *Coupling conditions for kinetic and drift diffusion semiconductor equations* in preparation.

[11] S.K. Loyalka, *Approximate Method in the Kinetic Theory* in Phys. Fluids 11, Vol. 14, 1971.

[12] A. Lukschin, H. Neunzert, J. Struckmeier, *Interim Report for the Hermes Project* DPH 6174/91, 1992.

COUPLED PHYSICAL MODELLING FOR THE ANALYSIS OF GROUNDWATER SYSTEMS

O. Kolditz & R. Ratke & W. Zielke

Institut für Strömungsmechanik, Universität Hannover
Appelstraße 9a, D - 30167 Hannover, Germany

H.-J. Diersch

Gesellschaft für Wasserwirtschaftliche Planung und Systemforschung
Waltersdorfer Straße 105, D - 12526 Berlin, Germany

SUMMARY

This work examines variable density flow and corresponding transport of solute as well as heat. The mathematical model comprises a set of three non-linear coupled partial differential equations to be solved for pressure, solute concentration and temperature. Especially if considering such non-linear problems, numerical simulators must be verified in a particular series of tests. Frequently used benchmarks for testing variable density models are Henry's and Elder's problems. We calculate these benchmarks using different codes - ROCKFLOW and FEFLOW. At that time the impact of different numerical schemes on the convergence and the accuracy of the approximate solutions is studied. Based on the verified simulators more complex problems are tackled. In this way we consider thermohaline convection and mixed convection in thermoelastic fractures. Finally an example of a 3-D Benard convection is presented.

INTRODUCTION

Saltwater intrusion and hydrothermal convection are classic examples of density-dependent problems in groundwater hydrology. The ecologic and economic importance of these natural phenomena has invited continuous research in the fields of aquifer contamination/remediation, waste disposal, and utilization of geothermal energy. Further geophysical application of density-dependent models is connected eg. with salinity circulations in stratified microtidal estuaries (Zielke et al. 1994 [1]), deep hydrothermal convection in sedimentary basins (Clauser & Neugebauer 1991 [2]), magmatic convection within the Earth's crust (Fradkov et al. 1993 [3]) and convection in Arctic and Antarctic leads and polynyas (Graduiertenkolleg "Komplexe Dynamische Systeme" 1994 [4]).

Theoretical and experimental studies on convective phenomena started already in the middle of our century. Horton & Rogers (1945) [5] and Wooding (1962) [6] firstly studied buoyancy-driven convection processes in porous media. Henry (1964) [7] developed a semi-analytical solution for the calculation of the advance of a saltwater front in a confined aquifer. Elder (1967) [8] carried out first numerical simulations of thermal convections in Hele-Shaw cells. Combarnous & Bories (1975) [9] reviewed the former theoretical and numerical investigations related to density-coupled transport processes in porous media. In the following Henry's and Elder's works have become two of the standard tests of variable density groundwater models.

This paper should provide an overview and a collection of benchmarks and examples for variable density problems in groundwater hydrology we considered in the past.

BASIC EQUATIONS

The equations governing flow, mass and heat transport in groundwater systems are derived from the basic conservation principles for mass, momentum and energy. We condense these balance equations here shortly. Details of the physical and mathematical modelling can be found eg. in [10], [11], [12], and [13]. The symbols are summarized in the Appendix.

fluid mass balance

$$\frac{\partial n\rho}{\partial t} + \nabla(n\rho\vec{v}) = 0 \tag{1}$$

solute mass balance

$$\frac{\partial nC}{\partial t} + \nabla(n\vec{v}C) - \nabla(n\hat{D}\nabla C) = 0 \tag{2}$$

momentum balance (extended Darcy's law):

$$\vec{q} = n\vec{v} = -\frac{\hat{k}}{\mu(C,T)}\left(\nabla p - \rho(C,T)\,\vec{g} \right) \tag{3}$$

heat balance

$$[nc\rho + (1-n)c^s\rho^s]\frac{\partial T}{\partial t} + nc\rho\vec{v}\cdot\nabla T - \nabla([n\hat{\lambda} + (1-n)\hat{\lambda}^s]\nabla T) = 0\,. \tag{4}$$

The equations (1) - (4) are coupled through the velocity and the mass density which depends on temperature and solute concentration. To close the set of equations we use following formulations of the state for density and viscosity of the liquid:

$$\rho = \rho_0\,[\,1 + \beta_C(C-C_0) - \beta_T(T-T_0)\,] \tag{5}$$

$$\mu(T,C) = \mu_0\frac{1 + 1.85\omega - 4.1\omega^2 + 44.5\omega^3}{1 + 0.7063\zeta - 0.04832\zeta^3} \quad ; \quad \omega = \frac{C}{\rho_0}\,,\ \zeta = \frac{T-T_0}{T_1}\,. \tag{6}$$

Often the so-called *Boussinesq-approximation* is employed for simplifying the model. It neglects density variations within the continuity equation of the fluid mass. Basicly, by this assumption the conservation of mass is substituted by the conservation of volume.

The dimensional analysis provides three numbers that control the coupled flow problem:

the Rayleigh-number: $Ra = \dfrac{kg\rho_0 L\beta_T \Delta T}{\mu\kappa}$, $\kappa = \dfrac{n\lambda + (1-n)\lambda^s}{c\rho}$ (7)

the Lewis-number: $Le = \dfrac{\kappa}{nD}$ and the buoyancy ratio: $N = \dfrac{\beta_C \Delta C}{\beta_T \Delta T}$.

On the numerical implementations of density-dependent models is reported in detail by [10], [11], [12], [14], and [15].

SALTWATER INTRUSION

Henry's Problem

Henry's problem describes the advance of a saltwater front in a confined aquifer which is initially charged with fresh water. Henry (1964) [7] developes a solution technique to this problem which can calculate the steady-state solute distribution. To this end he applies the Boussinesq-approximation which implies the existence of stream functions.

Henry (1964) [7] derives analytical expressions for the stream function and the concentration in the form of Fourier series. The resulting algebraic equations for the determination of the coefficients of the Fourier series must be obtained by numerical techniques. The "mystery" of Henry's solution is, that no numerical model so far has been reproduced closely his semi-analytical model [16]. Nevertheless, because of the absence of any other non-numerical technique for this kind of non-linear problems, Henry's solution has become one of the standard tests of variable density groundwater models (eg. [17], [18], and [19]).

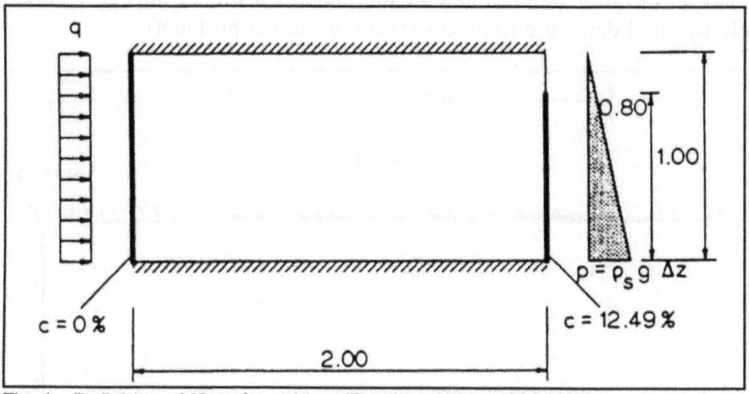

Fig. 1 Definition of Henry's problem (Fig. from Kröhn 1991 [12])

The idealized aquifer for the simulation of Henry's problem is shown in Fig. 1. The boundary conditions for flow consist of impermeable boundaries along the top and the bottom. Hydrostatic pressure is assumed along the vertical boundaries. For mass transport zero initial concentration is established within the aquifer. At the inland side (left) the concentration is zero which corresponds to the concentration of fresh water and at the coastal side (right) equal to the seawater concentration.

Computed concentration distributions obtained by Kröhn (1991) [12] in comparison to the results of other authors are shown in the Fig. 2.

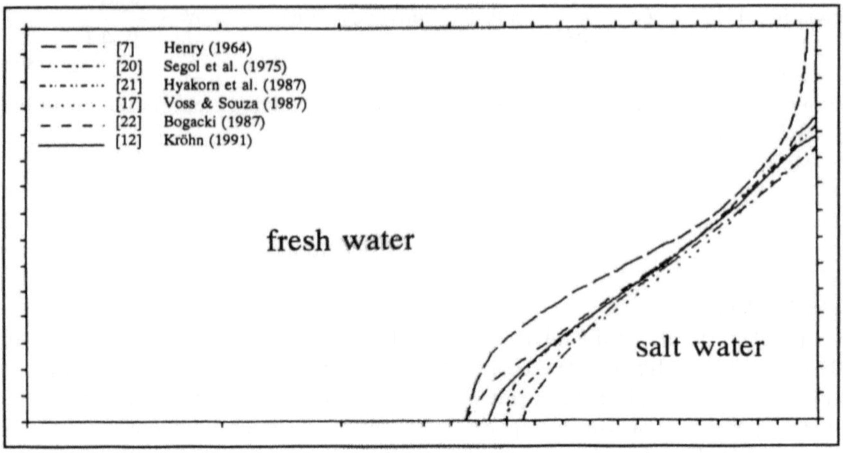

Fig. 2 Comparison of Henry's steady-state 50 % isohaline with different numerical solutions (Fig. from Kröhn 1991 [12])

Elder's problem

Elder (1967) [8] presents experimental and numerical studies concerning the convection in a Hele-Shaw cell produced by heating a part of the base of a cavity. Elder's problem can be treated as a thermal analogon for the saltwater intrusion into aquifers. The mass transport analogy is shown in Fig. 3. A permanent source of solute is located at the top while a zero concentration is maintained at the base of the closed rectangular aquifer. The advantage of this benchmark is to proof the accuracy of the numerical model in representing fluid flow driven purely by fluid density differences (free convection problem).

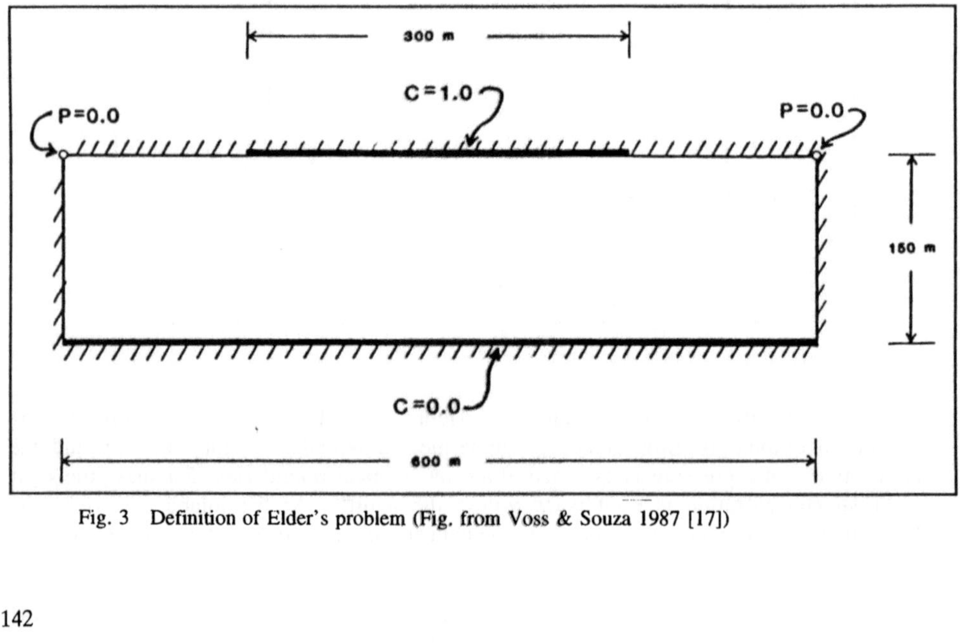

Fig. 3 Definition of Elder's problem (Fig. from Voss & Souza 1987 [17])

This example of a transient convection is frequently used for the cross-verification of numerical simulators ([17], [23], [24], and [25]). Voss & Souza (1987) [17] report on their numerical model for the simulation of Elder's (1967) [8] original problem. They obtain similar but not identical results. Especially at later computation times occur deviations in the salinity pattern. This may be caused by the different numerical approaches they used. Elder (1967) [8] and Voss & Souza (1987) [17] apply finite difference and finite element schemes, respectively. Further, they used slightly differing local discretizations. Encouraged by these previous works, Kolditz (1994b) [24] investigates the impact of the numerical approximation more in detail. In this way different finite element schemes (Galerkin-FEM, Upwind-FEM), local and temporal discretizations (predictor-corrector, semi-implicit, fully implicit) are tested. Furthermore, the influence of model assumptions concerning the approximation of the fluid mass balance is considered. Results of this study are presented in Fig. 4.

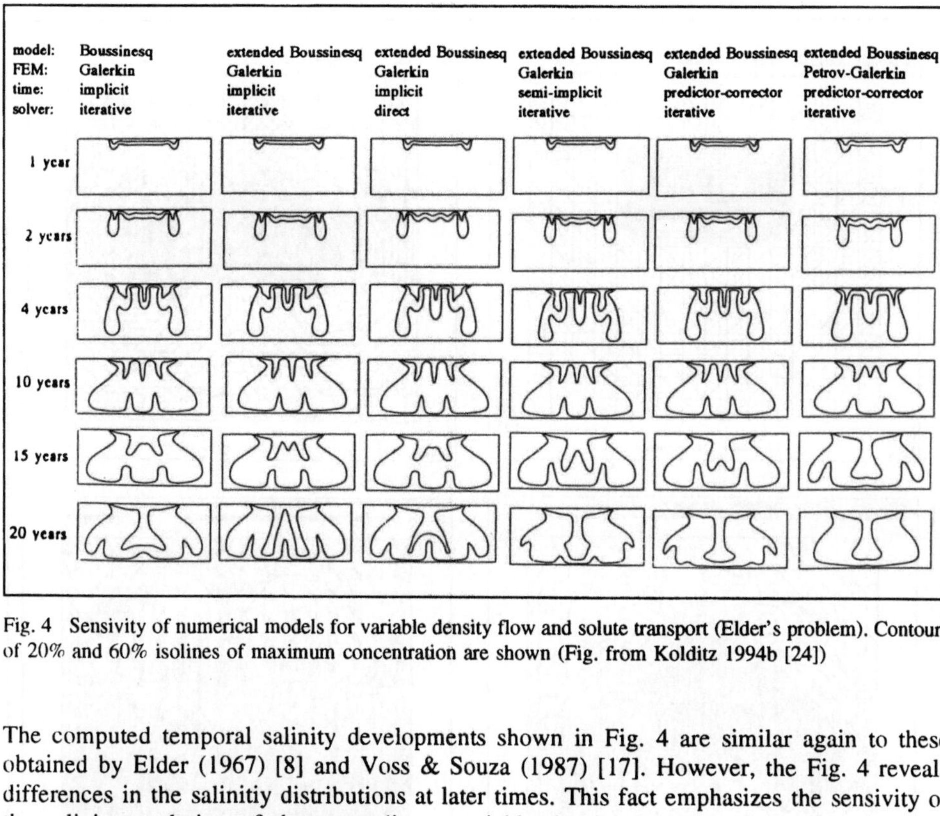

Fig. 4 Sensivity of numerical models for variable density flow and solute transport (Elder's problem). Contours of 20% and 60% isolines of maximum concentration are shown (Fig. from Kolditz 1994b [24])

The computed temporal salinity developments shown in Fig. 4 are similar again to these obtained by Elder (1967) [8] and Voss & Souza (1987) [17]. However, the Fig. 4 reveals differences in the salinitiy distributions at later times. This fact emphasizes the sensivity of the salinity evolution of these non-linear variable density processes from the numerical approximation. The results depend on the specific numerical realization of the mathematical model.

Sinking saltwater block
The convections described by Henry (1964) [7] and Elder (1967) [8] are maintained by permanent supply of substance or heat. Ratke (1995) [15] supposes another test where saltwater is initially captured in a block located within the aquifer (Fig. 5a). Then the saltwater block is released and it starts to sink due to the gravity. The sequenze of figures displays the developing convection within the impermeable bounded cavity (Fig. 5a-d). The last Fig. 5d shows a groundwater structure, adjusted by a layered density arrangement. This hetreogeneous groundwater system is further diffused untill the steady-state situation - organization of equal density distribution - is reached. Finally this test is a suited benchmark for the proof of the accuracy of the numerical scheme concerning the mass balance. The total mass of salt, contained within the closed aquifer, must be conserved during the whole process.

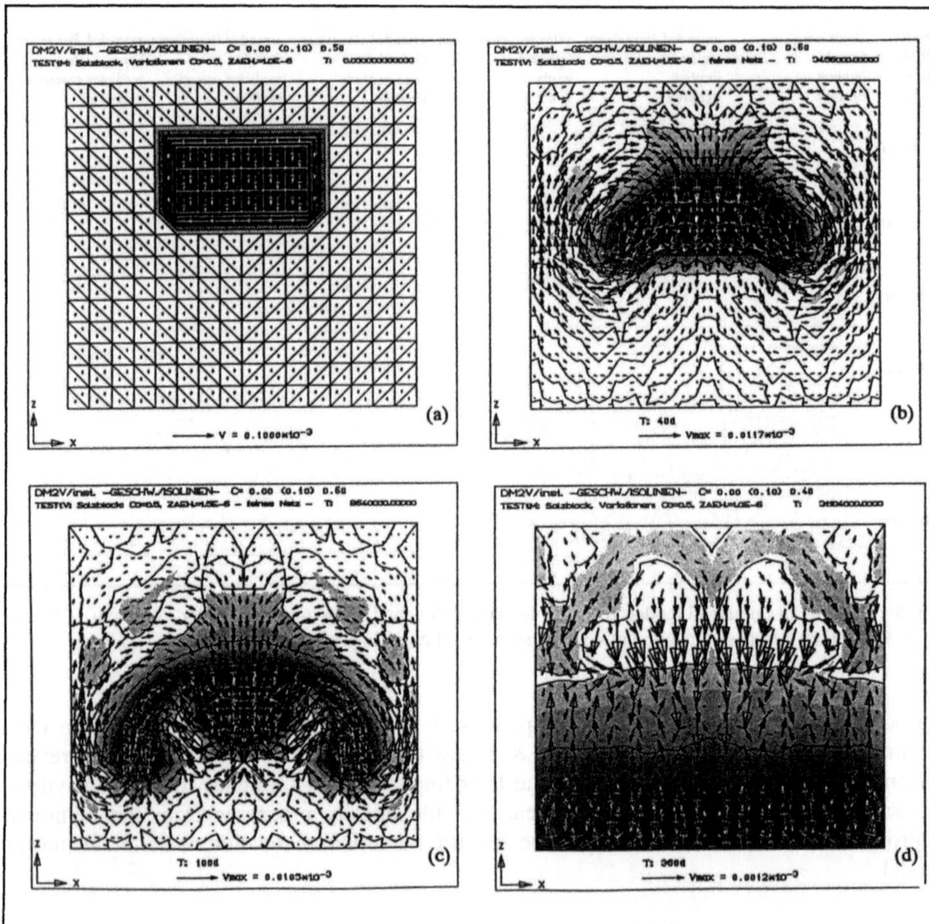

Fig. 5 Sinking saltwater block: Evolution of the salinity distribution calculated at (a) t=0d, (b) t=40d, (c) t=100d, and (d) t=360d (Fig. from Ratke 1995 [15])

Thermohaline processes in groundwater systems are connected with the presence of heterogeneous temperature and concentration fields. Thus, complex convections can arise from heat and salinity gradients acting simultaneously. Geophysical applications of thermohaline models can be found eg. in the field of geothermics and waste disposal in salt formations. In this way thermohaline effects gain importance for the production of mineralized thermal water and groundwater movement near salt domes.

Laying on the above study of Elder's problem, Kolditz (1994b)[24] defines a thermohaline expansion. In addition to the salinity differences a thermal gradient is superposed to the aquifer while permanently heating the base of the rectangular aquifer (compare with Fig. 3).

The Fig. 6 shows the developing temperature and the salinity distributions corresponding to a buoyancy ratio number $N = 4$. As can be seen, if comparing with Fig. 4 corresponding to the pure salinity convection, the thermal contribution delays the further intrusion of saltwater. This is caused by the contradicting orientations of buoyancy forces due to thermal and salinity gradients, respectively. Whereas heated water is rising to the top, weighty saltwater is sinking to the bottom of the aquifer.

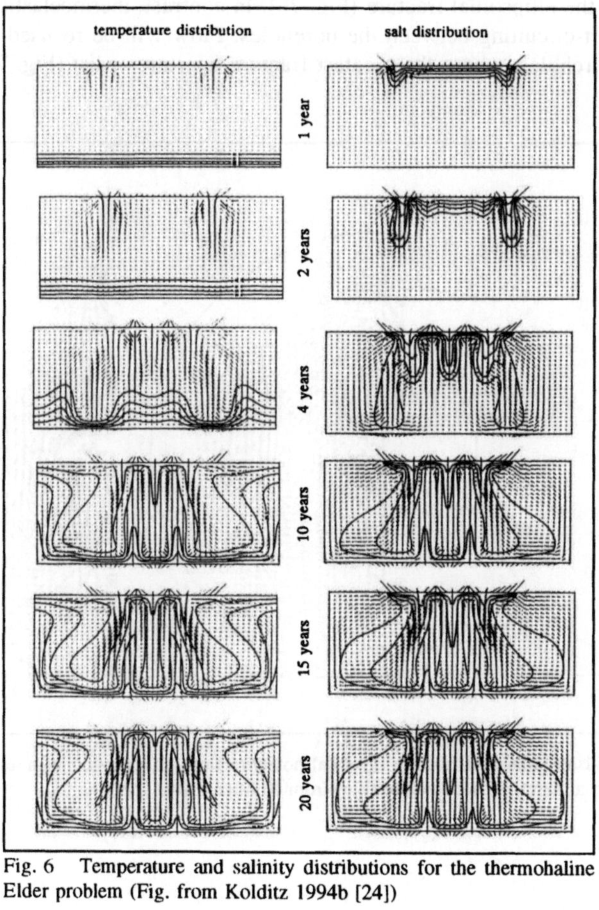

Fig. 6 Temperature and salinity distributions for the thermohaline Elder problem (Fig. from Kolditz 1994b [24])

Another case of a complex hydrothermal convection is connected with heat extraction from crystalline rocks. These hot dry rocks (HDR) receive increasing attention as an alternative source of energy. The basis of this concept is to create hydraulically large fractures in the rock matrix. Then a circulation system is established in such fractures by injecting cold water into one borehole and extracting the heated water through a second borehole. Large thermal gradients along these vertical fractures give rise to convection. The geometry of the fractures is continually altered by the interaction of poroelastic, thermoelastic and tectonic stresses during the heat extraction (Kohl 1992 [26]).

A set of the highly non-linear differential equations results for the mathematical description of flow and heat transfer in such geothermal reservoirs [27]. Kolditz & Diersch (1993) [28] use additionally relaxation methods and adaptive mesh refinement to simulate the coupled processes during heat extraction from the fractured reservoir.

The following ensemble of figures shows the velocity distribution within the fracture for different models. Neglecting density variations the flow is directed by the gradient of the hydraulic potential (Fig. 7a). Taking into account the density variations a large circulation developes within the ellipsoidal fracture (Fig. 7b). In contrast, thermoelastic effects give rise for hydraulic short-circuiting between the boreholes. Flow will be focused within a channel connecting the boreholes, where the greatest fracture apertures exist (Fig. 7c).

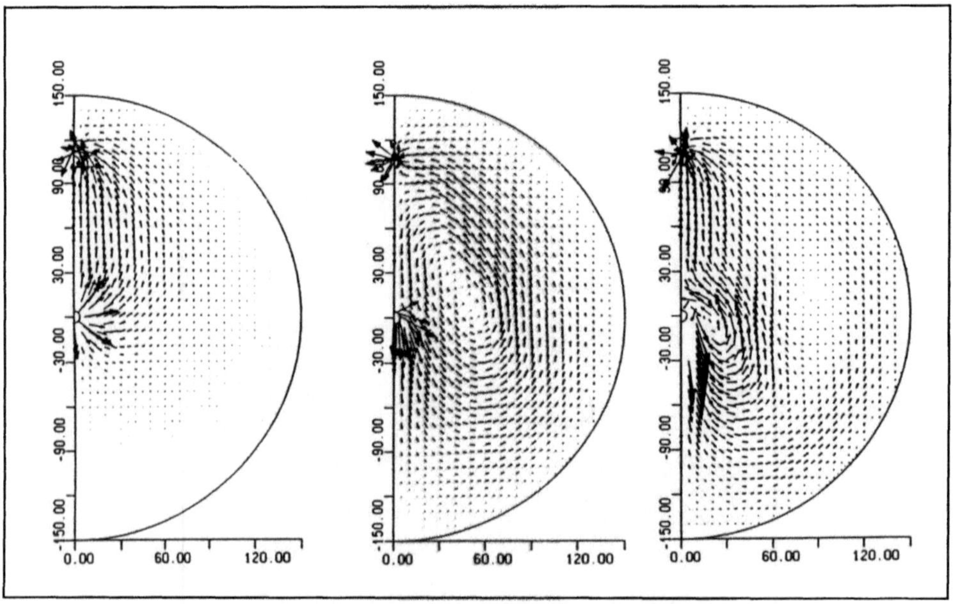

Fig. 7 Velocity fields of simulated fracture flow: (a) potential flow, (b) convection in an ellipsoidal fracture, and (c) convection in a thermoelastic fracture (Fig. from Kolditz 1994a [29])

There is both a theoretical and practical need of modelling density-coupled transport processes in three dimensions where free convection and transient influences dominate. Three-dimensional buoyancy-driven flows cover a complex task. Their computational analysis represents a challenging problem from various perspectives: Governing the convective instabilities arising from the strongly non-linear system and overcoming the numerical burden in simulating convective circulating pattern at high temporal and spatial resolutions.

While previous studies (eg. [30], [31], and [32]) devoted to stability criteria and the differences against the known 2-D processes a fully 3-D analysis of the multicellular pattern formation is only the subject of recent research. An appropriate test field concerns the 3-D Benard problem for a prorous layer where convective structures are caused by a heated or mass-intruded border face. This porous Benard problem can be considered as a 3-D extension of the Elder problem for different aspect ratios.

The first example refers to the 3-D cellular convection in a porous box with an aspect ratio of 4 equivalent to the above studied Elder problem in two dimensions. FEFLOW uses an one-step Newton Adams-Bashforth/trapezoid rule predictor-corrector FEM technique to tackle the problem. Fig. 8 shows the computed fingering and circulating pattern in the porous box where only a quarter of the domain was actually discretized.

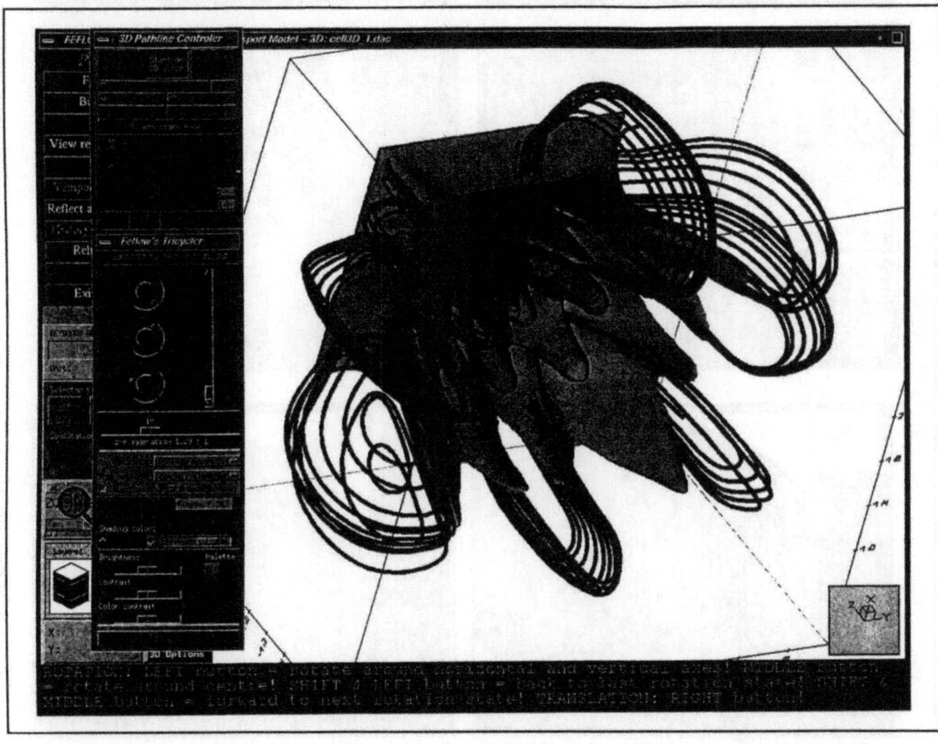

Fig. 8 Computed concentration isosurfaces and recirculating pathlines for the 3-D extended Elder problem at Ra=600 and dimensionless time 10^{-4}, the discretized quarter domain exists of 48.000 hexahedral elements with 51.701 nodes (Fig. from Diersch 1995 [33])

A greater numerical effort is necessary to simulate a 3-D multicellular example with an aspect ratio of 10 without any symmetric preassumptions. The porous layer was completely discretized by 250.000 hexahedral finite elements and an upwing FEM scheme based on an one-step Newton forward Euler/backward Euler predictor-corrector time stepping is used to compute successfully the transient development of the multicellular three-dimensional convective currents. Results at selected times are exhibited in Fig. 9.

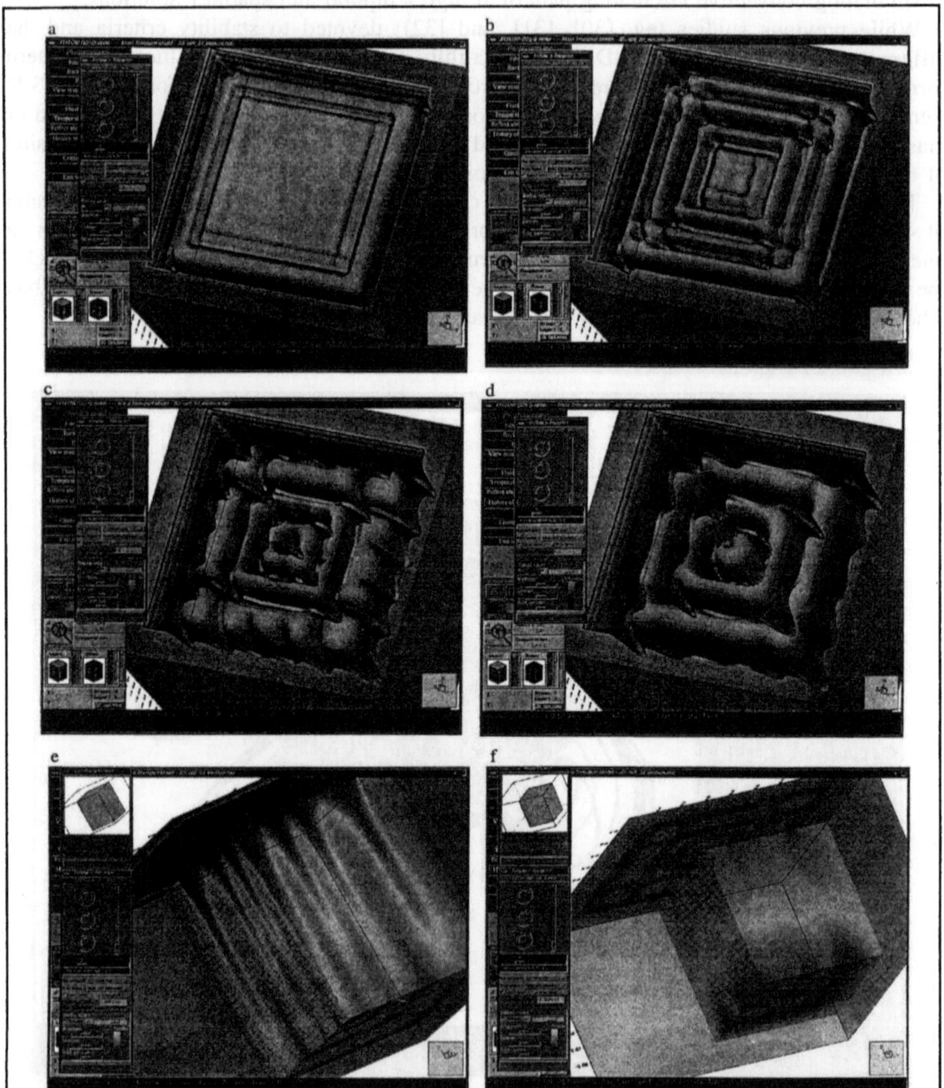

Fig. 9 Pattern formation for a porous Benard problem at Rayleigh number Ra=200 simulated by FEFLOW using 250.000 hexahedral elements with 257.761 nodes: Concentration isosurfaces at dimensionless times (a) 0.013, (b) 0.026, (c) 0.039, and (d) 0.078; (e) cross sectional concentration and (f) hydraulic head at time 0.039 (Fig. from Diersch 1995 [33]).

REFERENCES

[1] Zielke, W. & Stengel, T. & Schubert, R. & Malcherek, A. (1994): Impact of channel deepening on salinity distribution in a stratified estuary.- *Proc. Int. Symposium of Ecology and Engineering*, Kuala Lumpur.

[2] Clauser, C. & Neugebauer, H. J. (1991): Thermisch relevante Tiefenzirkulation in der Oberkruste unter dem Oberrheingraben ? Eingrenzung mit Hilfe hydrothermischer Modellrechnungen. - *Geol. Jb.*, E 48: 185-217.

[3] Fradkov, A. S. & Nauheimer, H. J. & Neugebauer, H. J. (1993): A weak non-linear thermo-diffusive-sedimentary convection system.- *Preprint*, Institute of Geodynamics-Physics of the Lithosphere, University Bonn, Germany.

[4] Graduiertenkolleg "Komplexe Dynamische Systeme" (1994): Mathematische Modelle für Polynyas und Leads.- *Report* Nr. 318, Institut für Dynamische Systeme, Universität Bremen.

[5] Horton, C. W. & Rogers, F. T. (1945): Convection currents in a porous medium.- *J. Appl. Phys.*, 16: 367-369.

[6] Wooding, R. A. (1962): Free convection of fluids in a vertical tube filled with porous material.- *J. Fluid Mech.*, 13: 126-144.

[7] Henry, H. R. (1964): Interfaces between salt water and fresh water in coastal aquifers.- *US Geological Survey Water-Supply Paper* 1613-C, Sea Water in Coastal Aquifers: C35-C70.

[8] Elder, J. W. (1967): Transient convection in a porous medium.- *J. Fluid Mech.*, 27(3): 609-623.

[9] Combarnous, M. A. & Bories, S. A. (1975): Hydrothermal convection in saturated porous media.- *Advances in Hydroscience*, Ven Te Chow (ed), 10: 231-307, Academic Press, New York.

[10] Diersch, H.-J. (1985): Modellierung und numerische Simulation geohydrodynamischer Transportprozesse.- *Habilitation*, Akademie der Wissenschaften der DDR, Berlin, 267S.

[11] Kolditz, O. (1990): Zur Modellierung und Simulation geothermischer Transportprozesse in untertägigen Zirkulationssystemen.- *Dissertation*, Institut für Mechanik, Akademie der Wissenschaften der DDR, Berlin, 120 S.

[12] Kröhn, K.-P. (1991): Simulation von Transportvorgängen im klüftigen Gestein mit der Methode der Finiten Elemente.- *Dissertation, Bericht* Nr. 29/1991, Institut für Strömungsmechanik, Universität Hannover.

[13] Häfner, F. & Sames, D. & Voigt, H.-D. (1992): Wärme- und Stofftransport - Mathematische Methoden.- Springer Verlag, Berlin - Heidelberg - New York, ISBN 3-540-54665-0, 626S.

[14] Diersch, H.-J. (1994): Interactive, graphics-based finite-element simulation system FEFLOW for modelling groundwater flow, contaminant mass and heat transport processes.- *Users's Manual*, Release 4.20, WASY - Gesellschaft für wasserwirtschaftliche Planung und Systemforschung mbH, Berlin.

[15] Ratke, R. (1995): Zur Lösung der Strömungs- und Transportgleichung bei veränderlicher Dichte.- *Technischer Bericht*, Institut für Strömungsmechanik, Universität Hannover, in preparation.

[16] Segol, G. (1994): Classic groundwater simulations - Proving and improving numerical models. - PTR Prentice Hall, 531p.

[17] Voss, C. I. & Souza, W. R. (1987): Variable density flow and solute transport simulation of regional aquifers containing a narrow feshwater-saltwater transition zone.- *Water Resources Research*, 23(10): 1851-1866.

[18] Senger, R. K. & Fogg, G. E. (1990): Stream functions and equivalent freshwater heads for modeling regional flow of variable-density groundwater.- *Water Resources Research*, 26(9): 2097-2106.

[19] Galeati, G. & Gambolati, G. & Neumann, S. P. (1992): Coupled and partially coupled Eulerian-Lagrangian model of freshwater-seawater mixing.- *Water Resources Research*, 28(1): 149-165.

[20] Segol, G. & Pinder, G. F. & Gray, W. G. (1975): A Galerkin-finite element technique for calculating the transient position of the saltwater front.- *Water Resources Research,* 11(2): 343-347.

[21] Huyakorn, P. S. & Andersen, P. F. & Mercer, J. W. & White, H. O. (1987): Saltwater intrusion in aquifers: development and testing of a three-dimensional finite element model.- *Water Resources Research,* 23(2): 293-312.

[22] Bogacki, W. (1987): Optimale Bewirtschaftung von Süß-Salzwasser Aquiferen.- *Dissertation,* Mitteilungen Nr. 65, Institut für Wasserbau und Wasserwirtschaft, Rheinisch-Westfälische Technische Hochschule, Aachen.

[23] Diersch, H.-J. (1981): Primitive variables finite element solutions of free convection flow in porous media.- *Zeitschr. Angew. Math. Mech.* (ZAMM), 61(7): 325-337.

[24] Kolditz, O. (1994b): Benchmarks for numerical groundwater simulations. - In: DIERSCH, H.-J. (1994) FE-FLOW User's Manual, Release 4.20, WASY - Gesellschaft für wasserwirtschaftliche Planung und Systemforschung mbH, Berlin.

[25] Kröhn, K.-P. (1994): Zur Modellierung der Grundwasserströmung mit variabler Dichte - Sensitivitätsstudie auf der Grundlage eines Laborexperiments von Elder.- *Fachbericht* zum Forschungsvorhaben 02E8301 des BMFT, Archiv-Nr. 111994, Bundesanstalt für Geowissenschaften und Rohstoffe, Hannover.

[26] Kohl, T. (1992): Modellsimulation gekoppelter Vorgänge beim Wärmeentzug aus heißem Tiefengestein. - *Dissertation,* Eidgenössische TH, Zürich,149S.

[27] Diersch, H.-J. & Kolditz, O. & Jesse, J. (1989): Finite element analysis of geothermal circulation processes in hot dry rock fractures.- *Zeitschrift angew. Math. Mech.* (ZAMM), vol. 69 (3), 139-153.

[28] Kolditz, O. & Diersch, H.-J. (1993): Quasi-steady-state strategy for numerical simulation of geothermal circulation in Hot Dry Rock fractures. - *Int. J. Non-Linear Mechanics,* 28(4): 467-481.

[29] Kolditz, O. (1994a): Modellierung von Strömungs- und Transportprozessen im geklüfteten Festgestein.- *NLfB-Bericht* Nr. 112281, Niedersächsisches Landesamt für Bodenforschung, Hannover, 202S., in Vorbereitung zur Publikation im *Geologischen Jahrbuch.*

[30] Holst, P. H. & Aziz, K. (1972): Transient three-dimensional natural convection in confined porous media.- *Int. J. Heat Mass Transfer,* 15: 73-90.

[31] Zebib, A. & Kassoy, D. R. (1978): Three-dimensional natural convection motion in a confined porous medium.- *Phys. Fluids,* 21: 1-3.

[32] Schubert, G. & Straus, J. M. (1979): Three-dimensional and multicellular steady and unsteady convection in fluid-saturated porous media at high Rayleigh numbers.- *J. Fluid Mech.,* 94: 25-38.

[33] Diersch, H.-J. (1995): Finite element analysis of three-dimensional transient free convection processes in porous media.- submitted to *IMACS-COST Conference* on "Three-dimensional complex flows", Sep. 13-15, Lausanne.

COUPLING OF BOUNDARY AND FINITE ELEMENTS IN AEROACOUSTIC CALCULATIONS

H. Schippers and J.A. Wensing[1]
National Aerospace Laboratory NLR
P.O. Box 90502, 1006 BM Amsterdam,The Netherlands

SUMMARY

A coupled system of finite elements and boundary elements is presented for the modelling of sound propagation in a nonuniform mean flow. Finite elements are used to discretize the acoustic pressure in a bounded computational domain including the nonuniform flow field. The radiation of acoustic waves in the exterior uniform flow field is modelled by a boundary integral representation formula. The interface condition on the shared boundary yields an integral equation which is solved using boundary elements. The coupling of finite elements and boundary elements is described in detail for sound propagation in a stationary fluid where the acoustic pressure is governed by the Helmholtz equation. The coupled computational system is applied to estimate scattering and refractive effects on the acoustic pressure waves through a boundary layer surrounding an aircraft's fuselage. The coupled mathematical formulation for three dimensional sound radiation around an engine in subsonic flow is discussed.

1. INTRODUCTION

Investigations on the acoustic field around commercial aircraft are motivated by the urge to reduce the noise levels inside and outside the aircraft. Examples are the prediction of sound pressure levels on the fuselage surface of propeller driven aircraft and the diffraction of fan noise from the nacelle inlets of turbofan engines including the evaluation of acoustic absorbing liners. The noise prediction requires the modelling of the sound pressure in a nonuniform mean flow. The governing aeroacoustic equations for the sound pressure follow from linearizing or perturbing the Navier-Stokes equations about a mean steady flow field. Since this flow field can be rather complicated, it is a prerequisite to develop numerical simulation tools for sound propagation in a nonuniform mean flow. Most of these tools are based on either finite element or finite difference methods in a bounded computational domain around the aircraft or a part of the aircraft. At the outer boundary of the computational domain one needs appropriate conditions for the radiation of acoustic waves. Here, one can apply local absorbing boundary conditions (see e.g. [2], [3], [4]) or global boundary conditions related to the use of integral operators. The drawback of the local far field boundary conditions is that they have to be applied at sufficiently distant from the radiating source and that they do not radiate exactly. Boundary conditions of this type depend on knowing which direction the waves are travelling, but in a general noise propagation problem, there are many waves travelling in many directions. Unless the outer boundary is so far away that the direction becomes radial. An alternative way is to use global boundary conditions which follow from the application of a boundary

[1]Present address Twente University, Enschede, The Netherlands

integral formula representing the acoustic pressure outside the computational domain. The integral formula follows from applying Green's theorems to the homogeneous acoustic equation. This formula involves as unknown distributions the pressure and its normal derivative at the outer boundary. At the outer boundary the pressure and its normal derivative are required to be continuous, which yields a boundary integral equation and a coupling with the sound pressure in the nonuniform flow field. The main feature of the boundary integral formula is that it handles the Sommerfeld radiation condition at infinity automatically.

In this paper the boundary integral formulation is proposed for the acoustic radiation and the coupling with the numerical solution in the bounded computational domain is discussed. The numerical solution follows from discretizing the governing aeroacoustic equation using piecewise linear finite elements on a triangular grid. The boundary integral equation is numerically solved using boundary elements. The pressure at the outer boundary is approximated by piecewise linear boundary elements, whereas the normal derivative is modelled by piecewise constant boundary elements. It is assumed that the boundary element grid coincides with the outer boundary of the finite element grid. This way a coupled computational system of finite elements and boundary elements is obtained. In section 2 the mathematical formulation is presented for sound propagation in a nonuniform mean flow. In section 3 the convergence of the coupled computational system of finite elements and boundary elements is discussed for the propagation of sound waves in a stationary fluid, i.e. in both the finite computational domain and the outer region the Helmholtz equation is solved. In section 4 the coupled system is applied to estimate scattering and refractive effects on the acoustic pressure waves through a boundary layer surrounding an aircraft's fuselage. The fuselage is represented by an infinitely long cylinder with a non-circular cross section. In section 5 some modelling aspects of three dimensional sound radiation around an engine in subsonic flow are discussed.

2. MATHEMATICAL MODELLING

The modelling of acoustic fields propagating in a nonuniform mean flow and the interaction of such fields with the mean flow itself and with reflecting boundaries can be described by the Lilley equation. When the summation convention is used this equation reads (see e.g. [1]):

$$\frac{D}{Dt}\left(\frac{D^2\Pi}{Dt^2} - \frac{\partial}{\partial x_i}c^2\frac{\partial \Pi}{\partial x_i}\right) + 2\frac{\partial v_j}{\partial x_i}\frac{\partial}{\partial x_j}c^2\frac{\partial \Pi}{\partial x_i} = -2\frac{\partial v_j}{\partial x_i}\frac{\partial v_k}{\partial x_j}\frac{\partial v_i}{\partial x_k} + \Psi \qquad (1)$$

where $\Pi = (1/\gamma)ln(p/p_0)$, p_0 some convenient constant reference pressure, v_i the velocity and c the speed of sound. Furthermore, $\frac{D}{Dt}$ is the substantial derivative

$$\frac{D}{Dt} = \frac{\partial}{\partial t} + v_i\frac{\partial}{\partial x_i}$$

and Ψ represents effects of entropy fluctuations and fluid viscosity. Equation (1) is a direct consequence of the laws of conservation of mass and momentum for a compressible fluid and is thus nonlinear. However, in many cases of interest (e.g., sound propagation in a fuselage boundary layer and sound generation in aircraft engines), it is reasonable to replace the velocity and the speed of sound by their mean values, $v_i := V_i$ and $c := \bar{c}$. For the prediction of fan noise from engines, it is argued in [1] that the propagation terms in the left hand side of equation (1) are determined mainly by the large region of steady flow and not by the smaller regions of unsteady flow in the vicinity of the fan. Furthermore, the effects due to entropy fluctuations and fluid viscosity may be neglected in many aerodynamic sound problems of interest, i.e. Ψ may be deleted. When these approximations are made, equation (1) reduces to the linear aeroacoustics equation

$$\frac{\bar{D}}{Dt}\left(\frac{\bar{D}^2\Pi}{Dt^2} - \frac{\partial}{\partial x_i}\bar{c}^2\frac{\partial\Pi}{\partial x_i}\right) + 2\frac{\partial V_j}{\partial x_i}\frac{\partial}{\partial x_j}\bar{c}^2\frac{\partial\Pi}{\partial x_i} = -2\frac{\partial V_j}{\partial x_i}\frac{\partial V_k}{\partial x_j}\frac{\partial V_i}{\partial x_k} \tag{2}$$

$$\frac{\bar{D}}{Dt} = \frac{\partial}{\partial t} + V_i\frac{\partial}{\partial x_i},$$

with a source term on the right hand side modelling the effects of nonlinear convective forces. At large distance from the sound source or from the reflecting object, equation (2) reduces to the homogeneous convective wave equation

$$\frac{\bar{D}^2\Pi}{Dt^2} - \bar{c}_\infty^2\frac{\partial}{\partial x_i}\frac{\partial\Pi}{\partial x_i} = 0, \tag{3}$$

wherein the velocity V_i may be replaced by the constant velocity at infinity.

3. COUPLING OF FINITE ELEMENTS AND BOUNDARY ELEMENTS

The application of a boundary integral representation formula for the acoustic radiation and the coupling with the numerical solution in a bounded domain is discussed for the problem of diffraction of waves around a circular two-dimensional object due to an incident plane wave. The acoustic waves are assumed to propagate harmonically in time with frequency ω in a stationary fluid. Then, equation (3) reduces to the Helmholtz equation

$$(\nabla^2 + k^2)P = 0, \tag{4}$$

with $k = \omega/\bar{c}_\infty$. Around the circular object a triangular boundary conforming grid is constructed. Equation (4) has to be satisfied in both the bounded computational domain and outside this domain. The total pressure field consists of an incoming and a scattered field, i.e.

$$P(y,z) = P^{inc}(y,z) + P^s(y,z).$$

Equation (4) has to be solved under the following boundary conditions:

1. P^s should consist of outgoing waves.
2. The normal component of the particle velocity on the circular boundary vanishes, i.e.

$$\frac{\partial P}{\partial n} = 0. \tag{5}$$

3. The acoustic pressure and its normal derivative are continuous at the outer boundary of the computational domain.

3.1 Coupled mathematical formulation

Equation (4) is solved in a weak sense. Let the computational domain be given by A_1 with boundaries S_i and S_e, where S_i corresponds with the boundary of the circular cylinder and S_e with the outer boundary. By the divergence theorem of Gauss, it follows that the acoustic pressure field P_1 in the Sobolev space $H^1(A_1)$ has to satisfy

$$\int_{A_1} (\nabla P_1 \cdot \nabla W - k^2 P_1 W) dA - \int_{S_i} \frac{\partial P_1}{\partial n} W dS - \int_{S_e} \frac{\partial P_1}{\partial n} W dS = 0, \tag{6}$$

for all test functions $W \in H^1(A_1)$. The boundary integral over S_i vanishes due to condition (5). Equation (6) is briefly written as

$$a(P_1, W) - < \frac{\partial P_1}{\partial n}, W >= 0, \tag{7}$$

where a is a functional of the first term of equation (6) and $<,>$ is an inner product over boundary S_e. Outside the computational domain the scattered pressure P_2^s is given by the boundary integral representation formula

$$P_2^s(\vec{r}) = \int_{S_e} \frac{\partial P_2^s(\vec{r}\,')}{\partial n} G(\vec{r}, \vec{r}\,') dS\,' - \int_{S_e} P_2^s(\vec{r}\,') \frac{\partial G}{\partial n\,'}(\vec{r}, \vec{r}\,') dS\,' \tag{8}$$

with $\vec{r} \in A_2$, the exterior domain. The fundamental solution G of the two-dimensional Helmholtz equation is given by

$$G(\vec{r}, \vec{r}\,') = \frac{i}{4} H_0^{(2)}(k \mid \vec{r} - \vec{r}\,' \mid),$$

where $H_0^{(2)}$ is the Hankel function of zeroth order and of second kind.

Whenever $\vec{r}$ is part of boundary S_e,the second integral of (8) becomes singular. By the jump relations of classical potential theory, it follows that

$$\frac{1}{2} P_2^s(\vec{r}) - \int_{S_e} \frac{\partial P_2^s(\vec{r}\,')}{\partial n} G(\vec{r}, \vec{r}\,') dS\,' + \int_{S_e} P_2^s(\vec{r}\,') \frac{\partial G}{\partial n\,'}(\vec{r}, \vec{r}\,') dS\,' = 0, \tag{9}$$

with $\vec{r} \in S_e$, which is briefly written in operator notation as

$$\frac{1}{2} P_2^s - V \frac{\partial P_2^s}{\partial n} + K P^s = 0. \tag{10}$$

The integral operators V and K correspond to potentials due to single layer and double layer distributions on the boundary S_e. This equation is also solved in a weak sense. For test functions ψ in the Sobolev space $H^{-1/2}(S_e)$, we require that

$$< \frac{1}{2} P_2^s, \psi > - < V \frac{\partial P_2^s}{\partial n}, \psi > + < K P_2^s, \psi >= 0, \tag{11}$$

with $P_2^s \in H^{1/2}(S_e)$ and $\partial P_2^s / \partial n \in H^{-1/2}(S_e)$.

The coupled mathematical formulation is achieved by means of the pressure field continuity equations on the interface boundary S_e:

$$\frac{\partial P_1}{\partial n} = \frac{\partial P_2}{\partial n} \quad \text{and} \quad P_1 = P_2 = P_2^{inc} + P_2^s \tag{12}$$

When these continuity equations are substituted into equations (7) and (11), the coupled problem to be solved reads: find the pressure $P_1 \in H^1(A_1)$ and its normal derivative at boundary S_e,

$$\sigma = \frac{\partial P_1}{\partial n} = \frac{\partial P_2}{\partial n} \in H^{-1/2}(S_e),$$

such that

$$a(P_1, W) - < \sigma, W >= 0 \tag{13}$$

and

$$\begin{aligned} < \frac{1}{2}P_1, \psi > \quad - < V\sigma, \psi > \quad + < KP_1, \psi >= \\ < \frac{1}{2}P_2^{inc}, \psi > \quad - < V\tfrac{\partial P_2^{inc}}{\partial n}, \psi > \quad + < KP_2^{inc}, \psi > \end{aligned} \tag{14}$$

for all test functions $W \in H^1(A_1)$ and $\psi \in H^{-1/2}(S_e)$. This coupling procedure has been described for Laplace's equation in reference [5].

3.2 Numerical approach

The coupled mathematical formulation (13) and (14) is numerically solved by a combination of finite elements and boundary elements. The pressure field P_1 in the inner area A_1 is discretized using piecewise linear shape functions. The pressure on the boundary S_e and its normal derivative σ are discretized using boundary elements. The boundary element mesh on S_e coincides with the outer boundary of the finite element mesh covering the inner area. The pressure is approximated by piecewise linear shape functions analogous to the pressure in the inner area and as a consequence the normal derivative σ is approximated by piecewise constant shape functions. According to Galerkin's method the test functions W are are chosen to correspond with the piecewise linear shape functions and the test functions ψ with piecewise constant functions on S_e.

The domain integrals of functional a in (13) are calculated with a second order Gaussian quadrature rule with four weight points. The boundary integrals in equation (13) are also calculated with Gaussian quadrature rules, i.e. a fourth order rule for the inner integral occurring in the boundary operators K and V and a second order rule for the outer integrals. An exception is made for the boundary element part due to V, when the contribution of an element to itself has to be determined. For this case the operator V yields a weakly singular integral, which is calculated analytically where the Hankel function is represented by a mathematical series for small arguments. Details of the calculation of the boundary integrals have been given in Ref. [8].

The discretized coupled system of equations is solved by block Gauss decomposition using a skyline solver for the finite element part. The bandwidth of the finite element matrix is minimized using the Gibbs-King ([6]) reordering algorithm for the node numbering. The computational model has been implemented on NLR's NEC SX3 supercomputer.

3.3 Convergence of the coupled system

For the case of Laplace's equation the convergence of the coupled system of finite element and boundary element equations was investigated in Ref. [5]. By mathematical

Table 1: Computational statistics of scattering of acoustic waves around a circular cylinder.

Grid	Coarse	Medium	Fine
Error	0.02760	0.00618	0.00165
Convergence rate	-	2.16	1.91
Number of finite elements	400	1440	5440
Number of boundary elements	80	160	320
Computation times (CPU-seconds)	13	55	307

analysis the following error estimate was derived

$$||P_1 - P_h||_{L^2(A_1)} + h||\sigma - \sigma_h||_{L^2(S_e)} \leq Ch^2, \tag{15}$$

where P_h and σ_h represent the numerical solution as obtained by the above approach. From table 1 it is observed that this estimate also holds for the Helmholtz equation. In this table the convergence rate of the L^2-error $||P_1 - P_h||_{L^2(S_i)}$ on the boundary S_i is presented for the case $kR = 10$ (R the radius of the cylinder). The results are obtained using nested computational grids by refining the mesh size uniformly.

4. NOISE PROPAGATION THROUGH A FUSELAGE BOUNDARY LAYER

The coupled system of finite elements and boundary elements is applied to estimate the scattering and refractive effects through a boundary layer surrounding an aircraft's fuselage. The fuselage is modelled by an infinitely long cylinder with a non-circular cross section, ranging from $x = -\infty$ to $x = +\infty$. In the present investigations only the effects of velocity gradients are considered. The effects of gradients in thermodynamic variables are neglected. Then, the solution of the Navier-Stokes equations is given by a unidirectional cylindrically sheared mean flow with:

$$\vec{V} = \vec{\imath}\, U(r(y,z)), \ \rho_0 = \text{constant}, \ p_0 = \text{constant}, \tag{16}$$

where $\vec{\imath}$ denotes the unit vector in x-direction. The coordinate variable r is an arbitrary function of the coordinates y and z, defining the cross section plane. The surfaces r = constant are coordinate surfaces in a cylindrical coordinate system where the mean velocity $U(r)$ remains constant. Changes in U occur only in the direction normal to these surfaces. Outside the boundary layer we have a uniform flow in the positive x-direction with constant velocity, i.e. $U(r) = U_\infty$ for $r > r_e$. Inside the boundary layer, i.e. $r_i < r < r_e$, the mean velocity $U(r)$ is a function of the coordinate r.

The governing equation for the acoustic pressure field follows from substituting (16) into (2). When the speed of sound is replaced by the constant value of the uniform outer flow, it follows that

$$\frac{\bar{D}}{Dt}\left(\frac{\bar{D}^2 p}{Dt^2} - \bar{c}_\infty^2 \nabla^2 p\right) + 2\bar{c}_\infty^2 \frac{dU}{dr}\frac{\partial^2 p}{\partial r \partial x} = 0 \tag{17}$$

with the substantial derivative

$$\frac{\bar{D}}{Dt} = \frac{\partial}{\partial t} + U(r)\frac{\partial}{\partial x}$$

and

$$\frac{d}{dr} = \frac{\nabla r}{|\nabla r|^2} \cdot \nabla.$$

The waves are assumed to propagate harmonically in time with frequency ω and with wave number α in axial direction, i.e.

$$p(t,x,y,z) = P(x,y)e^{\imath(\omega t+\alpha x)}.$$

By Fourier transformation equation (17) yields for P:

$$(\nabla^2 + \gamma^2(r))P - \frac{2\alpha}{(k+M(r)\alpha)}\frac{dM}{dr}\frac{\partial P}{\partial r} = 0 \tag{18}$$

with $M(r) = U(r)/\bar{c}_\infty$, $k = \omega/\bar{c}_\infty$ and

$$\gamma(r) = \sqrt{(k+(1+M(r))\alpha)}\sqrt{(k-(1-M(r))\alpha)}.$$

Outside the boundary layer equation (18) reduces to

$$(\nabla^2 + \gamma_0^2)P = 0, \tag{19}$$

with $\gamma_0 = \sqrt{(k+(1+M_\infty)\alpha)}\sqrt{(k-(1-M_\infty)\alpha)}$.

Equation (18) is solved in the weak sense as described in the previous section. For (18) the functional a becomes

$$a(P,W) = \int_{A_1}(\nabla P\cdot\nabla W - \gamma^2(r)PW + WF(r)\frac{\partial P}{\partial r})dA - \int_{S_e}\frac{\partial P}{\partial n}WdS, \tag{20}$$

with the function F given by

$$F(r) = \frac{2\alpha}{(k+M(r)\alpha)}\frac{dM}{dr}. \tag{21}$$

When this definition of a is compared with (6), it is observed that the domain integral of a is no longer symmetric with respect to P and W due to the presence of the term with $F(r)$. As a consequence, the resulting finite element matrix becomes also asymmetric. The boundary integral formulation is similar to (8), where the frequency k has to be replaced by γ_0.

The effects of the boundary layer shear flow on the noise propagation are investigated for a fuselage of a Fokker 50 like aircraft. The incoming pressure wave is prescribed by a plane wave with visual angle of incidence ϕ_v (i.e. ϕ_v is the angle between the direction of propagation and the cylinder axis, $\phi_v = 0^o$ corresponding to a wave propagating in the positive $x-$direction). Whenever, ϕ_v, k and M_∞ are prescribed, the actual values of the axial wave number α and the transverse wavenumber γ_0 can be calculated (see [8]).

Let the normalized boundary layer coordinate in radial direction be given by $\rho = (r-r_i)/\delta$, with δ the boundary layer thickness (i.e. $\delta = r_e - r_i$). For the modelling of the shear flow a turbulent velocity profile is considered according to the 1/7 power law, i.e.

$$M(\rho) = M_\infty \rho^{\frac{1}{7}}. \tag{22}$$

In ref. [9] approximations with respect to this profile have been investigated, a.o. a linear velocity profile and a piecewise linear profile based on the same displacement thickness.

It appears that the refraction of the acoustic waves strongly depends on the shape of the velocity profile. It was concluded that the velocity profile should be modelled as accurately as possible.

Computations are performed for cruise conditions, that are representative for a propfan driven aircraft. The fuselage is modelled by a cylinder with a cross section similar to a Fokker 50 fuselage. The plane waves impinge on the cylinder under angles of incidence of 90, 135 and 153 degrees. The computational grid is shown in figure 1, where the thickness of the boundary layer has been blown up for reasons of visualization. The influence of the turbulent boundary layer is assessed by computing the acoustic surface pressure. The results are presented in figures 2 to 4. It is obvious that for large angles of incidence a considerable modification is obtained. From figure 4 it is observed that a maximum attenuation is reached of about 8 dB. For angles of incidence larger than 90 degrees the acoustic waves bend away from the fuselage and as a result lower pressure amplitudes are found on the surface of the fuselage. For these angles of incidence the wave number can become complex in the boundary layer. At this cut-off point, the wave front is moving parallel to the fuselage. For real wave numbers, a small attenuation of the pressure results due to a decrease of the normal component of the velocity. For complex wave numbers, however, there is a strong exponential attenuation of the pressure.

5. SOUND PROPAGATION IN THREE DIMENSIONAL SUBSONIC FLOW

In this section some coupling aspects are discussed for the three dimensional sound radiation around an engine in subsonic flow. In and around the engine duct the propagation of the acoustic waves due to fan pressure fluctuations is governed by equation (2). The waves are assumed to propagate harmonically in time with frequency ω. Let $\alpha = \bar{c}/c_\infty$, $k = \omega/c_\infty$ and $\mu_i = V_i/c_\infty$. Then equation (2) can be rewritten as

$$(\imath k + \mu_i \frac{\partial}{\partial x_i})\chi - 2\frac{\partial \mu_j}{\partial x_i}\frac{\partial}{\partial x_j}\alpha^2\frac{\partial \Pi}{\partial x_i} = 2\frac{\partial \mu_j}{\partial x_i}\frac{\partial \mu_k}{\partial x_j}\frac{\partial \mu_i}{\partial x_k} \tag{23}$$

and

$$\frac{\partial}{\partial x_i}\alpha^2\frac{\partial \Pi}{\partial x_i} - (\imath k + \mu_i \frac{\partial}{\partial x_i})^2\Pi = \chi. \tag{24}$$

Remark that at some distant from the engine $|\chi| \to 0$, $\mu_1 \to M_\infty$, $\mu_2 \to 0$, $\mu_3 \to 0$ and $\alpha \to 1$, so that equation (24) reduces to the well known aeroacoustic equation in steady uniform flow

$$\frac{\partial}{\partial x_i}\frac{\partial \Pi}{\partial x_i} + k^2\Pi - 2\imath k M_\infty \frac{\partial \Pi}{\partial x_1} - M_\infty^2\frac{\partial^2 \Pi}{\partial x_1^2} = 0. \tag{25}$$

The coupling assumes that the acoustic waves are modelled inside the computational domain by (23) and (24), and outside the domain by (25). Since $|\chi| \to 0$ at some distance from the engine, the dominating equations at the outer boundary of the computational domain are given by (24) and (25). Hence the coupling terms have to follow from the boundary integrals of the weak formulation of equation (24) and the boundary integral formula that satisfies (25). By partial integration it follows that inside the computational domain Π has to satisfy

$$\begin{aligned}
\int_{A_1} W\chi dA &= \int_{A_1} (k^2 W\Pi - \frac{\partial W}{\partial x_i}(\alpha^2 \frac{\partial \Pi}{\partial x_i} - \mu_i\mu_j \frac{\partial \Pi}{\partial x_j}) \\
&\quad - 2\imath k W \mu_i \frac{\partial \Pi}{\partial x_i} + W\mu_i \frac{\partial \Pi}{\partial x_i}\frac{\partial \mu_j}{\partial x_j}) dA \\
&\quad + \int_{S_i} n_i(\delta_{ij}\alpha^2 - \mu_i\mu_j)\frac{\partial \Pi}{\partial x_j} W dS \\
&\quad + \int_{S_e} n_i(\delta_{ij}\alpha^2 - \mu_i\mu_j)\frac{\partial \Pi}{\partial x_j} W dS, \qquad (26)
\end{aligned}$$

for all test functions $W \in H^1(A_1)$. The last integral of (26) over the outer boundary reduces at some distant from the engine to

$$\int_{S_e} (n_i \frac{\partial \Pi}{\partial x_i} - n_1 M_\infty^2 \frac{\partial \Pi}{\partial x_1}) W dS.$$

Let the single layer distribution σ be given by

$$\sigma = n_i \frac{\partial \Pi}{\partial x_i} - n_1 M_\infty^2 \frac{\partial \Pi}{\partial x_1}. \qquad (27)$$

The coupling can be obtained by replacing the last integral of (26) by

$$\int_{S_e} \sigma W dS$$

and by solving σ from an appropriate boundary integral equation on S_e. In Ref. [7] the following boundary integral formula has been derived that satisfies (25)

$$\begin{aligned}
\Pi(\vec r) &= \int_{S_e} G(\vec r, \vec r\,')\sigma(\vec r\,')dS\,' \\
&\quad - \int_{S_e} 2\imath k M_\infty G(\vec r, \vec r\,')\Pi(\vec r\,') n_1' \\
&\quad + (n_i' \frac{\partial G}{\partial x_i'}(\vec r, \vec r\,') - n_1' M_\infty^2 \frac{\partial G}{\partial x_1'}(\vec r, \vec r\,'))\Pi(\vec r\,')dS\,', \qquad (28)
\end{aligned}$$

$\vec r \in A_2$, with σ given by (27) and the Green function G given by

$$G(\vec r, \vec r\,') = \frac{1}{4\pi \tilde r} e^{-\imath k(\tilde r + M_\infty(x-x'))/(1-M_\infty^2)} \qquad (29)$$

with

$$\tilde r = \sqrt{(x-x')^2 + (1-M_\infty^2)((y-y')^2 + (z-z')^2)}.$$

The continuity of Π at the outer boundary of the computational domain provides the coupling boundary integral equation for σ. It is noticed that this formulation involves the solution of σ instead of the normal derivative $n_i\partial\Pi/\partial x_i$, which was advocated for coupling in [7]. When the normal derivative has to be resolved, it is necessary to convert $\partial\Pi/\partial x_1$ in (27) into a linear combination of the normal derivative and the tangential derivatives of Π (See [7]), which complicates the numerical solution of the integral equation considerably.

The numerical solution of the coupled system can be obtained by using finite elements for χ and Π in equations (23) and (24), and boundary elements for for σ and $\chi_{|S_e}$ in the integral equation. Analogously to section 3.2, σ may be approximated by piecewise constant functions and $\chi_{|S_e}$ by piecewise linear functions.

6. CONCLUSIONS

The mathematical formulation for the propagation of sound waves in a nonuniform flow field has been discussed. The coupling of finite elements and boundary elements has been described. The finite elements are used to discretize the acoustic pressure in the nonuniform flow field, while boundary elements are applied to solve the boundary integral equation which follows from the interface condition at the outer boundary of the computational finite element domain. For three dimensional sound radiation in subsonic flow the coupling can be expressed explicitly in terms of the pressure at the boundary and a single layer distribution σ given by equation (27) describing a linear combination of the normal derivative and the axial derivative of the pressure. It is not necessary, as has been described in Ref. [7] to express the the axial derivative in terms of the pressure and its normal derivative. Only the pressure and the single layer distribution σ are retained in the present coupled mathematical formulation.

The convergence of the coupled system of finite elements and boundary elements has been investigated for the sound propagation in a stationary fluid where the acoustic pressure is governed by the Helmholtz equation. The convergence of the coupled system appears to be of second order when the computational mesh is refined uniformly. This observed rate of convergence for the Helmholtz equation is found to fit the theoretical error estimate of [5], which was derived for Laplace's equation.

The coupled computational system has been applied to estimate scattering and refractive effects on the acoustic pressure waves through a boundary layer surrounding an aircraft's fuselage. The propagation effects through the boundary layer can be significant and should be included in the noise predictions.

ACKNOWLEDGEMENT

Part of this research has been carried out in cooperation with Twente University, Department of Mechanical Engineering.
The authors would like to thank Professor H. Tijdeman (Twente University) and Dr. P. Sijtsma (National Aerospace Laboratory NLR) for fruitful discussions. The authors are grateful to Dr. H.H. Brouwer (National Aerospace Laboratory NLR) for bringing the aeroacoustic problems to their attention.

References

[1] GOLDSTEIN, M.E.; Aeroacoustics, McGraw-Hill Book co., New York, 1976.

[2] SIGMAN, R.K., MAJJIGI, R.K. and ZINN, B.T.; Determination of turbofan inlet acoustic using finite elements, AIAA Journal, Vol. 16, 1978, pp. 1139-1145.

[3] ENGQUIST, B. and MAJDA, A.; Absorbing boundary conditions for the numerical simulation of waves, Mathematics of Computation, Vol, 31, 1977, pp. 629-651.

[4] CANNING, F.X.; On the application of some radiation boundary conditions, IEEE Transactions on Antennas and Propagation, Vol. 38, 1990, pp. 740-745.

[5] COSTABEL, M., ERVIN, V.J., STEPHAN, E.P.; Experimental convergence rates for various couplings of boundary and finite elements, Math. Comput. Modelling, Vol. 15, No. 3-5, 1991, pp. 93-102.

[6] ACM-Trans. Math. Software, Vol.8, No.2, 1982, p.190.

[7] WU, T.E. and LEE, L.; A direct boundary integral formulation for acoustic radiation in a subsonic uniform flow, Journal of Sound and Vibration, Vol. 175, 1994, pp. 51-63.

[8] WENSING, J.A.; Computation of the acoustic field around an aircraft fuselage, NLR report TR 94235 L, National Aerospace Laboratory NLR, Amsterdam.

[9] SCHIPPERS, H. and WENSING, J.A.; A hybrid computational model for noise propagation through a fuselage boundary layer, NLR report TP 95125 L, National Aerospace Laboratory NLR, Amsterdam.

Fig. 1 Example of a computational grid around a Fokker 50-like fuselage

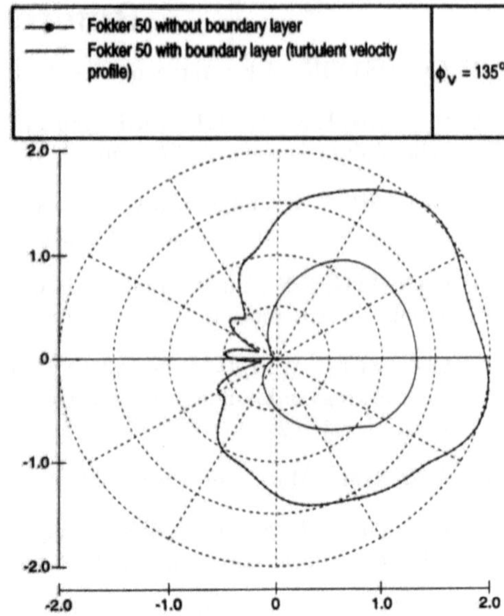

Fig. 3 Effects of the boundary layer on the pressure distribution (amplitude) around the cylinder with a cross section similar to a Fokker 50 fuselage ($k = 5.6$, $M_O = 0.8$, $\phi_V = 135°$)

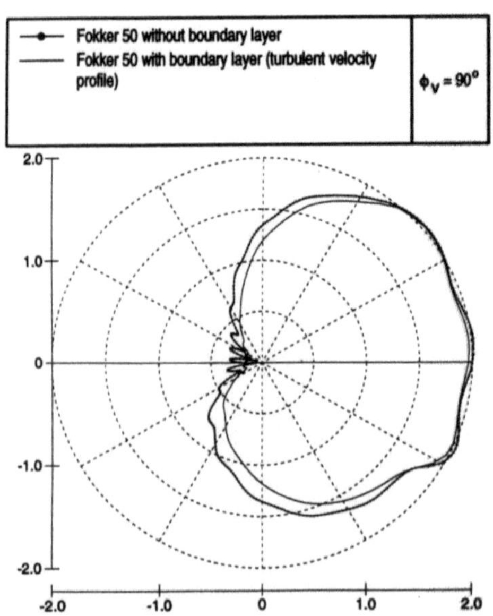

Fig. 2 Effects of the boundary layer on the pressure distribution (amplitude) around the cylinder with a cross section similar to a Fokker 50 fuselage ($k = 5.6$, $M_O = 0.8$, $\phi_V = 90°$)

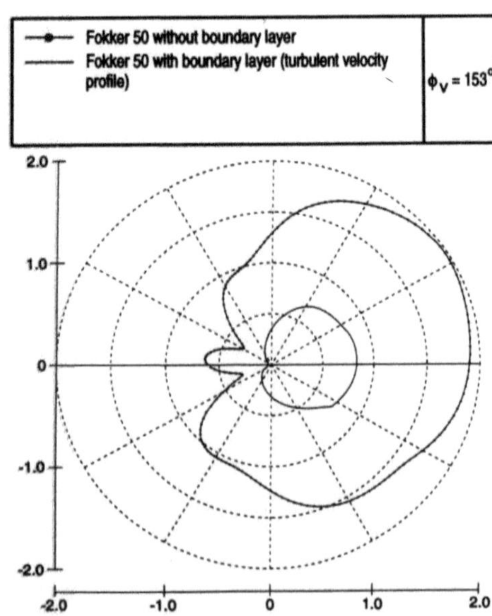

Fig. 4 Effects of the boundary layer on the pressure distribution (amplitude) around the cylinder with a cross section similar to a Fokker 50 fuselage ($k = 5.6$, $M_O = 0.8$, $\phi_V = 153°$)

Numerical Solution of the Neutron Diffusion Equation—Adaptive Concepts in Time and Space

Werner Schmid Frank Wagner

Mathematisches Institut der Technischen Universität München,
D-80290 München, Germany

(schmid@mathematik.tu-muenchen.de, wagnerf@mathematik.tu-muenchen.de)

Supported by grant 03-H07TUM of the German Federal Department of Research and Technology (BMFT). The first author is also supported by a Ph.D grant of SIEMENS AG.

Abstract

A central problem in the safety analysis of nuclear reactors is the determination of the neutron distribution in the reactor core. This is usually done by treating the neutron motion as a diffusion process and solving the diffusion equation numerically. Dividing the range of the energy variable into a certain number of intervals, the so-called energy groups, we obtain the multigroup neutron diffusion equations, a coupled system of either elliptic or parabolic partial differential equations (stationary or instationary). The use of hybrid mixed finite elements for their numerical solution is discussed and closely related adaptive concepts are presented. For a 2-D model problem some numerical results are given. Finally, the use of adaptive hybrid mixed finite element discretizations is considered for 3-D calculations on a rectangular geometry and an implementational concept is shortly presented.

1 The Multigroup Neutron Diffusion Equations

A significant tool in the safety analysis of nuclear reactors is the numerical simulation of the reactor core. Important physical quantities are the *neutron density* n and the *neutron flux* $\phi = n \cdot v$ with neutron velocity v. To describe the neutron distribution in the reactor core, the diffusion theory is often used instead of the more exact but expensive transport theory. The multigroup neutron diffusion equations reflect a neutron balance where the energy variable is split into G intervals, the *energy groups* $g = 1 \dots G$. (For a general description of reactor physics see e.g. Duderstadt/Hamilton [6].)

Usually, only a few energy groups, sometimes only two, are needed to obtain reliable results. Therefore, for the sake of simplicity, we will restrict ourselves to two groups. Let us first consider stationary diffusion without external sources and without upscattering, i.e., without scattering from the energy group 2 into the energy group 1. We have

$$\begin{aligned} -\nabla \cdot (D_1 \nabla \phi_1) + (\Sigma_{a1} + \Sigma_{12})\, \phi_1 &= \frac{1}{\lambda}\chi_1(\nu\Sigma_{f1}\phi_1 + \nu\Sigma_{f2}\phi_2) \\ -\nabla \cdot (D_2 \nabla \phi_2) + \Sigma_{a2}\phi_2 - \Sigma_{12}\phi_1 &= \frac{1}{\lambda}\chi_2(\nu\Sigma_{f1}\phi_1 + \nu\Sigma_{f2}\phi_2) \end{aligned}$$

or

$$(L - \frac{1}{\lambda}F) \cdot \Phi = 0, \tag{1}$$

where ϕ_g denotes the *neutron flux* of the energy group g, Σ_{ag} the respecting *absorption cross section*, Σ_{12} the *scattering cross section* from energy the group 1 of *fast neutrons* into the energy group 2 of *thermal neutrons* and $\nu\Sigma_{fg}$ the *fission cross section*.

(Out-)Diffusion and absorption are loss terms, fission acts as a neutron source and scattering is a loss term for the energy group 1 and a source term for the energy group 2.

The balance equation results in a generalized eigenvalue problem where

- the dominant eigenvalue λ (*multiplication constant* of the assembly)
- the associated eigenfunction Φ

are to be determined.

For a given geometry the long time behaviour, the power distribution and the burnup behaviour of the assembly can be deduced (see Schmid [11] for more details on the generalized eigenvalue problem).

$$V\frac{\partial \Phi}{\partial t} + (L - \frac{1}{\lambda}F)\Phi = -D\frac{\partial C}{\partial t}, \tag{2}$$

where V denotes the diagonal matrix of the *inverse neutron velocities* of the two groups, F the *fission operator* and C the *precursor concentrations* which are calculated simultaneously from

$$\frac{\partial C}{\partial t} + \lambda_C C = \frac{1}{\lambda}\beta[\nu\Sigma_{f1}\phi_1 + \nu\Sigma_{f2}\phi_2], \tag{3}$$

with *fission spectra* χ_{dg}, χ_{pg} for *delayed* and *prompt* neutrons, respectively, the *fraction* β of delayed neutrons and the *precursor decay constant* λ_C. (To simplify the notation, we only use a single precursor instead of six physically possible ones.)

2 Concepts for Time Step Control

The instationary problem (2) is interpreted as an *abstract Cauchy problem* in the Banach space $H := L^2([0,T], W)$:

$$V\frac{d}{dt}u + Au = f, \qquad u(0) = u_0,$$

with $u \in H$, $u_0 \in W$ and $f \in H$. This enables us to use concepts from ordinary differential equations for the time step control of this parabolic equation (see Bornemann [2]). For the two group diffusion equation we have (see Wagner [13] for more details):

W is chosen to be a suitable subspace of $[H^1(\Omega)]^2$ satisfying the corresponding boundary conditions and

$$A := L - \frac{1}{\lambda}F, \qquad f := -D\frac{d}{dt}C.$$

The initial value u_0 and the eigenvalue λ are the solution of the corresponding stationary problem (1).

We consider the transient problem in the time intervall $[t_0, t_1]$ with the stepsize $\tau = t_1 - t_0$. A discretization based on the *implicit Euler* scheme results in the following system of elliptic equations

$$[V + \tau A]u^1 = Vu^0 + \tau f, \quad u^0 = u(t_0). \tag{4}$$

An approximation of second order u^2 is obtained by an additive correction of the form $u^2 := u^1 + \eta_1$, where η_1 is the solution of

$$[V + \tau A]\eta_1 = \frac{1}{2}\tau \left[A(u^1 - u^0) + f(t_1) - f(t_0)\right]. \qquad (5)$$

Solving both systems of elliptic equations (4) and (5) *exactly*, the new time step τ_{new} would be calculated by

$$\tau_{new} := \sqrt{\frac{\text{TOL}}{\epsilon_1}} \cdot \tau,$$

where TOL is a prescribed tolerance and $\epsilon_1 := \|\eta_1\|$ is an estimate for the error in time.

u^1 and η_1 cannot be calculated exactly, though. We only get approximations $\hat{u}^1$ and $\hat{\eta}_1$ with perturbations $\delta_1 := \hat{u}^1 - u^1$, $\omega_1 := \hat{\eta}_1 - \eta_1$, respectively.

We require that these perturbations are below a certain threshold to assure that $\hat{u}_2$ is good enough with respect to TOL and that the resulting time step sequence is close to the unperturbed one. This means that the difference $\theta_1 := \hat{\epsilon}_1 - \epsilon_1$ should be small compared to ϵ_1 where $\hat{\epsilon}_1 := \|\hat{\eta}_1\| \doteq \|\hat{u}^1 - u(t_1)\|$.

A closer look at the calculations reveals

$$\begin{aligned}
\hat{u}^1 &= [V + \tau A]^{-1}(Vu^0 + \tau f) + \delta_1 = u^1 + \delta_1 \\
\hat{\eta}_1 &= \frac{1}{2}[V + \tau A]^{-1}\left(\tau A(\hat{u}^1 - u^0) + \tau f(t_1) - \tau f(t_0)\right) + \hat{\omega}_1 \\
&= \eta_1 + \frac{1}{2}[V + \tau A]^{-1}\tau A\delta_1 + \hat{\omega}_1 = \eta_1 + \omega_1 \\
\hat{u}^2 &= \hat{u}^1 + \hat{\eta}_1 = u^2 + \delta_1 + \frac{1}{2}[V + \tau A]^{-1}\tau A\delta_1 + \hat{\omega}_1 \\
&= u^2 + \delta_2.
\end{aligned}$$

If error estimates $[\delta_1]$ and $[\hat{\omega}_1]$ are provided for the perturbations $\|\delta_1\|$ and $\|\hat{\omega}_1\|$, their influence on the discretization in time can be estimated.

In general, δ_1 will dominate the perturbations. Therefore, we neglect $\hat{\omega}_1$ and, using the property $\|[V + \tau A_\tau]^{-1}\tau A_\tau\| < 1$, obtain the following estimates for $|\theta_1|$ and $\|\delta_2\|$ (cf. Bornemann [2, Chapter3.2]):

$$\begin{aligned}
[\theta_1] &:= \frac{1}{2}[\delta_1] \\
[\delta_2] &:= [\delta_1] + [\theta_1] = \frac{3}{2}[\delta_1]
\end{aligned}$$

We want our solution to satisfy

$$\hat{\epsilon}_1 + [\delta_2] < \text{TOL}, \qquad (6)$$

$$[\theta_1] < \frac{1}{4}\hat{\epsilon}_1, \qquad (7)$$

so that $\hat{\epsilon}_1$ is a reliable approximation for ϵ_1 and $\hat{u}^2$ is close to u^2.

We split (6) into the criterion $\hat{\epsilon}_1 < \rho\text{TOL}$ for the time discretization, resolved by choosing the new time step as

$$\tau_{new} := \sqrt{\frac{\rho\text{TOL}}{\hat{\epsilon}_1}} \cdot \tau,$$

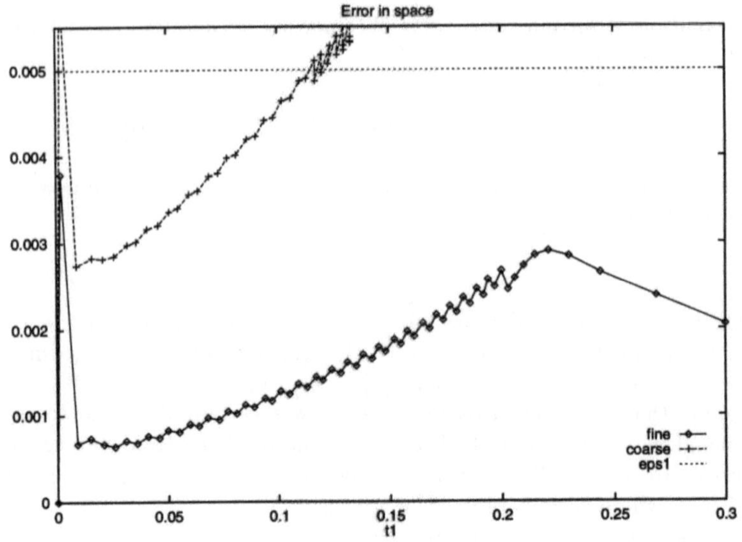

Figure 1: Discretization errors in space for elliptic solvers of different accuracies

and the criterion $[\delta_2] < (1-\rho)\mathrm{TOL}$ for the space discretization which is resolved by an elliptic solver fullfilling

$$[\delta_1] < \mathbf{eps}_1 := \frac{2}{3}(1-\rho)\mathrm{TOL}$$

for the solution $\hat{u}^1$ of (4).

Remark 1 *For $\Omega \subset \mathbf{R}^d$ we are discretizing $d+1$ dimensions in time and space. Therefore, $\rho = \frac{1}{d+1}$ is a possible choice.*

Remark 2 *If the elliptic solver is not close to $[\delta_1] < \mathbf{eps}_1$, e.g. when not using an adaptive solver, one could try to resolve the total tolerance more efficently by choosing the new time step as*

$$\tilde{\tau}_{\mathrm{new}} := \sqrt{\frac{TOL - [\delta_2]}{\hat{\epsilon}_1}} \cdot \tau.$$

A numerical example was computed using a cell-centered five-point stencil on a uniform grid with a multigrid solver and block Gauss-Seidel smoothing for a modification of the benchmark-problem TWIGL.

In Figure 1 we show the evolution of the space-discretization errors for elliptic solvers of different accuracies, denoted by *fine* and *coarse*, resp., obtained by using a different number of uniformly refined levels in the multigrid hierarchy. We observe an identic time step sequence for both solvers as long as the elliptic accuracy $\mathbf{eps}_1$ is satisfied. The whole solution process fails after the coarse solver is no longer able to solve with the demanded accuracy.

Therefore, we again emphasize the use of an adaptive method based on an efficient and reliable error estimator.

3 Mixed Hybrid Finite Elements

The physically interesting terms for the neutron diffusion equations are the neutron flux ϕ as well as the neutron current. Therefore, the use of mixed finite elements is adequate, because it offers the simultaneous calculation of the flux and the current. We do not consider the multigroup equations for the presentation of mixed finite elements and their implementation in hybrid form. For the sake of simplicity, we restrict to one equation and to the simplicial 2D case and refer to the excellent monography of Brezzi/Fortin [3, chapter V] for the general case. For the use of finite elements in the field of neutron diffusion and transport see e.g. Coulomb and Fedon-Magnaud [4], Del Valle et al. [5] and Schmidt [12].

First, let us give some notation: Let $\Omega \subset \mathbb{R}^2$ be a bounded polygonal domain with boundary Γ. Let a triangulation $\mathcal{T}_h$ be given where K denotes an element of $\mathcal{T}_h$, e_ν^K, $1 \leq \nu \leq 3$, the edges of K and $\mathcal{E}_h$ the set of edges e in $\mathcal{T}_h$.

We partly use the notation of Hoppe/Wohlmuth in [9].

3.1 Hybridization

We start with the following elliptic partial differential equation:

$$-\nabla \cdot (D\nabla\phi) + \Sigma\phi = S \text{ in } \Omega \tag{8}$$

with boundary conditions

$$\phi = 0 \text{ on } \Gamma_0 \quad \text{and} \quad D\frac{\partial\phi}{\partial n} = 0 \quad \text{on} \quad \Gamma_1; \quad (\Gamma_0 \cup \Gamma_1 = \Gamma).$$

Introducing the *neutron current* $\mathbf{j} := -D\nabla\phi$, the second order equation (8) can be formally written as the first order system

$$\begin{aligned} \tfrac{1}{D}\mathbf{j} + \nabla\phi &= 0 \\ \operatorname{div}(\mathbf{j}) + \Sigma\phi &= S \end{aligned} \tag{9}$$

with the corresponding boundary conditions.

Defining the spaces

$$\begin{aligned} H(\operatorname{div};\Omega) &:= \{\mathbf{q} \in [L^2(\Omega)]^2 \mid \operatorname{div}(\mathbf{q}) \in L^2(\Omega)\}, \\ H_{0,\Gamma_1}(\operatorname{div};\Omega) &:= \{\mathbf{q} \in H(\operatorname{div};\Omega) \mid \mathbf{q}\cdot\mathbf{n} = 0 \text{ on } \Gamma_1\}, \end{aligned}$$

the bilinear forms $a(\mathbf{p},\mathbf{q}) := \int_\Omega \frac{1}{D}\mathbf{p}\cdot\mathbf{q}\,dx$, $b(\mathbf{q},v) := \int_\Omega v \operatorname{div}(\mathbf{q})\,dx$, $c(u,v) := \int_\Omega \Sigma uv\,dx$, and the linear form $S(v) := \int_\Omega Sv\,dx$, the variational formulation of (9) is commonly referred to as the *mixed formulation* and is given as follows: find $(\mathbf{j},\phi) \in H_{0,\Gamma_1}(\operatorname{div};\Omega) \times L^2(\Omega)$, such that

$$\begin{aligned} a(\mathbf{j},\mathbf{q}) \; - \; b(\mathbf{q},\phi) &= 0 && \text{for } \mathbf{q} \in H_{0,\Gamma_1}(\operatorname{div};\Omega), \\ b(\mathbf{j},v) \; + \; c(\phi,v) &= S(v) && \text{for } v \in L^2(\Omega). \end{aligned} \tag{10}$$

The mixed finite element discretization of (8) is based on its mixed formulation (10) where the current space $H(\operatorname{div};\Omega)$ will be approximated by

$$RT_0(\Omega;\mathcal{T}_h) := \{\mathbf{q}_h \in H(\operatorname{div};\Omega) \mid \mathbf{q}_h|_K \in RT_0(K), K \in \mathcal{T}_h\},$$

where $RT_0(K)$ denotes the lowest order Raviart-Thomas element defined by

$$RT_0(K) := [P_0(K)]^2 + \mathbf{x}P_0(K), \quad \mathbf{x} := (x_1, x_2)^T.$$

By $P_k(K)$, $k \geq 0$, we denote the set of polynomials of degree $\leq k$ on K.

Note that any $\mathbf{q}_h|_K \in RT_0(K)$ is uniquely determined by the three degrees of freedom

$$\int\limits_{\partial K} (\mathbf{n} \cdot \mathbf{q}_h|_K) p_i \, d\sigma, \quad p_i \in R_0(\partial K), \ 1 \leq i \leq 3,$$

where $R_0(\partial K) := \{p \in L^2(\partial K) \mid p|_{e_\nu^K} \in P_0(e_\nu^K), 1 \leq \nu \leq 3\}$ and $\mathbf{n}$ is the outer normal to ∂K.

The conformity of the approximation is guaranteed by specifying the basis fields in such a way that continuity of the normal components

$$(\mathbf{n} \cdot \mathbf{q}_h)\,|_{e\cap K} = -\,(\mathbf{n}' \cdot \mathbf{q}_h)\,|_{e\cap K'}, \quad K \cap K' = e \in \mathcal{E}_h \cap \Omega, \tag{11}$$

is satisfied across interelement boundaries where $\mathbf{n}$ and $\mathbf{n}'$ are the outer normals on $e \cap \partial K$ and $e \cap \partial K'$, respectively.

Observing $\operatorname{div}(\mathbf{q}_h)|_K \in P_0(K)$, $K \in \mathcal{T}_h$, a natural choice for the approximation of the primal variable ϕ is to use piecewise constants leading to the ansatz space

$$W_0(\Omega; \mathcal{T}_h) := \left\{ v_h \in L^2(\Omega) \mid v_h|_K \in P_0(K), \ K \in \mathcal{T}_h \right\}.$$

Then, the mixed discretization requires the computation of the pair of unknowns $(\mathbf{j}_h, \phi_h) \in RT_0(\Omega; \mathcal{T}_h) \times W_0(\Omega; \mathcal{T}_h)$ such that

$$\begin{array}{rcrcll} a(\mathbf{j}_h, \mathbf{q}_h) & - & b(\mathbf{q}_h, \phi_h) & = & 0 & \text{for } \mathbf{q}_h \in RT_0(\Omega; \mathcal{T}_h), \\ b(\mathbf{j}_h, v_h) & + & c(\phi_h, v_h) & = & S(v_h) & \text{for } v_h \in W_0(\Omega; \mathcal{T}_h). \end{array}$$

The corresponding linear system is indefinit. In order to avoid the well-known difficulties in the solution of such problems, we use the technique of *mixed hybridization*. The idea is to eliminate the continuity constraints (11) for the normal components on the interelement boundaries from the ansatz space $RT_0(\Omega; \mathcal{T}_h)$. This leads to the nonconforming Raviart-Thomas approximation

$$RT_0^{-1}(\Omega; \mathcal{T}_h) := \left\{ \mathbf{q}_h \in (L^2(\Omega))^2 \mid \mathbf{q}_h|_K \in RT_0(K), \ K \in \mathcal{T}_h \right\}.$$

Note that the dimension of $RT_0^{-1}(\Omega; \mathcal{T}_h)$ exceeds that of its conforming counterpart $RT_0(\Omega; \mathcal{T}_h)$ by the number of interelement boundaries, since now two basis fields are associated with each interior edge. Instead, the continuity constraints are taken care of by Lagrangian multipliers from the *multiplier space*

$$M_0(\mathcal{E}_h) := \{\rho_h \in (L^2(\mathcal{E}_h)^2 \mid \rho_h|_e \in P_0(e), e \in \mathcal{E}_h \cap \Omega, \rho_h|_e = 0, e \in \mathcal{E}_h \cap \Gamma_0\}.$$

Thus, the nonconforming mixed hybrid discretization requires the computation of $(\mathbf{j}_h, \phi_h, \mu_h) \in RT_0^{-1}(\Omega; \mathcal{T}_h) \times W_0(\Omega; \mathcal{T}_h) \times M_0(\mathcal{E}_h)$ such that

$$\begin{array}{rcrcll} \hat{a}(\mathbf{j}_h, \mathbf{q}_h) & - & \hat{b}(\mathbf{q}_h, \phi_h) & = & -\hat{d}(\mu_h, \mathbf{q}_h) & \text{for } \mathbf{q}_h \in RT_0^{-1}(\Omega; \mathcal{T}_h), \\ \hat{b}(\mathbf{j}_h, v_h) & + & \hat{c}(\phi_h, v_h) & = & S(v_h) & \text{for } v_h \in W_0(\Omega; \mathcal{T}_h), \\ \hat{d}(\rho_h, \mathbf{j}_h) & & & = & 0 & \text{for } \rho_h \in M_0(\mathcal{E}_h), \end{array}$$

where the bilinear forms $\hat{a}$, $\hat{b}$, $\hat{c}$ and $\hat{d}$ are given by

$$\hat{a} := \sum_{K \in \mathcal{T}_h} a(\cdot|_K, \cdot|_K), \ \hat{b} := \sum_{K \in \mathcal{T}_h} b(\cdot|_K, \cdot|_K), \ \hat{c} := c$$

$$\hat{d}(\mu_h, \mathbf{q}_h) := \sum_{K \in \mathcal{T}_h} \int_{\partial K} \mu_h n \cdot \mathbf{q}_h \, d\sigma, \quad \mu_h \in M_0(\mathcal{E}_h), \ \mathbf{q}_h \in RT_0^{-1}(\Omega; \mathcal{T}_h).$$

Written in operator (or matrix) terms we get

$$\begin{bmatrix} \hat{A} & -\hat{B}^T & \hat{D}^T \\ \hat{B} & \hat{C} & 0 \\ \hat{D} & 0 & 0 \end{bmatrix} \begin{bmatrix} \mathbf{j}_h \\ \phi_h \\ \mu_h \end{bmatrix} = \begin{bmatrix} 0 \\ S \\ 0 \end{bmatrix}.$$

The distinctive feature of mixed hybridization is to eliminate the discrete flux $\mathbf{j}_h$, i.e., to use *static condensation*, and to take advantage of the equivalence of the resulting variational equations in $(\phi_h, \mu_h) \in W_0(\Omega; \mathcal{T}_h) \times M_0(\mathcal{E}_h)$

$$\begin{bmatrix} \hat{C} + \hat{B}\hat{A}^{-1}\hat{B}^T & -\hat{B}\hat{A}^{-1}\hat{D}^T \\ \hat{D}\hat{A}^{-1}\hat{B}^T & -\hat{D}\hat{A}^{-1}\hat{D}^T \end{bmatrix} \begin{bmatrix} \phi_h \\ \mu_h \end{bmatrix} = \begin{bmatrix} \tilde{S} \\ 0 \end{bmatrix} \tag{12}$$

with an extended nonconforming approximation where the lowest order nonconforming Crouzeix-Raviart ansatz space is augmented with bubbles.

Remark 3 *Arnold/Brezzi [1, Lemma 2.4] have shown that for 2D simplicial triangulations the mixed hybrid finite element approximation defines a nonconforming approximation φ_h that is, up to L^2-projections of the right hand side onto $W_0(\Omega; \mathcal{T}_h)$, equivalent to an approximation by the standard nonconforming Crouzeix-Raviart elements augmented by bubbles. If the exact solution satisfies $\phi \in H^2(\Omega)$ and $S \in H^1(\Omega)$, this approximation is of order $O(h^2)$ in contrast to an order of $O(h)$ w.r.t $\|.\|_0$ for the piecewise constant approximation ϕ_h.*

3.2 An L^2-Error Estimator for 2D-Simplicial Triangulation

Adaptivity by local refinement can be realized using an efficient and reliable a posteriori error estimator for the L^2-norm of the total error in the primal variable ϕ_h. Such an estimator has been developed in Hoppe/Wohlmuth [9] and can be derived from a saturation assumption motivated by the superconvergence result for mixed hybridization mentioned in Remark 3.

Let $CR(\Omega; \mathcal{T}_h)$ denote the standard nonconforming Crouzeix-Raviart ansatz space for the given simplicial triangulation of Ω and let $B(\Omega; \mathcal{T}_h)$ denote the space of cubic bubble functions (locally cubic polynomials vanishing on the boundary of the elements). For an exact definition of these spaces see [9, Chapter 2].

The solution $(\phi_h, \mu_h) \in W_0(\Omega; \mathcal{T}_h) \times M_0(\mathcal{E}_h)$ of (12) defines a *nonconforming interpoland* $\varphi_h \in CR(\Omega; \mathcal{T}_h)$, $\Pi_h \varphi_h := \mu_h$ to the Langrangian multiplier μ_h, where Π_h is the L_2-projection onto $M_0(\mathcal{E}_h)$. Supported by the superconvergence result, we assume the *saturation assumption*

$$\|\phi - \varphi_h\|_0 \leq \beta \|\phi - \phi_h\|_0, \quad 0 \leq \beta < 1.$$

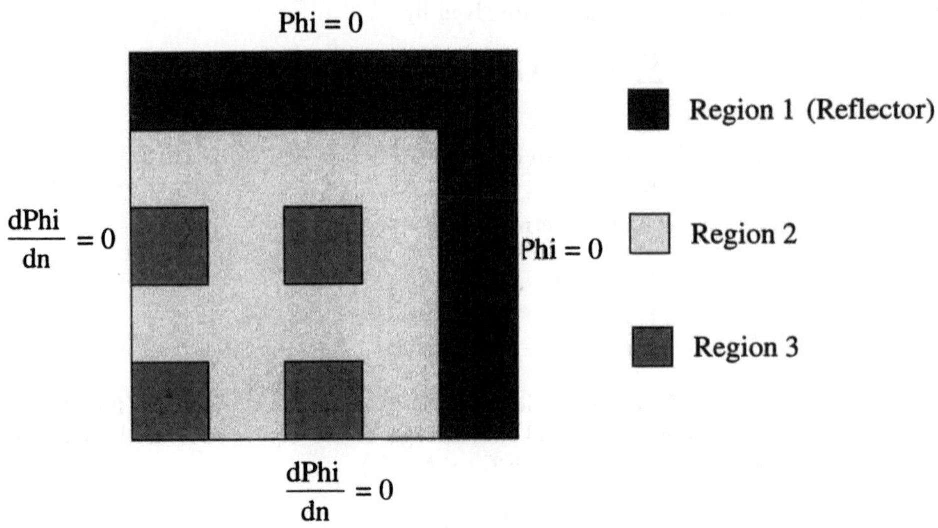

Figure 2: Simplified geometry for the test problem

The unique approximation $\psi_h \in NC(\Omega;\mathcal{T}_h) := CR(\Omega;\mathcal{T}_h) \bigoplus B(\Omega;\mathcal{T}_h)$ to ϕ is equivalent to (ϕ_h, μ_h) (see Remark 3).

Given an iterative approximation $\tilde{\psi}_h \in NC(\Omega;\mathcal{T}_h)$ of ψ_h, we define

$$(\tilde{\phi}_h, \tilde{\mu}_h) \in W_0(\Omega;\mathcal{T}_h) \times M_0(\mathcal{E}_h) \text{ and } \tilde{\varphi}_h \in CR(\Omega;\mathcal{T}_h)$$

as

$$\tilde{\phi}_h := P_h\tilde{\psi}_h, \; \tilde{\mu}_h := \Pi_h\tilde{\psi}_h, \; \Pi_h\tilde{\varphi}_h := \tilde{\mu}_h,$$

where P_h and Π_h are the L_2-projections onto the corresponding spaces.

The following error estimation for the primal variable ϕ_h holds:

$$\begin{array}{c} \frac{1}{1+\beta}\|\tilde{\phi}_h - \tilde{\varphi}_h\|_0 - \sqrt{\frac{10}{3}}\|\psi_h - \tilde{\psi}_h\|_0 \\ \leq \|\phi - \tilde{\phi}_h\|_0 \leq \\ \frac{1}{1-\beta}\|\tilde{\phi}_h - \tilde{\varphi}_h\|_0 + \sqrt{\frac{10}{3}\frac{1+\beta}{1-\beta}}\|\psi_h - \tilde{\psi}_h\|_0 \,. \end{array} \tag{13}$$

We remark that $\|\psi_h - \tilde{\psi}_h\|_0$ in (13) represents the iteration error which can be controlled by the iterative solution process. Therefore, we may use the cheaply computable local contributions $\|\tilde{\phi}_h - \tilde{\varphi}_h\|_{0,K}$ as indicators for local refinement of a given triangulation.

The following numerical test in case of the two group diffusion equations was performed based on the algorithm in [9]. We adapted it to the two group diffusion equations by using a block Gauss-Seidel scheme as an outer iteration. Local refinement was performed using the error estimate of the equation for the thermal energy group 2. The data for this test problem are taken from a realistic simulation of a light water reactor with simplified geometry (Figure 2). They correspond to a cross section through a 3D quarter core and consist of two differently enriched regions (regions 2,3) and the pressurized water reflector (region 1) with constant coefficients in every region.

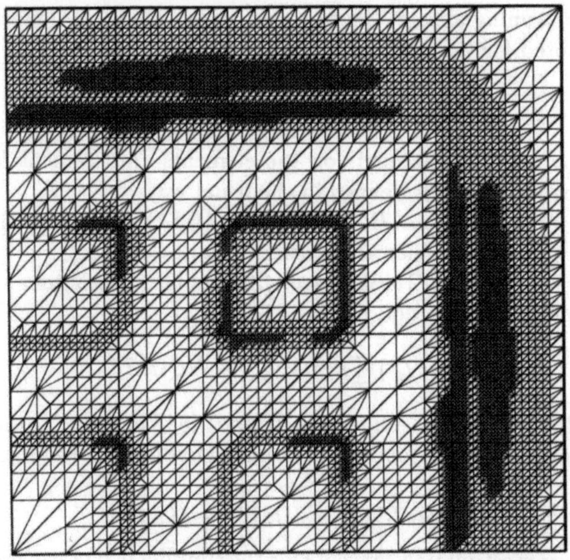

Figure 3: Adaptive grid after five local refinements

Figure 3 shows that the structures of the given assembly are well resolved. This encouraged us to transfer this concept to a 3D rectangular geometry. The next section will resume our considerations for that case.

4 Concepts for 3D-Computations

We assume a rectangular triangulation $\mathcal{T}_h^*$ to be given for $\Omega \in \mathbb{R}^3$, a polyhedral bounded domain. By $\mathcal{F}_h$ we denote set of faces for this triangulation and $f_\nu^K, 1 \leq \nu \leq 6$, are the faces of an element $K \in \mathcal{T}_h^*$.

4.1 Rectangular Geometry in 3D

For calculations in reactor simulation, typically a rectangular geometry is used. Therefore, we must apply the concepts mentioned in the preceding section to rectangular geometries in 3D.

Instead of Crouzeix-Raviart elements we refer to nonconforming finite elements that apply to 2D and 3D rectangular geometries developed by Rannacher/Turek [10]. For the 3D reference element $\tilde{K} := [-1,1]^3$ we use the local ansatz space

$$\tilde{Q}(\tilde{K}) := \text{span}\{1, x, y, z, x^2 - y^2, y^2 - z^2\}.$$

The corresponding global finite element space $H_h(\Omega; \mathcal{T}_h^*)$ is constructed to be piecewise in this local ansatz space and continuous with respect to the weighted first moments on the element faces.

Another nonconforming ansatz, the so-called *nodal scheme of lowest order* (see Hennart [7]) is constructed for the local ansatz space

$$\tilde{N}(\tilde{K}) := \text{span}\{1, x, y, z, x^2, y^2, z^2\},$$

uniquely defined by the first moments on the elements and element faces. The global finite element space $\tilde{N}_h(\Omega; \mathcal{T}_h^*)$ is defined in the usual way.

The space of bubble functions $B(\Omega; \mathcal{T}_h^*) \subset \tilde{N}_h(\Omega; \mathcal{T}_h^*)$ consists of the functions with vanishing first moments on the element faces.

Augmenting the Rannacher/Turek space with bubbles, we obtain a space equivalent to the nodal scheme of lowest order, i.e.,

$$H_h(\Omega; \mathcal{T}_h^*) \bigoplus B(\Omega; \mathcal{T}_h^*) \equiv \tilde{N}_h(\Omega; \mathcal{T}_h^*).$$

On the other hand, we have the result from Hennart [8, Theorem 5]) that in our case the nodal scheme of lowest order is equivalent to a mixed hybrid finite element approximation, i.e.,

$$\tilde{N}_h(\Omega; \mathcal{T}_h^*) \equiv W_0(\Omega; \mathcal{T}_h^*) \times M_0(\mathcal{F}_h).$$

Therefore, the error estimation concepts for the simplicial 2D case, presented in section 3.2, are transfered to the rectangular 3D case, using Rannacher/Turek nonconforming finite elements instead of Crouzeix-Raviart elements.

Remark 4 *The detailed theoretical and numerical analysis of this error estimator for rectangular geometries in 2D and 3D is a current research project of the authors, especially its numerical validation. For an adaptive refinement on a rectangular geometry slave nodes, introduced in section 4.3, are needed. Their explicit consideration within a possibly nonuniform refinement will be another area of our future work.*

Remark 5 *Both Rannacher/Turek and nodal nonconforming space can be applied to 2D rectangular geometries as well. The construction of the corresponding spaces in 2D is straightforward.*

4.2 Two Group Calculations

For a more efficent coupling of the two equations we propose, due to the identic nature of all equations in the system, to compute local 2×2 blocks with the unknown of both energy groups simultaneously. We are convinced that the local coupling between the energy groups via scattering and fission renders it especially favourable to closely link the two unknowns with identical "localization".

This enables the efficient use of shift strategies for the generalized eigenvalue problem and can easily be included in the iterative solution of the linear subproblems. Two possibilities for the use of this block-concept within iterative solvers are block-oriented smoothers within multigrid which has been done with good results for the 2D case (cf. [11]) and the use of block-preconditioners (like block-ILU) for cg-like iterations. First numerical tests in 3D for the latter have been promising.

A closer examination of these blockwise calculations with respect to multilevel and cg-like iterations with multilevel preconditioning is in progress.

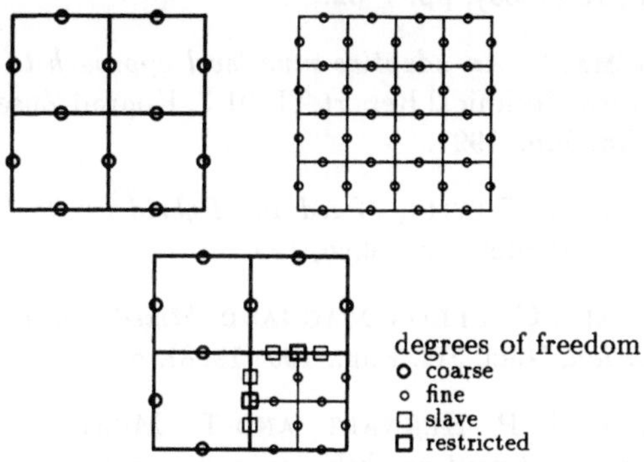

Figure 4: A rectangular 2D situation with slave nodes

4.3 Slave Nodes

In rectangular geometries there exists nothing like a *green closure* known from simplicial geometries to preserve conformity of the triangulation. To handle the neighbourhood of refined and unrefined elements, we introduce *slave nodes*:

For coarse degrees of freedom associated with a face shared by a refined and an unrefined element, we construct a new corresponding base function with a *restricted* support.

In computations, we use the corresponding fine degrees of freedom associated with the refinement faces of such a face as an *auxiliary tool* for those restricted degrees of freedom, e.g. as *slaves* of the *restricted degrees of freedom*.

Consequences are that these slaves are interpolated values from restricted degrees of freedom and they behave differently during the assembly of a global stiffness matrix.

In Figure 4 we show a typical situation with slave nodes for a rectangular geometry in 2D.

5 Conclusion

We have briefly discussed the derivation of the two group neutron diffusion equations. A short discussion of time step control for the transient equations indicated the possible benefits of adaptive concepts. Mixed hybrid finite elements have been considered as a possibility for the numerical solution of these equations. For 2D simplicial triangulation we discussed adaptivity concepts and showed that they are promising for the use in this area. Concepts for a transfer to 3D rectangular geometries have been shortly presented. Adaptive calculations for 3D reactangular grids are to be done in the near future for stationary and instationary neutron diffusion to reveal if the promising 2D results can be achieved there, too.

References

[1] D. N. ARNOLD AND F. BREZZI, *Mixed and nonconforming finite element methods: Implementation, postprocessing and error estimates*, M^{2}AN Math. Modelling and

Numer. Anal., 19 (1985), pp. 7–32.

[2] F. A. BORNEMANN, *An adaptive multilevel approach to parabolic equations in two space dimensions*, Technical Report TR 91-7, Konrad-Zuse-Zentrum für Informationstechnik, Berlin, June 1991.

[3] F. BREZZI AND M. FORTIN, *Mixed and Hybrid Finite Element Methods*, Springer, New York Berlin Heidelberg Tokyo, 1991.

[4] F. COULOMB AND C. FEDON-MAGNAUD, *Mixed and mixed-hybrid elements for the diffusion equation*, Nucl. Sci. Eng., 100 (1988), pp. 218–225.

[5] E. DEL VALLE, J. P. HENNART, AND D. MEADE, *Finite element formulations of nodal schemes for neutron diffusion and transport problems*, Nucl. Sci. Eng., 92 (1986), pp. 204–211.

[6] J. J. DUDERSTADT AND L. J. HAMILTON, *Nuclear Reactor Analysis*, J. Wiley & Sons, New York, 1976.

[7] J. P. HENNART, *A general family of nodal schemes*, SIAM J. Sci. Stat. Comput., 7 (1986), pp. 264–287.

[8] J. P. HENNART AND E. DEL VALLE, *On the relationship between nodal schemes and mixed-hybrid finite elements*, Numer. Methods Partial Differential Equations, 9 (1993), pp. 411–430.

[9] R. H. W. HOPPE AND B. WOHLMUTH, *Adaptive multilevel techniques for mixed finite element discretizations of elliptic boundary value problems*, TUM-M 9411, Mathematical Institute, Technical University Munich, Munich, June 1994. (submitted to SIAM J. Numer. Anal.).

[10] R. RANNACHER AND S. TUREK, *Simple nonconforming quadrilateral stokes element*, Numer. Methods Partial Differential Equations, 8 (1992), pp. 97–111.

[11] W. SCHMID, *Multigrid Eigenvalue Methods for the Neutron Multigroup Diffusion Equations*, TUM-M 94-07, Mathematical Institute, Technical University Munich, Munich, April 1994.

[12] F. SCHMIDT, *Solution of the neutron diffusion equation by the finite element method*, Kerntechnik, 4 (1992), pp. 223–227.

[13] F. WAGNER, *Time step control for transient multigroup diffusion equations.* in preparation, 1995.

A NUMERICAL SCHEME FOR STRESS WAVES AT A FLUID-SOLID INTERFACE

U. Specht, Ch. Dickopp and J. Ballmann
Lehr- und Forschungsgebiet für Mechanik
RWTH Aachen
TH-Postfach
D-52056 Aachen

SUMMARY

In this paper a numerical scheme for wave processes at a solid-fluid interface is presented which are governed by nonlinear hyperbolic systems of PDE's. One example of application is cavitation damaging in liquids by shock waves emanating from collapsing bubbles.
The liquid is modeled by the Euler or Navier-Stokes equations for compressible fluids and the solid by the equations of elastoplasticity.
In the scheme a Finite Element solver for the liquid side is coupled with a Godunov-type method for elastic-plastic waves in solids. Applying the contact conditions an interative convergent scheme is obtained. Numerical results are presented for a plane problem with a cylindrical shock wave developing in the liquid.

1. INTRODUCTION

Damage of marine propellers gave the reason to Lord Rayleigh to investigate the pressure field around a collapsing bubble [1]. Even earlier Besant treated the problem within a mathematical framework [2]. Since then many theoretical and experimental investigations have been done. In experiments it was detected by Vogel et al. [3], Tomita and Shima [4], Ran and Katz [5] and Lauterborn and Hentschel [6] that shock waves appear around a collapsing bubble and that liquid jets can occur if a bubble collapses near a solid wall. The development of a jet was first simulated numerically by Plesset and Chapman under the assumption of an incompressible liquid and a homogeneous pressure distribution in the bubble [7].
Westenberger assumed an inviscid compressible liquid and simulated the collapse of a spherical bubble by the method of characteristics [8]. It was shown that the pressure field in the bubble is truly non-homogeneous and that shocks appear preferably in the liquid around the bubble. Grant and Lush [9] considered the one-dimensional problem of a liquid column impacting on an elastic-plastic solid.
In order to get a clearer explanation of the damaging mechanism one should treat the full multi-phase and multi-dimensional problem, admitting compressibility, non-vanishing viscosity and heat conductivity, evaporation or condensation, and assuming

non-homogeneous fields in the bubble and in the liquid. The interaction of a fluid with the solid must also be treated as a multi-dimensional problem since surface waves may play an important role.
In this paper we concentrate on the appearance of a shock wave in the liquid and its following interaction with the liquid-solid interface. Furthermore we assume the problem as two-dimensional with a cylindrical shock wave. For the fluid we use a Finite Element Navier-Stokes solver; for the elastic-plastic solid a Godunov-type method is employed. Because of the shock-capturing capability we also use a Godunov-type method for the fluid-solid interface scheme.

2. GOVERNING EQUATIONS

2.1. Liquid phase

The fluid is modeled by the Navier-Stokes equations for compressible heat conducting fluids. For a two-dimensional problem in the x, y-plane they can be written in conservation form as

$$\mathbf{U}_{,t} + (\mathbf{F} - \mathbf{F}_v)_{,x} + (\mathbf{G} - \mathbf{G}_v)_{,y} = \mathbf{0} \tag{1}$$

with

$$\left.\begin{aligned} \mathbf{U} &= \begin{pmatrix} \rho_l \\ \rho_l U \\ \rho_l V \\ e \end{pmatrix} \quad \mathbf{F} = \begin{pmatrix} \rho_l U \\ \rho_l U^2 + p \\ \rho_l UV \\ U(e+p) \end{pmatrix} \quad \mathbf{G} = \begin{pmatrix} \rho_l V \\ \rho_l UV \\ \rho_l V^2 + p \\ V(e+p) \end{pmatrix} \\ \mathbf{F}_v &= \begin{pmatrix} 0 \\ \tau_{xx} \\ \tau_{xy} \\ U\tau_{xx} + V\tau_{xy} - q_x \end{pmatrix} \quad \mathbf{G}_v = \begin{pmatrix} 0 \\ \tau_{xy} \\ \tau_{yy} \\ U\tau_{xy} + V\tau_{yy} - q_y \end{pmatrix}. \end{aligned}\right\} \tag{2}$$

ρ_l denotes the density, U and V the velocity components in x and y direction, p the pressure, e the total energy. A comma stands for partial differentiation (i.e., $f_{,x} = \partial f/\partial x$). Assuming Stokes' hypothesis for the viscous stresses and Fourier's law of heat conduction, then

$$\tau_{xx} = \frac{2}{3}\mu(2U_{,x} - V_{,y}), \qquad \tau_{xy} = \mu(U_{,y} + V_{,x}), \qquad \tau_{yy} = \frac{2}{3}\mu(2V_{,y} - U_{,x}) \tag{3}$$

and

$$q_x = -kT_{,x}, \qquad q_y = -kT_{,y}. \tag{4}$$

μ denotes the coefficient of viscosity, T the temperature and k the thermal conductivity. Assuming water for the liquid phase the system of equations (1) can be completed by the equation of state proposed by Tait

$$\left(\frac{\rho_l}{\rho_0}\right)^n = \frac{p+B}{p_0+B}, \tag{5}$$

where n and B may be functions of the temperature. For the total energy of the liquid the relation

$$e = \rho_l \epsilon + \frac{1}{2}\rho_l(U^2 + V^2) \tag{6}$$

holds, where ϵ is the internal energy, assuming a constant specific heat c, $\epsilon = cT$.

2.2. Solid phase

The governing dynamic equations for an elastic-plastic solid under plane strain conditions can be written in the following matrix form

$$\mathbf{A}\mathbf{u}_{,t} = \mathbf{f}_{,x} + \mathbf{g}_{,y} \tag{7}$$

with

$$\left.\begin{aligned} &\mathbf{A} = \operatorname{diag}\left(\rho_s, \rho_s, \frac{1}{3K}, \frac{1+h}{2\mu_{el}}, \frac{1+h}{2\mu_{el}}, \frac{1+h}{\mu_{el}}\right) \\ &\mathbf{u} = \begin{pmatrix} u \\ v \\ \sigma_x + \sigma_y + \sigma_z \\ \sigma_x - \sigma_z \\ \sigma_y - \sigma_z \\ \tau \end{pmatrix}, \quad \mathbf{f} = \begin{pmatrix} \sigma_x \\ \tau \\ u \\ u \\ 0 \\ v \end{pmatrix}, \quad \mathbf{g} = \begin{pmatrix} \tau \\ \sigma_y \\ v \\ 0 \\ v \\ u \end{pmatrix}. \end{aligned}\right\} \tag{8}$$

ρ_s denotes the density, u and v the particle velocities, μ_{el} the elastic shear modulus and K the bulk modulus. σ_x, σ_y, σ_z and τ are the non-vanishing components of the stress tensor. h is the so-called plastic factor defined by $h = \mu_{el}/\mu_p(\kappa) - 1$, where μ_p denotes the plastic shear modulus. The system of equations (7) is completed by the von Mises yield condition

$$(\sigma_x - \sigma_z)^2 + (\sigma_y - \sigma_z)^2 - (\sigma_x - \sigma_z)(\sigma_y - \sigma_z) + 3\tau^2 \leq 3\kappa^2. \tag{9}$$

κ is called the yield stress. If the inequality (9) is satisfied, the solid behaves linear-elastic, otherwise elastic-plastic. Since we assume $\mu_p \leq \mu_{el}$, h is never negative, so the system (7) is hyperbolic. There exist four possible wave velocities $c_1 < c_f < c_2 < c_s$ with

$$c_1^2 = \frac{1}{\rho_s}\left(K + \frac{4}{3}\mu_{el}\right), \quad c_f^2 = \frac{1}{\rho_s}\left(K + \frac{4}{3}\frac{\mu_{el}}{1+h}\right), \quad c_2^2 = \frac{\mu_{el}}{\rho_s}, \quad c_s^2 = \frac{\mu_{el}}{\rho_s(1+h)}. \tag{10}$$

3. NUMERICAL METHODS

3.1. Liquid phase

For the fluid we use a Finite Element Navier-Stokes solver proposed by Dickopp and Ballmann [10]; They modified the flux-corrected transport (FCT) approach of Löhner at al. [11], since Tait's equation of state is very sensitive to density changes. The technique of FCT was originally introduced by Boris and Book [12, 13, 14].

3.2. Solid phase

For the elastic-plastic solid we use the two step scheme of Lin and Ballmann [15] which is based on Godunov's [16] and Zwas' [17] methods. The first step is to calculate the fluxes

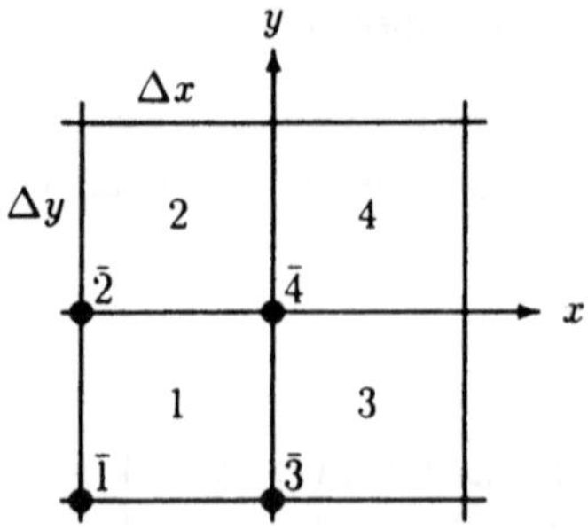

Figure 1: Cells and grids in the domain

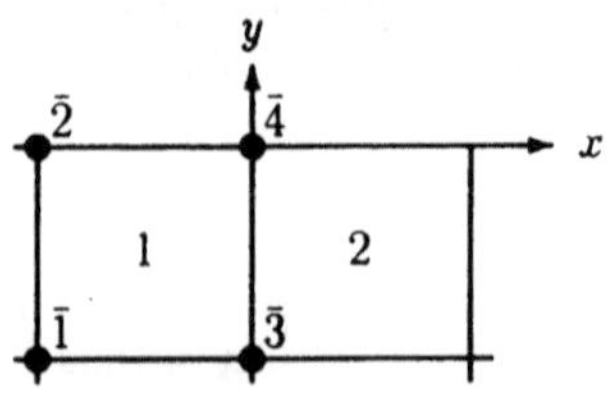

Figure 2: Cells and grids at the boundary

at a grid point, e.g. point $\bar{4}$ in Fig. 1:

$$\sum_{i=1}^{4} \int_{\mathbf{u}_i^0}^{\mathbf{u}_{\bar{4}}^{1/2}} \mathbf{A}\, d\mathbf{u} = \frac{\Delta t}{\Delta x}(\mathbf{f}_4^0 + \mathbf{f}_3^0 - \mathbf{f}_2^0 - \mathbf{f}_1^0) + \frac{\Delta t}{\Delta y}(\mathbf{g}_4^0 - \mathbf{g}_3^0 + \mathbf{g}_2^0 - \mathbf{g}_1^0). \tag{11}$$

Having obtained all the fluxes at the vertices of a cell, the values at the cell center are updated, e.g. for the cell center 1:

$$\int_{\mathbf{u}_1^0}^{\mathbf{u}_1^1} \mathbf{A}\, d\mathbf{u} = \frac{\Delta t}{2\Delta x}(\mathbf{f}_{\bar{4}}^{1/2} + \mathbf{f}_{\bar{3}}^{1/2} - \mathbf{f}_{\bar{2}}^{1/2} - \mathbf{f}_{\bar{1}}^{1/2}) + \frac{\Delta t}{2\Delta y}(\mathbf{g}_{\bar{4}}^{1/2} - \mathbf{g}_{\bar{3}}^{1/2} + \mathbf{g}_{\bar{2}}^{1/2} - \mathbf{g}_{\bar{1}}^{1/2}). \tag{12}$$

A scheme similar to equation (11) can be applied at the boundary, e.g. for point $\bar{4}$ the scheme used is (see Fig. 2)

$$\sum_{i=1}^{2} \int_{\mathbf{u}_i^0}^{\mathbf{u}_{\bar{4}}^{1/2}} \mathbf{A}\, d\mathbf{u} = \frac{\Delta t}{\Delta x}(\mathbf{f}_2^0 - \mathbf{f}_1^0) + \frac{\Delta t}{\Delta y}(2\mathbf{g}_{\bar{4}}^{1/2} - \mathbf{g}_1^0 - \mathbf{g}_2^0). \tag{13}$$

Prescribing the tractions τ and σ_y one can calculate the first two components of $\mathbf{u}_{\bar{4}}^{1/2}$, say u and v. Having obtained the velocities the last four equations can be solved for the stresses. In the plastic case an iterative method has to be applied.
Calculating the fluxes by equation (11) or (13) integrals of the form

$$\int_{\tau_1^0}^{\tau_{\bar{4}}^{1/2}} \frac{1 + h(\kappa)}{\mu_{el}}\, d\tau \tag{14}$$

have to be evaluated. Therefore an elastic-plastic stress loading path has to be chosen. Following Lin and Ballmann [15] and assuming linear work-hardening material the integral (14) can be split up into two parts:

$$\int_{\tau_1^0}^{\tau_{\bar{4}}^{1/2}} \frac{1 + h(\kappa)}{\mu_{el}}\, d\tau = \frac{1}{\mu_{el}} \int_{\tau_1^0}^{\tau_1^*} d\tau + \frac{1}{\mu_p} \int_{\tau_1^*}^{\tau_{\bar{4}}^{1/2}} d\tau, \tag{15}$$

wherein

$$\tau_1^* = \frac{\tau_{\bar{4}}^{1/2}}{\kappa_{\bar{4}}^{1/2}}\kappa_1^0. \tag{16}$$

4. COUPLING FLUID-SOLID

The main problem of fluid-solid interaction is to couple domains of solution of nonlinear PDE's with different types of waves having different wave speeds.
For an inviscid fluid a possible local wave scheme at the interface is presented in Fig. 3. There it is assumed that the velocity of the shear wave in the solid is larger than the speed of sound in the fluid and the velocity of the fluid is subsonic.

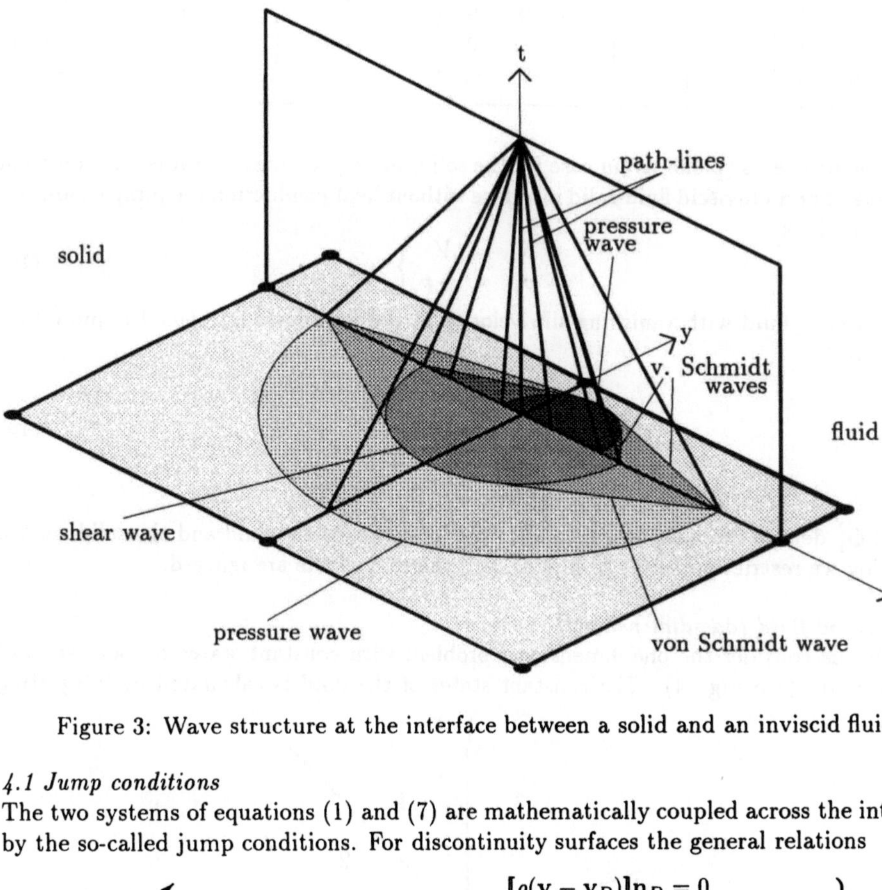

Figure 3: Wave structure at the interface between a solid and an inviscid fluid

4.1 Jump conditions

The two systems of equations (1) and (7) are mathematically coupled across the interface by the so-called jump conditions. For discontinuity surfaces the general relations

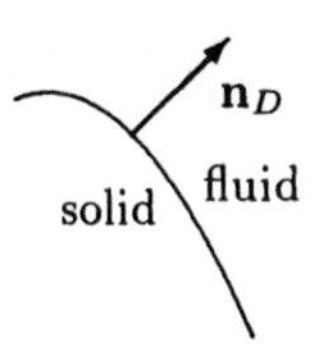

$$\left.\begin{aligned}
&[\rho(\mathbf{v}-\mathbf{v}_D)]\mathbf{n}_D = 0\\
&[\rho\mathbf{v}\circ(\mathbf{v}-\mathbf{v}_D)-\mathbf{T}]\mathbf{n}_D = \mathbf{0}\\
&\left[\rho\left(\epsilon+\frac{\mathbf{v}^2}{2}\right)(\mathbf{v}-\mathbf{v}_D)-\mathbf{v}\mathbf{T}+\mathbf{q}\right]\mathbf{n}_D = 0
\end{aligned}\right\} \tag{17}$$

hold [18]. In equation (17) jumps across a discontinuity are denoted by the standard bracket notation. $\mathbf{v}$ denotes the velocity (subscript D refers to the discontinuity surface), $\mathbf{T}$ denotes the stress tensor, $\mathbf{q}$ is the heat flux, and $\mathbf{n}_D$ is the unit normal vector to the interface. The symbol $\circ$ stands for the dyadic product.
The solid-fluid interface is a contact discontinuity, where the relative normal velocity $(\mathbf{v} - \mathbf{v}_D)\mathbf{n}_D$ vanishes on both sides, e.g. $v = V$ in terms of equations (1) and (7). Therefore the second equation of the system (17) gives $[\mathbf{T}]\mathbf{n}_D = \mathbf{0}$. The stress tensors $\mathbf{T}$

Table 1: Stress tensors for different continua

Solid	Inviscid fluid	Viscous fluid
$\begin{pmatrix} \sigma_x & \tau & 0 \\ \tau & \sigma_y & 0 \\ 0 & 0 & \sigma_z \end{pmatrix}$	$\begin{pmatrix} -p & 0 & 0 \\ 0 & -p & 0 \\ 0 & 0 & -p \end{pmatrix}$	$\begin{pmatrix} \tau_{xx} - p & \tau_{xy} & 0 \\ \tau_{xy} & \tau_{yy} - p & 0 \\ 0 & 0 & -p \end{pmatrix}$

for plane problems (plane strain case for the solid) are represented in Table 1 for different continua. For an inviscid fluid-solid interface without heat conduction the jump conditions are reduced to:

$$\left.\begin{aligned} v &= V \\ \sigma_y &= -p \end{aligned}\right\}. \qquad (18)$$

For a viscous fluid with vanishing slip velocity at the fluid-solid interface the jump conditios are

$$\left.\begin{aligned} u &= U \\ v &= V \\ \sigma_y &= \tau_{yy} - p \, . \\ \tau &= \tau_{xy} \\ q_y &= Q_y \end{aligned}\right\} \qquad (19)$$

q_y and Q_y denote the heat flux normal to the interface for the fluid and the solid. In the following we restrict ourselves to liquids, and thermal effects are ignored.

4.2 Inviscid fluid (one-dimensional)
First let us consider the one-dimensional problem with constant states on both sides of the interface (see Fig. 4). The constant states of the fluid is calculated by integrating

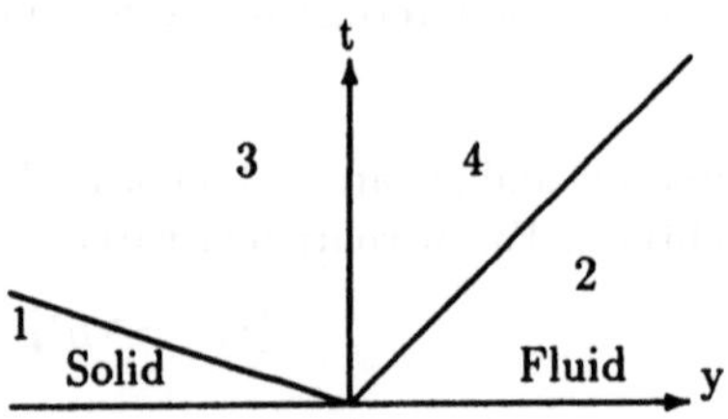

Figure 4: One-dimensional wave structure

the equations for the conservative variables ρ and ρV in each finite element. Integration is carried out along the characteristic lines in the $y-t$-plane. This means for the solid integrating the equations

$$d\sigma_y = \rho_s c(\kappa)\, dv, \qquad \text{along } \frac{dy}{dt} = -c(\kappa), \tag{20}$$

and for the inviscid fluid

$$dp = \rho_l a\, dV, \qquad \text{along } \frac{dy}{dt} = a + V, \tag{21}$$

where $c(\kappa)$ is determined by equations (9) and (10) and in the liquid a is the speed of sound, defined by $a^2 = n(\partial p/\partial \rho)_T = np/\rho$. The results are the following algebraic equations for the unknowns v_3, σ_{y3}, V_4 and p_4: For the solid

$$v_3 - v_1 = \int_{\sigma_{y1}}^{\sigma_{y3}} \frac{1}{\rho_s c(\kappa)}\, d\sigma_y \tag{22}$$

and for the liquid

$$V_4 - V_2 = \int_{p_2}^{p_4} \frac{1}{\rho_l a}\, dp. \tag{23}$$

This system of equations is completed by the jump conditions (18). Using the proposed elastic-plastic stress loading path and the equations for $a(p,\rho)$ and $\rho(p)$ in the liquid the integrals can be calculated exactly. The system can be solved using an iterative method prescribing σ_{y3} for equation (22) and V_4 for equation (23).

4.3 Inviscid fluid (two-dimensional)

For the two-dimensional coupling we use the boundary scheme (13) on both sides of the

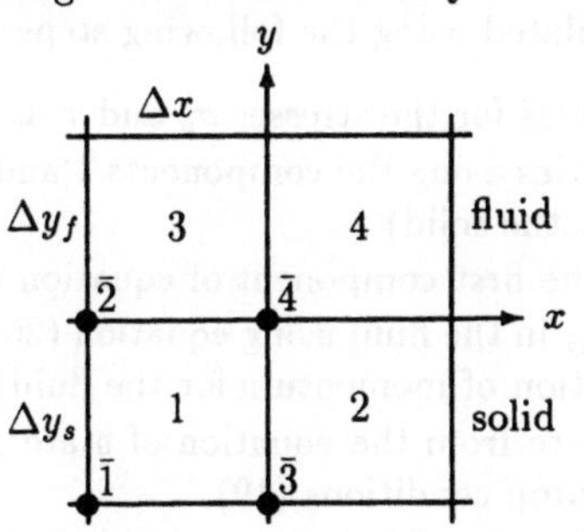

Figure 5: Cells and grids for the interface

interface. With the notations of Fig. 5 this leads for the solid to

$$\sum_{i=1}^{2} \int_{\mathbf{u}_i^0}^{\mathbf{u}_4^{1/2}} \mathbf{A}\, d\mathbf{u} = \frac{\Delta t}{\Delta x}(\mathbf{f}_2^0 - \mathbf{f}_1^0) + \frac{\Delta t}{\Delta y_s}(2\mathbf{g}_{\bar{4}}^{1/2} - \mathbf{g}_1^0 - \mathbf{g}_2^0). \tag{24}$$

and for the fluid to

$$\mathbf{U}_{\bar{4}}^{1/2} = \frac{1}{2}(\mathbf{U}_3^0 + \mathbf{U}_4^0) + \frac{\Delta t}{2\Delta x}(\mathbf{F}_3^0 - \mathbf{F}_4^0) + \frac{\Delta t}{2\Delta y_f}(2\mathbf{G}_{\bar{4}}^{1/2} - \mathbf{G}_3^0 - \mathbf{G}_4^0). \tag{25}$$

Two different ways have been chosen to couple the two systems of equations (24) and (25). The first method is:

1. Assuming a starting value for the normal stress σ_y at the solid side on the interface
2. Calculating the normal velocity from the second component of equation (24) (conservation of momentum in y-direction for the solid)
3. Calculating ρ_l using the first component of equation (25) (conservation of mass for the fluid)
4. Calculating the pressure from the equation of state (5) leads to an improved normal stress σ_y for the solid
5. Iteration until the relative error of the density is less than an error bound ϵ.

The second method is based on the one-dimensional method proposed in the previous section, applied to the cells 1,3 and 2,4. Therefore we know the second component g_2 of the flux vector $\mathbf{g}$ and the first component G_1 of the flux vector $\mathbf{G}$. The corresponding components of the vectors $\mathbf{G}_{\bar{4}}^{1/2}$ and $\mathbf{g}_{\bar{4}}^{1/2}$ are determined by the mean-value of the two values of the one-dimensional problem.

4.4 Viscous fluid

For the case of a viscous fluid the equation (25) has to be replaced by

$$\left.\begin{aligned}\mathbf{U}_{\bar{4}}^{1/2} &= \frac{1}{2}(\mathbf{U}_3^0 + \mathbf{U}_{\bar{4}}^0) + \frac{\Delta t}{2\Delta x}(\mathbf{F}_3^0 - \mathbf{F}_{v3}^0 - \mathbf{F}_{\bar{4}}^0 + \mathbf{F}_{v4}^0) \\ &+ \frac{\Delta t}{2\Delta y_f}\left(2(\mathbf{G}_{\bar{4}}^{1/2} - \mathbf{G}_{v\bar{4}}^{1/2}) - \mathbf{G}_3^0 + \mathbf{G}_{v3}^0 - \mathbf{G}_{\bar{4}}^0 + \mathbf{G}_{v4}^0\right).\end{aligned}\right\} \qquad (26)$$

The normal fluxes are calculated using the following steps:

1. Assuming starting values for the stresses σ_y and τ at the solid side on the interface
2. Calculating the velocities using the components 1 and 2 of equation (24) (conservation of momentum for the solid)
3. Calculating ρ_l using the first component of equation (26)
4. Calculating τ_{xy} and τ_{yy} in the fluid using equation (26) and the velocity components from step 2 (conservation of momentum for the fluid)
5. Calculating the pressure from the equation of state (5) leads to improved stresses σ_y and τ due to the jump conditions (19)
6. Iterate until the relative error of the density is less than an error bound ϵ.

The most important difference to the scheme for the inviscid fluid is the procedure mentioned in step 4. The viscous stresses cannot be calculated directly from equation (26). Under the assumption that the second and the third components are known the equation has to be solved for τ_{xy} and τ_{yy}.

5. NUMERICAL EXAMPLE

The numerical calculation starts with a cylindrical high pressure region in the fluid (water). The cylinder has an infinite length normal to the $x-y$-plane, its diameter is

0.02m and the distance of its center from the solid is 0.0225m. The initial conditions are: Fluid and solid (aluminum) at rest, the pressure in the high pressure region $p_1 = 10^8 N/m^2$ and $p_0 = 10^5 N/m^2$ everywhere else. The stresses in the solid are $\sigma_x = \sigma_y = -p_0, \tau = 0$. The plastic shear modulus is chosen according to the experiments of Lipkin and Clifton [19] as $\mu_p = 0.8\mu_{el}$. In Fig. 6 the results for the pressure distribution in the fluid and the distribution of the maximum shear stress τ_{max} in the solid are shown after 22μsec, where τ_{max} is defined by

$$\tau_{max}^2 = \frac{1}{3}\left((\sigma_x - \sigma_z)^2 + (\sigma_y - \sigma_z)^2 - (\sigma_x - \sigma_z)(\sigma_y - \sigma_z)\right) + \tau^2. \qquad (27)$$

The expected effects like the surface wave and the von Schmidt waves [20] in the fluid

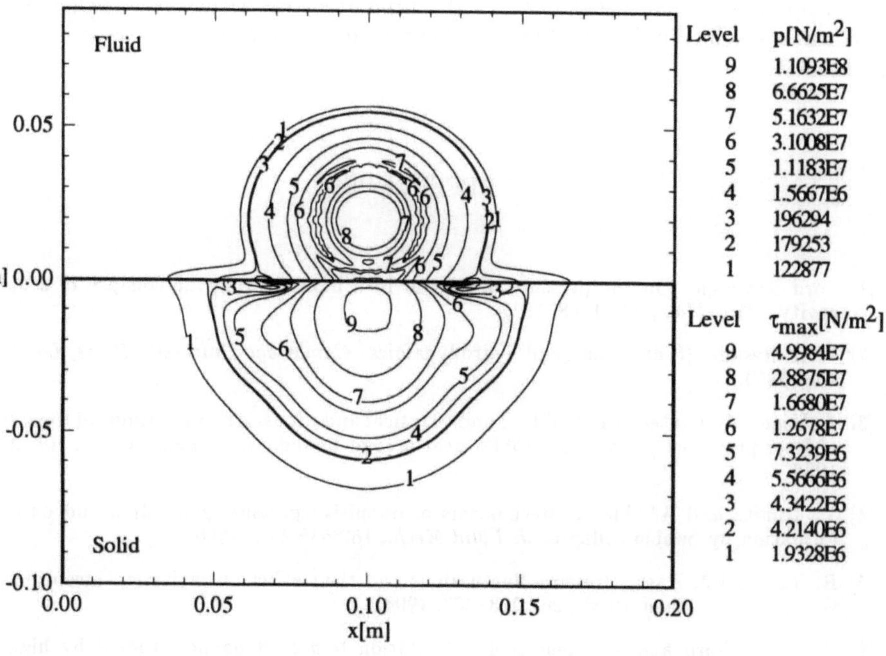

Figure 6: Contour lines of pressure (fluid) and the max. shear stress (solid) after 22μsec

can be clearly recognized. If the pressure wave in the fluid travels with the same speed as the Stoneley wave in the solid resonance effects can occur causing further damage of the solid. In this numerical example a small plastic zone (not shown in the figure) appeared near the present maximum.

6. CONCLUSIONS AND FURTHER DEVELOPMENTS

The numerical experiments have shown, that the scheme for the fluid-solid coupling

resolves interfacial effects well.
Further developements have to be made. First one has to consider the three phase problem, including the collapsing bubble. The next step will be implementing the rotational symmetry, to treat a collapsing bubble with an initially spherical shape, because the aim of this work, which is still in progress, is to give a clearer explanation of the cavitation damaging mechanism, which is not completely understood up to now.

ACKNOWLEDGEMENT

U. Specht gratefully acknowledges the support of the Deutsche Forschungsgemeinschaft as a scholarship holder of the Graduiertenkolleg "Analyse und Konstruktion in der Mathematik".

REFERENCES

[1] Lord Rayleigh. On the pressure developed in a liquid during the collapse of a spherical cavity. *Phil. Mag.*, 34:94–98, 1917.

[2] W.H. Besant. Hydrostatics and hydrodynamics. *Cambridge University Press, London*, art. 158, 1859.

[3] A. Vogel, W. Lauterborn, and R. Timm. Optical and acoustic investigations of the dynamics of laser-produced cavitation bubbles near a solid boundary. *J. Fluid Mech.*, 206:299–338, 1989.

[4] Y. Tomita and A. Shima. Mechanisms of impulsive pressure generation and damage pit formation by bubble collapse. *J. Fluid Mech.*, 169:535–564, 1986.

[5] B. Ran and J. Katz. Pressure fluctuations and their effect on cavitation inception within water jets. *J. Fluid Mech.*, 262:223–263, 1994.

[6] W. Lauterborn and W. Hentschel. Cavitation bubble dynamics studied by high speed photography and holography: Part one. *Ultrasonics*, 23:260–268, 1985.

[7] M.S. Plesset and R.B. Chapman. Collapse of an initially spherical vapour cavity in the neighbourhood of a solid boundary. *J. Fluid Mech.*, 47(2):283–290, 1971.

[8] H. Westenberger. *Zur Dynamik des Impulstransportes in blasigen, kompressiblen Füssigkeiten.* PhD thesis, RWTH Aachen, 1987.

[9] M.McD. Grant and P.A. Lush. Liquid impact on a bilinear elastic-plastic solid and its role in cavitation erosion. *J.Fluid Mech.*, 176:237–252, 1987.

[10] C. Dickopp and J. Ballmann. Wave processes in a three phase system of a collapsing bubble near a compliant wall. *to appear in: Proc. 5th Int.Conf. on Hyperbolic Problems*, 1994.

[11] R. Löhner, K. Morgan, J. Peraire, and M. Vahdati. Finite element flux-corrected transport (FEM-FCT) for the Euler and Navier-Stokes equations. *Finite Elements in Fluids*, 7:105–121, 1987.

[12] J.P. Boris and D.L. Book. Flux-corrected transport. I. SHASTA, a fluid transport algorithm that works. *J. Comp. Phys.*, 11:38–69, 1973.

[13] D.L. Book, J.P Boris, and K. Hain. Flux-corrected transport. II. Generalizations of the method. *J. Comp. Phys.*, 18:248–283, 1975.

[14] J.P. Boris and D.L. Book. Flux-corrected transport. III. Minimal-error FCT algorithm. *J. Comp. Phys.*, 20:397–431, 1976.

[15] X. Lin and J. Ballmann. Numerical method for elastic-plastic waves in cracked solids, part 2: Plane strain problem. *Archive of Applied Mechanics*, 63:283–295, 1993.

[16] S.K. Godunov. A finite difference method for numerical computation of discontinuous solutions of the equations of fluid dynamics. *Mat. Sbornik*, 47:271–306, 1959.

[17] B. Eilon, D. Gottlieb, and G. Zwas. Numerical stabilizers and computing time for second-order accurate schemes. *J. Comp. Phys.*, 9:387–397, 1972.

[18] W. Arrenbrecht and J. Ballmann. Unstetigkeitsflächen in der Kontinuumsmechanik. In P.L. Butzer and Fehér, editors, *E.B. Christoffel*, pages 449–460. Birkhäuser Verlag, Basel, 1981.

[19] J. Lipkin and R.J. Clifton. Plastic waves of combined stresses due to longitudinal impact of a pretorqued tube. *Journal of Applied Mechanics*, pages 1107–1112, December 1970.

[20] O. v. Schmidt. Über Knallwellenausbreitung in Flüssigkeiten und festen Körpern. *Zeitschr. f. techn. Physik*, 19:554–560, 1938.

Solution of the coupled Navier-Stokes equations

C. Vuik and A. Segal
Faculty of Technical Mathematics and Informatics,
Delft University of Technology,
Mekelweg 4, 2628 CD Delft, The Netherlands,
e-mail: c.vuik@math.tudelft.nl a.segal@math.tudelft.nl

Abstract

In this paper the incompressible Navier-Stokes equations are discretized by the finite element method. After linearization large, sparse systems of linear equations have to be solved. A well known problem is the occurrence of zero elements on the main diagonal. We describe ordering techniques of the grid points and the unknowns to avoid this problem, so direct and iterative methods can be used without pivoting. It appears that in three-dimensional problems the iterative methods are much better than the direct or penalty methods.

1 Introduction

There are several methods to solve the incompressible Navier-Stokes equations. In this paper we restrict ourselves to the three-dimensional case and use the velocities and the pressure as unknowns. The coupled equations, three momentum equations and the continuity equation, are discretized by the finite element method. After linearization we obtain a large, sparse system of linear equations. Due to zero elements on the main diagonal, break down of direct and iterative methods may occur.

In Section 2, we describe and investigate some orderings, where the zero main diagonal elements become non zero elements during the decomposition of the matrix. In our numerical experiments we observe no break down of the direct method without pivoting.

Thereafter, in Section 3, we specify two incomplete LU decompositions, which are used as preconditioners for Krylov subspace methods. Some theoretical results are given.

Finally the described methods are compared numerically in Section 4. It appears that already for small problem sizes the preconditioned Krylov subspace methods are much better than the direct or penalty methods. For large problem sizes it is impossible to use direct or penalty methods due to excessive memory requirements. Furthermore for medium problem sizes the CPU time for a direct or penalty method is orders of magnitudes larger than that for the iterative method.

2 Statement of the problem and ordering techniques

In this section we consider the discretization of the incompressible Navier-Stokes equations by finite element methods. Furthermore the consequences of the incompressibility condition are investigated. The general form of the stationary Navier-Stokes equations may be written as

$$-div\ \boldsymbol{\sigma} + \mathbf{u} \cdot \nabla \mathbf{u} = \mathbf{f} \tag{1}$$

where $\boldsymbol{\sigma}$ is the stress tensor:

$$\boldsymbol{\sigma} = -p\mathbf{I} + \mu(\nabla \mathbf{u}^T + \nabla \mathbf{u})\,. \tag{2}$$

μ denotes the viscosity, $\mathbf{f}$ is some external force and $\mathbf{u} \cdot \nabla \mathbf{u}$ represents the convective terms. The continuity equation is given by

$$div\ \mathbf{u} = 0\,. \tag{3}$$

Equations (1) to (3) are solved by a standard finite element method based upon the Galerkin formulation as can be found in for example Cuvelier et al [2]. In this paper we restrict ourselves to the three-dimensional case. The elements used are the triquadratic isoparametric Crouzeix Raviart hexahedrons. This means that the velocity is approximated by a quadratic polynomial in each direction in the reference element and the pressure is approximated by a linear polynomial, which is discontinuous over the element boundaries. Since the convective terms are non-linear it is necessary to use some linearization scheme. To that end we use the solution of the Stokes equations (i.e. the Navier-Stokes equations without convective terms) as initial guess, proceed with a number of so-called Picard iterations and finally apply Newton linearization. In each step of the iteration process, it is necessary to solve a large system of linear equations. Formally this system of equations can be written as:

$$\begin{pmatrix} A & B^T \\ B & 0 \end{pmatrix} \begin{pmatrix} u \\ p \end{pmatrix} = \begin{pmatrix} f \\ 0 \end{pmatrix}, \tag{4}$$

where A denotes the discretization of the stress tensor and the linearized convection terms, $B^T p$ the discretization of the pressure gradient and $Bu = 0$ the discretization of the continuity equation. The vector u represents the velocity unknowns and the vector p the unknown pressures.

The so-called *zero-block* at the main diagonal of the second equation of (4) is caused by the fact that the pressure is not present in the continuity equation, whereas the equation itself is coupled to the pressure unknowns. This property is inherent to incompressible Navier-Stokes equations and is independent of the type of discretization. Due to this *zero-block* a so-called saddle point problem arises. This is one of the major problems in the solution of the incompressible Navier-Stokes problems.

The system of linear equations (4) may be solved by either a direct method (LU-decomposition) or an iterative method. Let us first focus on the direct methods.

The ordering as suggested in equation (4), that is, first all velocity unknowns and then all pressure unknowns, appears to be very uneconomical, since the corresponding profile is very large. In the sequel we shall denote this ordering by p-last ordering. A much better ordering is achieved by ordering all unknowns per point. If a good nodal point numbering is present this gives by far most the smallest profile. Unfortunately the *zero-block* complicates things considerably. Due to boundary conditions it may be possible that the first diagonal element in the matrix is a zero and as a consequence straight-forward LU-decomposition is not possible. In that case it is necessary to use a kind of pivoting strategy, which influences the structure of the matrix and makes the estimation of the size of the matrix a difficult task. So in fact one would like to have some a priori numbering of nodal points and unknowns that prevents the presence of a zero diagonal element during the elimination process and moreover, produces a kind of optimal profile. This numbering must be such that the first unknowns are velocity degrees of freedom independent of the type of boundary conditions. Furthermore it must have a very small influence at the local band width.

Let us, for the sake of the argument, consider the simple rectangular domain of Figure 1, subdivided into triangles. Furthermore we assume that the velocity is prescribed at the complete boundary. If the nodes are numbered in a natural way, from left to right and line-wise from below to the upper boundary, it is clear that one of the first unknowns is the pressure in the first element. The band width is determined by one stroke of elements only. One may expect that if we first number the velocities and then the pressures, then during the elimination procedure the pressure diagonal elements become non-zero. In order to avoid increase of the profile a clever numbering would be to renumber the unknowns such that per stroke of elements first all velocities are numbered and then all pressures. In this way, due to fill in, the elimination may change the pressure diagonal elements to non-zero elements and the actual band width is hardly changed. Such a renumbering will be called pressure last renumbering per level or p-last per level.

This renumbering is simple for a rectangular region. In a general irregular shaped region an automatic procedure is necessary to define the equivalent of the strokes of elements. To

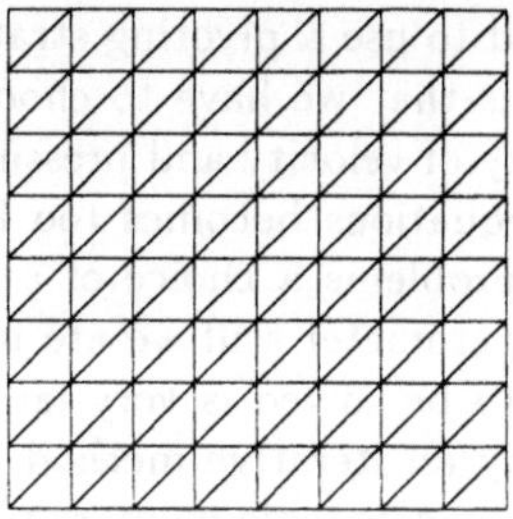

Figure 1: Rectangular domain, subdivided into triangles

that end we assume that the nodal points have been renumbered before, in order to get a small profile for example by the standard reversed Cuthill-McKee renumbering algorithm. Next we define a level structure which is very similar to the Cuthill-McKee structure. We start with node 1 and find all neighbours of this node. Let node i_1 have the highest number of this set of neighbours. Then level 1 is defined as the set of nodes 1 to i_1. The next level is found by considering all new neighbours of level 1. Let node i_2 have the highest number of this new set. Then the next level is defined as the i_1+1 to i_2. This process is repeated until all neighbours are part of a level. With respect to the start it might be necessary to combine levels 1 and 2. Per level we number first the velocities and then the pressure degrees of freedom. In this way we get a nearly optimal numbering, which may be applied in combination with Gaussian elimination. To our knowledge this type of numbering has not yet been published before. Experiments have shown that indeed this numbering is suitable for the solution of the equations by a direct method.

In the literature one often tries to solve the problems due to the *zero-block* by segregating the computation of pressure and velocity. Well known methods are for example the penalty function method, pressure correction and the method of divergence-free vector fields.

In terms of finite element methods the penalty approach is very popular. In this approach the continuity equation is disturbed by a small pressure term:

$$\varepsilon p + div\mathbf{u} = 0. \tag{5}$$

This allows us to express the pressure explicitly in the velocity and to substitute this pressure in the momentum equations. In this way an equation for the velocity only remains. A clear advantage of this approach is the reduction of the size of the system of equations and the fact that there is no need to use a pivoting strategy. An important disadvantage of the penalty function method is that we have to choose the parameter ε in a suitable way. If ε is too large, the accuracy of velocity and pressure is insufficient, if ε is too small, the condition of the system of equations becomes too large because $1/\varepsilon$ appears in the equations. For many practical problems a choice of $\varepsilon = 10^{-6}$ is reasonable. However, even if we have a good penalty parameter still we are faced with a problem. The condition of the system of equations to be solved is very large and as a consequence it is not possible to solve the equations by an iterative method. Especially for three-dimensional problems direct methods lead to unacceptable memory requirements and computing times.

The pressure correction method is typically a method for time-dependent problems. Its application in case of finite element methods is somewhat more complex than in case of finite volumes. We will not go into details.

The method of divergence-free vector fields is quite complicated to program, especially in three dimensions. For that reason it is not commonly used.

A method that tries to overcome some of the disadvantages of the penalty function method is the so-called Uzawa scheme. In fact this is a type of iteration method that is based on the same ideas as the penalty function method. The main difference is that the parameter ε has a moderate value of order 10^{-1}. Although this method does not have the ill-conditioned property of the penalty method, in practice the convergence is not very fast. For that reason we shall investigate an iterative method that is based on the original equations.

3 Incomplete LU decompositions

In this section we consider the iterative solution of the linear system, that has been given in the previous section. We solve this system with iterative methods of Krylov subspace type combined with preconditioners based on incomplete LU decompositions [1]. First we compare two different representations of the coupled linear system. Thereafter two different ILU preconditioners are defined. Finally, some existence results are proved for the ILU decompositions.

From Section 2 it appears that when we use the p-last ordering we have to solve the following coupled system of equations:

$$\tilde{M}\begin{pmatrix} u \\ p \end{pmatrix} = \begin{pmatrix} A & B^T \\ B & 0 \end{pmatrix}\begin{pmatrix} u \\ p \end{pmatrix} = \begin{pmatrix} f \\ 0 \end{pmatrix}. \tag{6}$$

Another representation of this equation is:

$$M\begin{pmatrix} u \\ p \end{pmatrix} = \begin{pmatrix} A & B^T \\ -B & 0 \end{pmatrix}\begin{pmatrix} u \\ p \end{pmatrix} = \begin{pmatrix} f \\ 0 \end{pmatrix}. \tag{7}$$

An advantage of (6) is that for the Stokes problem $\tilde{M}$ is a symmetric matrix. For Navier-Stokes A is non-symmetric, so $\tilde{M}$ is also non-symmetric. Disadvantages of (6) are: $\tilde{M}$ is never positive definite and a symmetric incomplete LL^T decomposition breaks down. First we consider the positive definiteness properties of $\tilde{M}$ and M. We have:

$$\begin{pmatrix} x^T & y^T \end{pmatrix} \begin{pmatrix} A & B^T \\ B & 0 \end{pmatrix} \begin{pmatrix} x \\ y \end{pmatrix} = x^T A x + 2x^T B y, \tag{8}$$

and

$$\begin{pmatrix} x^T & y^T \end{pmatrix} \begin{pmatrix} A & B^T \\ -B & 0 \end{pmatrix} \begin{pmatrix} x \\ y \end{pmatrix} = x^T A x. \tag{9}$$

For a given x the vector y can always be chosen in such a way that the right-hand side of (8) is negative. On the other hand for every choice of x and y the sign of the right-hand side of (9) only depends on the properties of A. So if A is positive definite then M is positive semi-definite.

We now consider two incomplete LU decompositions: the classic ILU decomposition [3] and ILUD, where only the diagonal is changed. Both versions are used. It appears that ILU is more robust, but more expensive, whereas ILUD is easier to analyse. The matrix M is decomposed into the following matrices: L a lower triangular matrix, D a diagonal matrix, U an upper triangular matrix. Furthermore $diag(L) = diag(U) = D$. The decomposition is made such that $\hat{L}\hat{U} \approx M$ where $\hat{L} = LD^{-1/2}$ and $\hat{U} = D^{-1/2}U$. The following rules are used:

<u>ILU</u>

1. $l_{i,j} = u_{j,i} = 0$ if there is no connection between i and j,
2. $(\hat{L}\hat{U})_{i,j} = m_{i,j}$ if there is a connection between i and j.

<u>ILUD</u>

1. $l_{i,j} = m_{i,j}$, and $u_{j,i} = m_{j,i}$ for $i > j$,
2. $(\hat{L}\hat{U})_{i,i} = m_{i,i}$.

Note that for both preconditioners $d_i > 0$ is necessary in order to form $\hat{L}$ and $\hat{U}$.

In the remaining part of this section, we give some existence results for the ILUD preconditioner. Therefore we consider the second rule from ILUD:
$(\hat{L}\hat{U})_{i,i} = (LD^{-1/2}D^{-1/2}U)_{i,i} = d_i + \sum_{j=1}^{i-1} \frac{l_{i,j} * u_{j,i}}{d_j} = m_{i,i}$.
Combination with rule 1 leads to:

$$d_i = m_{i,i} - \sum_{j=1}^{i-1} \frac{m_{i,j} * m_{j,i}}{d_j}. \tag{10}$$

Suppose $M \in \mathbb{R}^{n \times n}$ and $A \in \mathbb{R}^{n_1 \times n_1}$ then we have the following theorem:

Theorem 1 *If the ILUD decomposition of A exists and the norm of every column of B^T is non-zero, then the ILUD decomposition of M exists and $d_i > 0$ for $i \in [1, n]$.*

Proof: From the assumptions it follows that the ILUD decomposition of A exists and thus $d_j > 0$ for $j = 1, ..., n_1$. For $i \in (n_1, n]$ we have $m_{i,j} = -m_{j,i}$ and $m_{i,i} = 0$. This together with (10) implies that $d_i = \sum_{j=1}^{i-1} \frac{m_{i,j}^2}{d_j}$. Since the norm of a column of B^T is non-zero we have $\sum_{j=1}^{i-1} m_{i,j}^2 > 0$. Combined with $d_k > 0$ for $k < i$ it follows that

$$d_i \geq (min_{1 \leq k \leq i-1} \frac{1}{d_k}) \sum_{j=1}^{i-1} m_{i,j}^2 > 0.$$

□

Remark:
The assumption on B is satisfied in many practical applications, but the assumption on A is not always satisfied. If the ILUD preconditioner is applied to $\tilde{M}$ then $d_j > 0$ for $j \leq n_1$, but $d_{n_1+1} < 0$ so it is impossible to form $\hat{L}$ and $\hat{U}$.

Suppose another ordering is used, for instance the p-last per level ordering. Then there exists a permutation matrix P such that M is given by

$$M = P^T \begin{pmatrix} A & B^T \\ -B & 0 \end{pmatrix} P.$$

The equations $m_{i,j} = -m_{j,i}$ and $m_{i,i} = 0$ again hold for a row which corresponds with a pressure unknown.

Theorem 2 *Suppose that the ILUD decomposition exists for $j < i$ (so $d_j > 0$) and the ith row corresponds with a pressure unknown. If there is one $k < i$ such that $m_{i,k} \neq 0$ then $d_i > 0$.*

Proof: It follows again from (10) that

$$d_i = \sum_{j=1}^{i-1} \frac{m_{i,j}^2}{d_j}.$$

Since $d_j > 0$ for $j < i$ and $m_{i,k}^2 > 0$ for at least one $k < i$ we obtain $d_i > 0$. □

This implies that the ILUD decomposition does not break down ($d_i \neq 0$) in a pressure row, if every pressure unknown is preceded by a velocity unknown with a non-zero connection. Both theorems hold for Stokes and Navier-Stokes problems.

4 Numerical experiments

In this section we give some numerical experiments with the preconditioners given in Section 3. We start with the solution of the Stokes equations, where we compare the iterative method with a direct and a penalty method. Thereafter we solve the Navier-Stokes equations on a three-dimensional Backward Facing Step problem.

4.1 The Stokes equations

We consider the Stokes equations on a cube. In Table 1 we give the number of unknowns and the size of the matrices for the coupled system. Only non zero elements are stored and the rows and columns corresponding to essential boundary conditions have been removed. The ratio between these two numbers gives the average number of non-zero elements per row. It appears that this ratio is relatively large (≈ 180). This has two important implications: the CPU time for a matrix vector multiplication is large with respect to the CPU time for a vector update, and a large part of the fill-in is used in the classical ILU decomposition, so we expect a fast convergence of the preconditioned iterative method.

Table 1: The size of the problem with respect to the grid-size

number of elements	number of unknowns	non-zero entries of the matrix	ratio
$3 \times 3 \times 3$	483	75 000	155
$6 \times 6 \times 6$	4 857	840 000	173
$12 \times 12 \times 12$	43 400	7 800 000	180

In Table 2 we summarise the results for three different methods: a penalty method, a direct method and an iterative method (GMRES with ILUD). The last two methods are applied to the coupled problem (7). We observe no break down of the direct and iterative method if the p-last per level ordering technique is used (see Section 2). Comparing the different solution methods, it appears that using the iterative method leads to a large decrease in CPU time and memory requirements. The CPU time is measured in seconds on an HP 735 workstation.

Now we give a theoretical consideration of the work and memory requirements. It appears that for the penalty and direct method the memory required is proportional to n_1^5, where n_1 denotes the number of grid points in the x_1-direction. The amount of work for these methods depends on n_1^7. For the iterative method the values are n_1^3 and n_1^4 respectively. In Table 3 we compare these expressions with the measurements. For the iterative

Table 2: The CPU time and the memory requirements for the various methods

	$3 \times 3 \times 3$		$6 \times 6 \times 6$	
method	non-zero entries	CPU time	non-zero entries	CPU time
direct	160,000	1.3	5,000,000	237
penalty	100,000	0.47	3,400,000	132
iterative	75,000	0.08	840,000	2.74

Table 3: The ratio for the memory and work requirements for $n_1 = 6$ and $n_1 = 3$. Between brackets the theoretical expected values

method	memory	work
direct	31.7 (32)	182 (128)
penalty	34 (32)	247 (128)
iterative	11 (8)	34 (16)

method we were able to obtain also the ratios for $n_1 = 12$ and $n_1 = 6$. They are given by 9.4 (memory) and 24 (work). We see a reasonable correspondence between theory and experiment. Furthermore the differences between the direct and penalty method and the iterative method increases enormously for increasing grid size.

In this example both preconditioners are used. It appears that ILU is more robust than ILUD and it leads to less iterations of the preconditioned GMRES method. However the construction of the ILU decomposition takes a lot more work and doubles the memory required. For this reason if the ILUD decomposition does not breakdown ILUD is preferred, because the extra memory is negligible and the total CPU time is, in general, less than that for ILU.

4.2 The Navier-Stokes equations

In this section we solve the Navier-Stokes equations on a three-dimensional Backward Facing Step problem. The geometry is given in Figure 2. At the left boundary surface we use a Dirichlet inflow boundary condition and at the right boundary surface we use an outflow boundary condition: $\sigma_{nn} = 0$ and $\sigma_{\mathbf{nt}} = \mathbf{0}$. At all other boundaries we use a no slip condition. In this case we have to solve a non linear problem. The strategy to do this is given in Section 2. Initially we solve the corresponding Stokes equations, thereafter some Picard iterations are done and finally some Newton Raphson iterations are used.

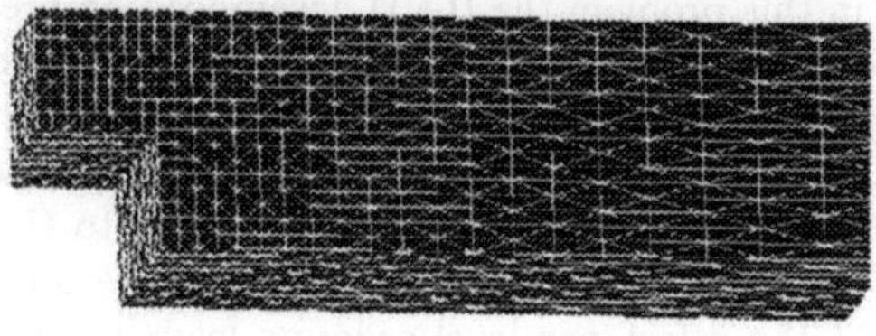

Figure 2: The geometry of the three-dimensional Backward Facing Step problem

The ILU/ILUD decompositions are based on the current coefficient matrix, so a new decomposition is made in every outer iteration. As Krylov subspace methods we use GMRES [4] and Bi-CGSTAB [5]. When we restrict ourselves to Stokes and Picard outer iterations then for both methods 9 outer iterations are needed. The total number of inner iterations is 153 for GMRES and 122 for Bi-CGSTAB. The total CPU time (including building of the matrices and decompositions) is 351 s for GMRES and 385 s for Bi-CGSTAB. Note that Bi-CGSTAB uses less iterations, but one iteration of Bi-CGSTAB is approximately two times as expensive as an iteration of GMRES. This explains the bigger CPU time for Bi-CGSTAB. The optimal CPU time for GMRES is not unexpected since it is known that if the number of iterations is small and a matrix vector product is expensive (which means a large number of non zero elements per row) then GMRES is the best method (see [6]).

We have experimented with different strategies to solve the non linear equations. It appears that the number of inner iterations in a Newton Raphson step is slightly more than in a Picard step. However in general less outer iterations are needed if Newton Raphsón is used. For this reason we use the Stokes equations in the first iteration, Picard in the second iteration and Newton Raphson in the following iterations. In general 5 or 6 outer iterations are sufficient to reduce the initial error with a factor of 10^{-4}. We stop the inner iteration if $\|r_k\|_2/\|r_0\|_2 < eps$. If Picard iterations are used $eps = 10^{-1}$ is sufficient. If Newton Raphson steps are used it may be better to use $eps = 10^{-2}$, because then the outer iterations converge quadratically, whereas if $eps = 10^{-1}$ is used Newton Raphson has a linear convergence behaviour.

Finally we observe that in this problem the ILUD decomposition breaks down, so we only use the ILU preconditioner. It appears that break down of the ILUD decomposition always happens in the velocity part, which is in agreement with the theory given in Section 3. The construction of the ILU decomposition is expensive. The CPU-time is comparable with the CPU time to build the coefficient matrix. With respect to the ordering techniques we observe no break down of the ILU decomposition using the p-last, or p-last per level ordering. It appears that the p-last per level ordering leads to less inner iterations and CPU time than the p-last ordering.

We end this section with some results for the BFS problem on an $8 \times 16 \times 28$ grid. The number of unknowns is equal to 8×10^4 and the number of non zero entries of the matrix is equal to 1.4×10^7. The CPU time to build the matrix is 2 min., whereas the CPU time to build the ILU preconditioner is 3 min. Using ILU, GMRES, p-last per level and $eps = 10^{-2}$, 5 outer iterations are needed. The total CPU time is 80 min. and the total number of inner iterations is 350.

5 Conclusions

In this paper the incompressible Navier-Stokes equations are solved. We consider the momentum equations coupled with the continuity equation.

Ordering techniques are described to prevent break down of the LU decomposition. From our numerical experiments it appears that direct methods can be used with the p-last per level ordering. The CPU time and memory requirements for the direct method are comparable to that of the penalty approach.

Thereafter the proposed orderings are combined with preconditioned Krylov subspace methods. It appears from the theory that the zero main diagonal elements no longer lead to break down or bad convergence. Furthermore, it appears that the iterative methods are much better, than direct or penalty methods.

ACKNOWLEDGEMENT

The authors would like to thank C.G.M. Kassels for software support.

References

[1] R. Barrett, M. Berry, T.F. Chan, J. Demmel, J. Donato, J. Dongarra, V. Eijkhout, R. Pozo, C. Romine, and H. van der Vorst. *Templates for the solution of linear systems.* SIAM, Philadelphia, 1994.

[2] C. Cuvelier, A. Segal, and A.A. van Steenhoven. *Finite element methods and Navier-Stokes equations.* Reidel Publishing Company, Dordrecht, Holland, 1986.

[3] J.A. Meijerink and H.A. Van der Vorst. An iterative solution method for linear systems of which the coefficient matrix is a symmetric M-matrix. *Math. Comp.*, 31:148–162, 1977.

[4] Y. Saad and M.H. Schultz. GMRES: a generalized minimal residual algorithm for solving non-symmetric linear systems. *SIAM J. Sci. Stat. Comp.*, 7:856–869, 1986.

[5] H.A. Van der Vorst. Bi-CGSTAB: a fast and smoothly converging variant of Bi-CG for solution of non-symmetric linear systems. *SIAM J. Sci. Stat. Comp.*, 13:631–644, 1992.

[6] C. Vuik. Further experiences with GMRESR. *Supercomputer*, 55:13–27, 1993.

Numerical Simulation of Temperature Distribution and Seam Forming in Narrow Gap Welding

D. Weiß [1] and G. Lube [2]

[1] Otto von Guericke University of Magdeburg, Institute of Thermodynamics, P.O.Box 4120, D-39106 Magdeburg, Germany

[2] Georg August University of Göttingen, Institute of Numerical and Applied Mathematics, Lotzestr. 16–18, D-37083 Göttingen, Germany

Summary

The knowledge of the temperature distribution of the workpiece during arc welding is necessary in order to ascertain changes in the material structure as well as the strength of the joint. The simulation has not only to consider the heat effect of the arc but also the weld pool deformation influenced by arc pressure and gravitation. A coupled nonlinear system based on the description of energy transport and minimization of surface energy has been employed. The system is solved using a stabilized finite element method of Galerkin/least-squares type. Numerical results for examples are given. Furthermore we discuss the robustness of the numerical method.

1. Mathematical model

The numerical simulation of temperature distribution and weld pool deformation during narrow gap arc welding is considered. Knowledge of the temperature field and the seam geometry is the basis for process optimization. From these results one can ascertain the quality of the joint, its strength and the material structure.

Modelling of the welding process has to include the energy input and acting forces as the most important effects of the arc. Most of the known models take into consideration pool convection, surface deformation and different energetic parts of arc, radiation, passing drops and evaporation by assumption of Gauss distributed surface sources [6]. A strong surface deformation appears in the case of narrow gap welding. We consider, as an example, narrow gap welding in inclined or vertical position, respectively. In this case, an intensity distribution $\dot{q}_{arc} = \dot{q}(x^2 + y^2)$ of Gaussian type independent of z does not take into account that the energy input takes place where the transitional resistance is minimal (cf. Fig. 1). This resistance is determined primarily by the distance between electrode and pool surface as well by the conductivity of the plasma. The conductivity is influenced by evaporation of the metal on the pool surface. The distribution function Φ is given by

$$\Phi(x,y,z,T) = \exp\left(-\frac{R^2 - R^2_{min}}{R^2_{max} - R^2_{min}}\right) \tag{1}$$

if the improved electrical conductivity is incorporated by a reduced transitional resistance R. As a first attempt, a linear relation was formulated for the dependence of R on T

$$R(x,y,z,T) \sim \begin{cases} L & \text{if } T < T_{liq} \\ L \cdot \left(1 - \vartheta \frac{T - T_{liq}}{T_{evap} - T_{liq}}\right) & \text{otherwise} \end{cases} . \tag{2}$$

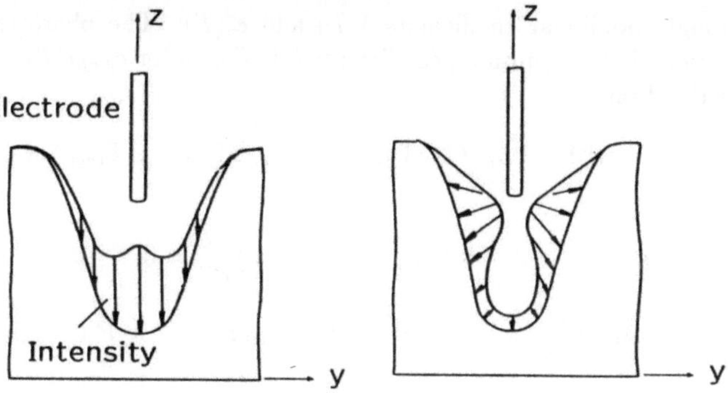

Fig. 1: Energy input on pool surface
a) Gauß distribution b) Distribution in dependence on resistance

ϑ is a constant weighting factor modelling the ionization conditions ($0 \leq \vartheta < 1$). Furthermore, L denotes the distance between the electrode and the pool surface.
The maximal values of the pressure $p_{arc,max}$ and heat flux density $\dot{q}_{arc,max}$ in equations (3) and (4) are determined in such a way that the corresponding surface integrals on the domain of the arc influence are equal to the total force or the total energy input, respectively :

$$p_{arc}(x,y,z,T) = p_{arc,max}\Phi(x,y,z,T) \cdot (\vec{n}_{arc} \cdot \vec{n}_{surf})(x,y,z) \tag{3}$$

$$\dot{q}_{arc}(x,y,z,T) = \dot{q}_{arc,max}\Phi(x,y,z,T) \cdot (\vec{n}_{arc} \cdot \vec{n}_{surf})(x,y,z)\,. \tag{4}$$

The mathematical model incorporates the following additional assumptions :

- The domain is a plate of constant thickness stretching to infinity in the direction of welding.
- The energy input and transport are time-independent.
- The boundary of the pool underside is a pool support or solid material.
- Electro-magnetic and chemical effects are neglected.

The energy transport in the domain Ω with

$$\Omega = \Omega(T) := \{P = (x,y,z) \in \mathcal{R}^3 : -L_{x_1} < x < L_{x_2}, 0 < y < L_y,\ 0 < z < Z(T)\}$$

is governed by the stationary Fourier-Kirchhoff equation using an appropriate co-ordinate system (cf. Fig. 2). The corresponding boundary value problem reads

$$-\nabla \cdot (\lambda(T)\nabla T) + c_p(T)\varrho\vec{v} \cdot \nabla T = \dot{q}_{vol} := 0 \qquad \forall\, P \in \Omega = \Omega(T).$$

with strongly nonlinear coefficients $\lambda(T)$ and $c_p(T)$. The phase transformation solid – liquid is modelled via jumping coefficients (cf. Fig. 3 for $c_{p,eff}(T)$). Boundary conditions are prescribed on

$$\partial\Omega = \Gamma_{top}(T) \cup \Gamma_{front} \cup \Gamma_{side} \cup \Gamma_{symm} \cup \Gamma_{back} \cup \Gamma_{bottom}$$

as

$$T = T_{front} \qquad \forall\, P \in \Gamma_{front}; \qquad -\lambda(T)\frac{\partial T}{\partial \vec{n}} = 0 \qquad \forall\, P \in \Gamma_{symm} \cup \Gamma_{back}$$

$$-\lambda(T)\frac{\partial T}{\partial \vec{n}} = -\alpha_{eff}(P,T)(T - T_\infty(P)) \qquad \forall\, P \in \Gamma_{bottom} \cup \Gamma_{side}$$

$$-\lambda(T)\frac{\partial T}{\partial \vec{n}} = \begin{cases} \dot{q}_{input}(P,T), & \text{if } r = \sqrt{x^2+y^2} \le \tilde{L}_{arc} \\ -\alpha_{eff}(P,T)(T - T_\infty(P)), & \text{if } r > \tilde{L}_{arc} \end{cases} \qquad \forall\, P \in \Gamma_{top} = \Gamma_{top}(T)$$

with

$$\begin{aligned} \dot{q}_{input}(x,y,z,T) &= \dot{q}_{arc,max}\Phi(P,Z,T)\,(\vec{n}_{arc}\cdot\vec{n}_{surf})\,(P,Z) \\ &\quad -\alpha_{eff}(P,T)(T - T_\infty(P)) - \dot{q}_{evap}. \end{aligned}$$

Note that Φ is the nonlinear distribution function as defined in (1), (2).
The pool surface $\Gamma_{top} = \Gamma_{top}(T)$, or equivalently $\hat{z} = L_z - z$ is determined via minimization of the surface energy :

$$\int_\Omega \left\{\sigma(T)\left(\left(1+|\nabla\hat{z}|^2\right)^{1/2} - 1\right) - p_{arc}\hat{z} + \tfrac{1}{2}\varrho g\hat{z}^2\right\} dxdy \;\Rightarrow\; \text{Min}\,, \tag{5}$$

more precisely, it is the solution of the following boundary value problem (cf. [8] for details):

$$-\nabla\cdot\left(\frac{\sigma(T)\nabla Z}{\sqrt{1+|\nabla Z|^2}}\right) + c\,Z = f(Q,Z,T) \qquad \forall\, Q \in G$$

in

$$G := \{Q = (x,y) \in \mathcal{R}^2 : -L_{x_1} < x < L_{x_2},\ 0 < y < L_y\}$$

with

$$c = \varrho g \cos\theta_{plate},$$

$$\begin{aligned} f = f(Q,Z,T) &= -p_{arc,max}(Z)\Phi(Q,Z,T)\cdot(\vec{n}_{arc}\cdot\vec{n}_{surf})\,(Q,Z) \\ &\quad +\varrho g\,(L_z\cos\theta_{plate} + x\sin\theta_{plate}) + p_0(Z) \end{aligned}$$

and boundary conditions

$$Z = L_z \qquad \forall\, Q \in \partial G :\ y > 0,\ x < L_{x_2},$$

$$\frac{\partial Z}{\partial \vec{n}} = 0 \qquad \forall\, Q \in \partial G :\ y = 0; \qquad \frac{\partial Z}{\partial \vec{n}} = 0 \qquad \forall\, Q \in \partial G :\ x = L_{x_2}.$$

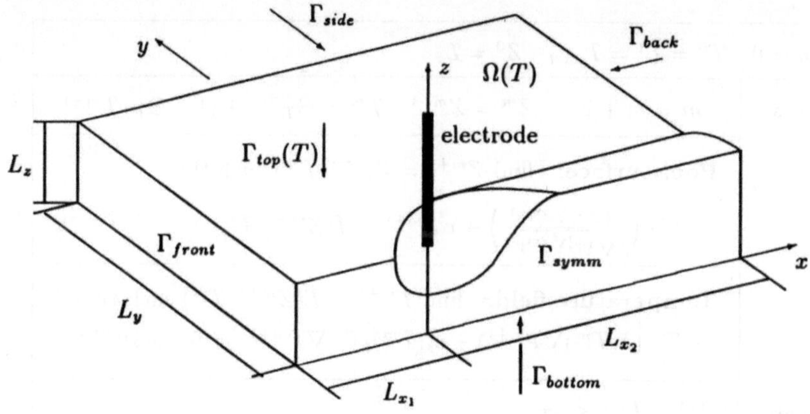

Fig. 2: Domain Ω for the energy transport problem with the free surface Γ_{top}

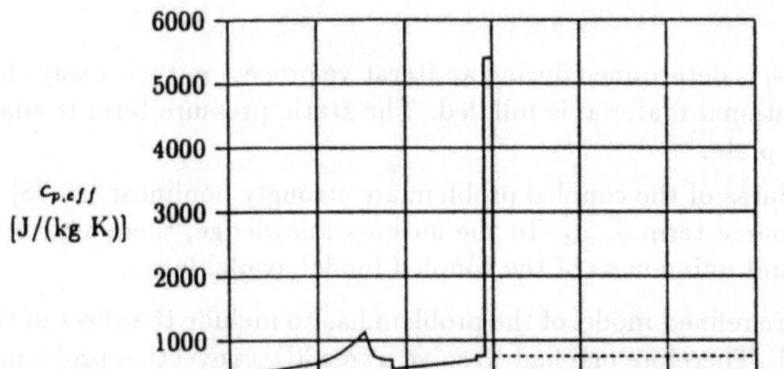

Fig. 3: Effective heat capacity $c_{p,eff}(T)$

$m = 0 \quad T^0 = T^1 = T_{start} \quad Z^0 = L_z$	
repeat	$m = m+1 \qquad Z^m = Z^{m-1} \quad T^m = \Theta_T T^m + (1-\Theta_T)\, T^{m-1}$
	Pool surface: find $Z^{m+1} = Z(Z^m, T^m)$ such that $-\nabla \cdot \left(\frac{\sigma(T^m) \nabla Z^{m+1}}{\sqrt{1+\lvert\nabla Z^m\rvert^2}} \right) + c\, Z^{m+1} = f(Z^{m+1}, T^m)$
	Temperature field: find $T^{m+1} = T(Z^{m+1}, T^m)$ such that $-\nabla \cdot (\lambda(T^m) \nabla T^{m+1}) + c_p(T^m) \varrho \vec{v} \cdot \nabla T^{m+1} = 0 \quad \text{in}\, \Omega(T^m)$
until $\lVert T^{m+1} - T^m \rVert < \varepsilon_T$	
Cooling conditions: $t_{8/5}(y,z)$	
Seam shape: $Z^{m+1}(y)\big\rvert_{x=L_{x_2}}$	

Box: The iteration scheme for the coupled problem

The free constant p_0 is determined during an iterative process in such a way that the mass balance of the additional material is fulfilled. The static pressure term is adapted to the case of an inclined plate.

Note that several datas of the coupled problem are strongly nonlinear (cf. [8] for details), in particular the source term $\dot{q}_{input}$. To the authors knowledge, there are no results concerning existence and uniqueness of the coupled model available.

Remark. A more refined model of the problem has to include the effect of the flow field in the melting pool. Therefore one had to solve a free 3D–convection problem with a free surface and strongly nonlinear source terms. Such a model seems so far too complicated. So we restrict ourselves to the mathematical model derived above. The influence of the convective energy transport is modelled via increased values of $\lambda(T)$ and $\vec{v}(T)$ for $T > T_{liq}$ in the melting zone leading to a nonlinear problem with jumping coefficients (cf. [8] for details). □

2. Numerical solution

We solve the described coupled nonlinear system using the simple iteration method (or method of "frozen coefficients") together with underrelaxation. $t_{8/5}(y,z)$ denotes the pointwise cooling time of the plate from 800^oC to 500^oC. From knowledge of both, the $t_{8/5}$–time and the seam shape $Z(y)$, one can ascertain the quality of the joint, its strength and the material structure (cf. Box). The resulting linear problems are of diffusion–

convection or diffusion-reaction type where u stands for the unknown functions T or Z, respectively :

$$Lu \equiv -\nabla \cdot (a\nabla u) + \vec{b} \cdot \nabla u + cu = f \quad \text{in } \Omega \tag{6}$$

$$u = g \quad \text{on } \Gamma_1, \qquad a\frac{\partial u}{\partial \vec{n}} + \alpha(u - u_\infty) = 0 \quad \text{on } \Gamma_2 . \tag{7}$$

The corresponding weak formulation reads:

Find $u \in V^g := \{v \in W^{1,2}(\Omega) : v|_{\Gamma_1} = g\}$ such that

$$a(u, \varphi) = f(\varphi) \quad \forall\, \varphi \in V^0, \tag{8}$$

with

$$\begin{aligned} a(u,\varphi) &= \int_\Omega (a\nabla u)\cdot \nabla\varphi \, d\Omega + \int_\Omega \vec{b}\cdot\nabla u \,\varphi\, d\Omega + \int_\Omega cu\,\varphi\, d\Omega + \int_{\Gamma_2} \alpha u\varphi\, d\Gamma \\ f(\varphi) &= \int_\Omega f\,\varphi\, d\Omega + \int_{\Gamma_2} \alpha u_\infty \varphi\, d\Gamma \end{aligned} \tag{9}$$

The standard Galerkin finite element method with piecewise linear ansatz and test functions φ on a tetrahedral mesh may suffer from spurious oscillations in the transport dominated case unless the mesh is sufficiently refined. As a remedy, in the Galerkin/least-squares formulation stabilizing terms depending on the residual $Lu - f$ are added to the standard Galerkin formulation. This corresponds on a regular mesh to the choice of discrete test functions:

$$\tilde{\varphi}_h = \varphi_h + \delta L\varphi_h. \tag{10}$$

The resulting discrete problem reads as follows :

Find $u_h \in V_h^g := \{v \in V^g,\ v|_K \in P_1(K)\ \forall\ K\}$ such that

$$a(u_h, \varphi_h) + \sum_K \delta_K \int_K (Lu_h - f)\, L\varphi_h \, dK = f(\varphi_h) \quad \forall\, \varphi_h \in V_h^0 . \tag{11}$$

The method is stable and high order accurate due to the residual form of equation (11). An error analysis of the Galerkin / Least– squares method including anisotropic mesh refinement in boundary layers is given in [1]. A good adaptibility to locally changing diffusive, convective and reactive influences can be noted [2,3]. Note that, in case of $\vec{b} = 0$, we recover the standard Galerkin method for the diffusion–reaction equation.

The approach in [3] allows a simple calculation of the upwind parameter δ_K. h_K is here the element length in the flow direction :

$$\delta_K = \frac{h_K}{2|\vec{b}_K|}\frac{Pe_K}{\sqrt{1+Pe_K^2}}, \qquad Pe_K = \frac{h_K|\vec{b}_K|}{a_K}, \qquad h_K = \frac{2|\vec{b}_K|}{\sum_{i=1}^4 |\vec{b}_K \nabla\varphi_i|}. \tag{12}$$

Remark : The numerical analysis of stabilized Galerkin methods to nonlinear diffusion–convection–reaction problems seems to be an open problem. The special case of nonlinear diffusion coefficients is considered in [5]. Nonlinearities in the lower order terms are studied

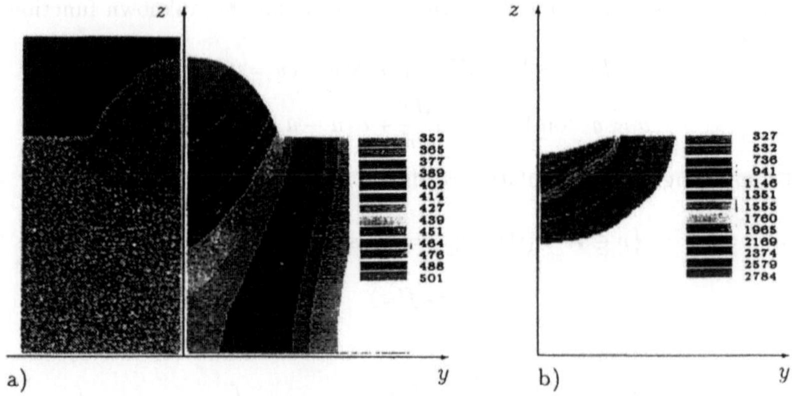

Fig. 4: Temperature distribution and surface formation for build–up welding in cross section to welding direction (temperature in [^oC])

in [4]. □

The tetrahedral mesh has been adapted to the size of the analytical solution of a simplified model. Furthermore, the position of a tetrahedron with respect to the solidification front has to be carefully considered due to the jumping coefficients of the Fourier–Kirchhoff equation.

For the solution of the linear algebraic systems, we employ a SOR–method with Jacobi preconditioning, downwind numbering of the nodes and a dynamical adaption of the relaxation parameter (essentially depending on the discrete residual). In the range up to 50.000 finite elements, no advantage could be found using more sophisticated solvers as BiCGstab or GMRES. Furthermore, the application of domain decomposition methods is in progress and will considerably enlarge the range of solvable discrete problems.

3. Numerical results

The test of the model performance took place in comparison with the experiment. The numerical solution has been tested with reference to the seam shape for build-up welding in horizontal position. The strong reinforcement represents here a special requirement to the algorithm. Fig. 4 a) shows the comparison of the calculated seam shape after the solidification ($x = 40 \cdot 10^{-3}$m) with the experimental result. In Fig. 4 b) the result is shown at the position of the electrode ($x = 0$ m). The temperature distribution and the weld pool deformation correspond with the experiment well.

The temperature distribution for the narrow gap welding in vertical position is shown in Fig. 5. The distribution has been measured by means of thermography on the back side of the plate in the $x - y$-plane. The theoretical and experimental 2D-field are opposed in Fig. 5a). The transient temperature field for $y = z = 0$ (Fig. 5b) confirms the quality

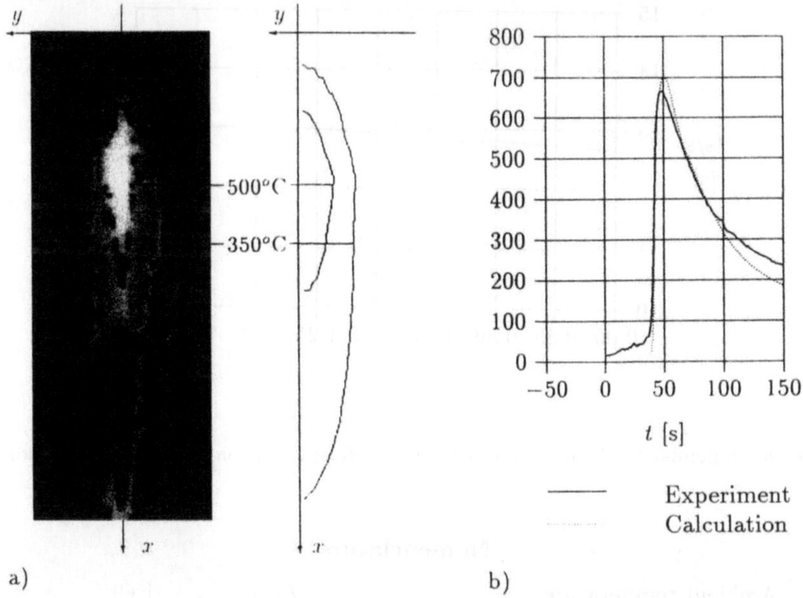

Fig. 5: Comparison between a) measured and b) calculated temperature distribution on the back side of the plate for narrow gap welding (temperature in [^oC])

of the solution. With the presented model we have a description of the process, which is in accordance to the specific conditions of the energy transport in the case of narrow gap welding. The choosen formulation of the arc effect allows the application to different deep depressed pool surfaces and different welding positions. Further numerical results and comparison of the computed results to experimental data are given in [8].

The report [9] contains results for a somewhat simplified 2D model concerning the robustness of the proposed algorithm with respect to different numerical and model parameters. Of particular interest is the necessity of the stabilization of the basic Galerkin method. Undershoots of the numerical solution (as a consequence of unphysical oscillations) were obtained in the range from $\delta^* = 0.0$ (standard Galerkin method) with 232^0C to $\delta^* = 0.5$. In Fig. 6 we present the dependence of the $t_{8/5}$−time on the weighting factor δ^*. In the range $\delta^* \in [1.0, 2.0]$ a nearly constant value of the $t_{8/5}$−time can be observed. For practical application of the algorithm, future work is necessary to ascertain the dependence of the model parameters on the technological parameters as the result of experimental and theoretical investigations.

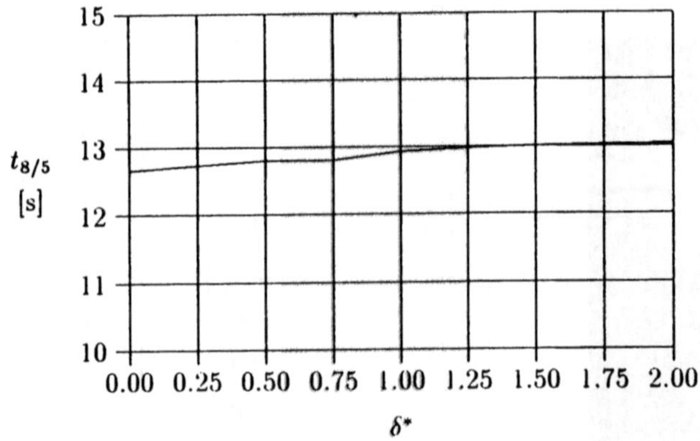

Fig. 6: Dependence of the cooling time $t_{8/5} = t_{8/5}(y = 0)$ on the weighting factor δ^*

Nomenclature

Ambient temperature	T_∞	[°C]
Arc length (average)	$\tilde{L}_{arc}$	[m]
Coefficient of heat transfer (conv. & rad.)	α_{eff}	[W/(m² K)]
Cooling time from 800°C to 500°C	$t_{8/5}$	[s]
Density	ϱ	[kg/m³]
Finite element	K	
Gravitational acceleration	g	[m/s²]
Heat capacity	c_p	[J/(kg K)]
Heat flux density	$\dot{q}$	[W/m²]
Heat source density (per volume)	$\dot{q}_{vol}$	[W/m³]
Inclination of the plate in $x-z$-plane	θ_{plate}	[°]
Normal unit vector	$\vec{n}$	
Plate lengths	L_{x_1}, L_{x_2}, L_y	[m]
Plate thickness	L_z	[m]
Temperature	T	[°C]
Thermal conductivity	λ	[W/(m K)]
Surface deformation	Z	[m]
Surface tension	σ	[N/m]
Velocity	$\vec{v}$	[m/s]

Acknowledgements

The research of the first author as a former member of the Graduiertenkolleg "Modellierung, Berechnung und Identifikation mechanischer Systeme" at the Magdeburg University was supported by the German Research Society and the Land Sachsen-Anhalt.

References

[1] Apel, Th., Lube, G.: Anisotropic mesh refinement in stabilized Galerkin methods, submitted to Numer. Math..

[2] Franca, L.P., Frey, S.L., Hughes, T.J.R.: Stabilized Finite Element Methods, I. Application to the Advective-diffusive Model, *CMAME* 95 (1992) 253-276.

[3] Lube, G.: Stabilized Galerkin Finite Element Methods for Convection Dominated and Incompressible Flow Problems, Banach Center Publications, Vol.29, 85-104, Inst. of Math., Polish Academy of Science, Warzawa 1994.

[4] Lube, G.: Streamline diffusion finite element method for quasilinear elliptic problems, Numer. Math. 61 (1992), 335–357.

[5] Pousin, J., Rappaz, J.: Consistence, stability, a–priori and a–posteriori errors for Petrov–Galerkin methods applied to nonlinear problems, Numer. Math. 69 (1994) 2, 213–227.

[6] Radaj, D. : *Heat Effects of Welding*, Springer-Verlag Berlin, Heidelberg and New York, 1991.

[7] Weiß, D., Franz, U. and Schmidt, J. : Model of Deformation and Temperature Distribution in Weld Pools during Arc Welding, in *Proceedings of the 2nd Int. Seminar Numerical Analysis of Weldability (Graz, Austria, May 1993)*, in press.

[8] Weiß, D.: Finite–Elemente Simulation von Temperaturfeld und Schmelzbaddeformation beim Lichtbogenschweißen, Thesis, Univ. Magdeburg 1994.

[9] Weiß, D.: Einfluß von Modellparametern auf das Lösungsverhalten bei der Berechnung von Schweißtemperaturfeldern mittels FEM, Univ. Magdeburg, Techn. Report (unpublished), Nov. 1994.

List of Participants:

Artlich, Stefan TU Hamburg-Harburg, Arbeitsbereich Verfahrenstechnik I, Denickestr. 15, 21071 Hamburg,

Axelsson, Owe Department of Mathematics, University Nijmegen, Toernooiveld 1, NL-6500 GL Nijmegen, The Netherlands,

Bächtold, Martin Institut für Quantenelektronik, HPT-H9, ETH Zürich, CH-8093 Zürich, Switzerland,

Biesenbach, Helmut Institut für Angewandte Mathematik der Universität Bonn, Wegeler Str. 6, 53115 Bonn,

Blum, Heribert Fachbereich Mathematik, Universität Dortmund, Postfach 50 05 00, 44221 Dortmund,

Brink, Ulrich Institut f. Baumechanik u. Num. Mechanik, Universität Hannover, Appelstr. 9A, 30167 Hannover,

Brokate, Martin Institut für Informatik und Praktische Mathematik, Universität Kiel, Olshausenstr. 40-60, 24098 Kiel,

Bungartz, Hans-J. Institut f. Informatik, TU München, 80290 München,

Bunse-Gerstner, Angelika Universität Bremen, Fachbereich 3 / Mathematik und Informatik, Postfach 330 440, 28334 Bremen,

Burmeister, Jens Institut für Informatik und Praktische Mathematik, Universität Kiel, Olshausenstr. 40-60, 24098 Kiel,

Buthmann, Ute Institut für Informatik und Praktische Mathematik, Universität Kiel, Olshausenstr. 40-60, 24098 Kiel,

Buzdin, A. Mathematische Fakultät, Universität Kaliningrad, Al. Nevsky Str. 14, 236041 Kaliningrad Russland,

Carstensen, Carsten Department of Mathematics, Heriot-Watt University, Edingburgh EH14 4AS, UK,

Chernigovskii, S. Institut f. Theoretische Astrophysik, Im Neuenheimer Feld 561, 69120 Heidelberg,

Coclici, Cristian Mathemat. Inst. A, Universität Stuttgart, Pfaffenwaldring 57, 70569 Stuttgart,

Dütsch, Heribert Lehrstuhl f. Strömungsmechanik, FAU Erlangen-Nürnberg, Cauerstr. 4, 91058 Erlangen,

Eckstein, Knut ISD, Pfaffenwaldring 27, 70550 Stuttgart,

Faermann, Birgit Institut für Informatik und Praktische Mathematik, Universität Kiel, Olshausenstr. 40-60, 24098 Kiel,

Fassbender, Heike Universität Bremen, Fachbereich 3 / Mathematik und Informatik, Postfach 330 440, 28334 Bremen,

Feistauer, Milislav Faculty of Mathematics and Physics, Charles University Prague, Malostranske nam. 25, 118 00 Praha 1, Czech Republic,

Ferket, Peter Department of Mathematics and Computer Science, Eindhoven University of Technology, P.O.Box 513, NL-5600 MB Eindhoven,

Feuchter, Dirk Institut für Computeranwendungen III, Universität Stuttgart, Pfaffenwaldring 27, 70550 Stuttgart,

Funken, Stefan A. Institut für Angewandte Mathematik, Universität Hannover, Welfengarten 1, 30167 Hannover,

Giebermann, Klaus Universität Karlsruhe, Institut f. Praktische Mathematik, Englerstr. 2, 76128 Karlsruhe,

Graham, Ivan G. University of Bath, School of Mathematical Sciences, Claverton Down, Bath BA2 7AY, United Kingdom,

Griebel, Michael Institut für Informatik der Technischen Universität München, Arcisstr. 21, Postfach 20 24 20, 80290 München,

Groh, Ulrich Techn. Universität Chemnitz-Zwickau, Fachb. Mathematik, Postf. 964, 09009 Chemnitz,

Haack, Carsten Techn. Universität Hamburg-Harburg, Arbeitsbereich Meerestechnik II, Eißendorfer Str. 42, 21071 Hamburg,

Hackbusch, Wolfgang Institut für Informatik und Praktische Mathematik, Universität Kiel, Olshausenstr. 40-60, 24098 Kiel,

Hanke, Michael Humboldt-Universität Berlin, Institut f. Angewandte Mathematik, Unter den Linden 6, 10099 Berlin,

Heinrich, Bernd FB Mathematik, Techn. Universität Chemnitz-Zwickau, Reichenhainer Str. 41, 09009 Chemnitz,

Heise, Bodo Joh.-Kepler-Universität Linz, Institut f. Mathematik, Altenberger Str. 69, A-4040 Linz,

Heisig, Michael Institut für Computeranwendungen III, Universität Stuttgart, Pfaffenwaldring 27, 70550 Stuttgart,

Heppner, Ingo Institut für Computeranwendungen III, Universität Stuttgart, Pfaffenwaldring 27, 70550 Stuttgart,

Hinkelmann, Reinhard Institut für Strömungsmechanik, Universität Hannover, 30167 Hannover,

Hömberg, Dietmar Weierstraß-Institut f., Angewandte Analysis u. Stochastik, Mohrenstr. 39, 10117 Berlin,

Hoffmann, Wolfgang Institut für Computeranwendungen III, Universität Stuttgart, Pfaffenwaldring 27, 70550 Stuttgart,

Ihlenburg, Frank R.-Wagner-Str. 2 N45, 23968 Wismar,

Johannsen, Klaus Institut für Computeranwendungen III, Universität Stuttgart, Pfaffenwaldring 27, 70550 Stuttgart,

Junkherr, Jörg Institut für Informatik und Praktische Mathematik, Universität Kiel, Olshausenstr. 40-60, 24098 Kiel,

Kaiser, H. Chr. Weierstraß-Institut f. Angew. Analysis, u. Stochastik (WIAS), Postf. 13 04, 10117 Berlin,

Kasper, Harald Institut f. Strömungsmechanik, Universität Hannover, Appelstr. 9A, 30167 Hannover,

Katzer, Edgar Institut für Analysis, Universität Magdeburg, Postfach 4120, 39016 Magdeburg,

Klar, Axel FB Mathematik, Universitt Kaiserslautern, 67663 Kaiserslautern,

Klawonn, Axel Institut fuer Numerische Mathematik, Westfälische Wilhelms-Universitaet, Einsteinstr. 62, D-48149 Münster, Germany,

Klein, Rupert Institut f. Techn. Mechanik, RWTH Aachen, Templergraben 64, 52056 Aachen,

Kolditz, O. Institut f. Strömungsmechanik, Uni. Hannover, Appelstr. 9a, 30167 Hannover,

Lage, Christian Institut für Informatik und Praktische Mathematik, Universität Kiel, Olshausenstr. 40-60, 24098 Kiel,

Lagendijk, Vincent Institut f. Wasserbau u. Wasserwirtschaft, RWTH Aachen, Mies-van-der-Rohe-Str. 1, 52074 Aachen,

Lang, Stefan Institut für Computeranwendungen III, Universität Stuttgart, Pfaffenwaldring 27, 70550 Stuttgart,

Langer, Andreas Institut f. Mathematik, Universität Dortmund, 44221 Dortmund,

Langer, Ulrich Institut f. Mathematik, Johannes Kepler Universität Linz, Altenberger Str. 69, A-4040 Linz, Austria,

Lege, Thomas Bundesanstalt f. Geowissenschaften u. Rohstoffe, Stillweg 2, 30655 Hannover,

Lube, Gert Mathematisches Institut, Universität Göttingen, Lotzestr. 16-18, 37073 Göttingen,

Liebau, Frank Technische Universität Berlin, Fachbereich 3 - Mathematik, Sekr. MA 6-2, Straße des 17.Juni 135, 10623 Berlin,

Maar, Bernd Institut für Computeranwendungen III, Universität Stuttgart, Pfaffenwaldring 27, 70550 Stuttgart,

Mazurkevich, George Institut für Computeranwendungen III, Universität Stuttgart, Pfaffenwaldring 27, 70550 Stuttgart,

Mordhorst, Uwe Rechenzentrum der Universität Kiel, Olshausenstr. 40, 24098 Kiel,

Mund, Dr. Inst. f. Angew. Mathematik, Universität Hannover, Welfengarten 1, 30167 Hannover,

Munz, C.-D. INR Kernforschungszentrum Karlsruhe GmbH, Postfach 3640, 76021 Karlsruhe,

Neumeyer, Thomas SCM, Techn. Universität München, 80290 München,

Neuß, Nicolas Institut für Wissenschaftliches Rechnen, Im Neuenheimer Feld 294, 69120 Heidelberg,

Noelle, Sebastian Institut f. Angewandte Mathematik, Universität Bonn, Wegelerstr. 10, 53115 Bonn,

Ohlhoff, Antje Universität Bremen, Institut für Dynamische Systeme, Bibliotheksstraße 1, Postfach 330440, 28334 Bremen 33,

Paul, Rainer Institut für Informatik und Praktische Mathematik, Universität Kiel, Olshausenstr. 40-60, 24098 Kiel,

Ratke, Rainer Institut für Strömungsmechanik und Elektr. Rechnen im Bauwesen, Universität Hannover, Appelstr. 9 A, 30167 Hannover,

Rentz-Reichert, Henrik Institut für Computeranwendungen III, Universität Stuttgart, Pfaffenwaldring 27, 70550 Stuttgart,

Reusken, Arnold Eindhoven University of Technology, Dept. of Mathematics and Computing Science, Postbus 513, NL - 5600 MB Eindhoven,

Rusten, Torgeir SINTEF, P.O.Box 124 Blindern, N-0314 Oslo, Norway,

Sauter, Stefan Institut für Informatik und Praktische Mathematik, Universität Kiel, Olshausenstr. 40-60, 24098 Kiel,

Schippers, Harmen National Aerospace Laboratory NLR, P.O.Box 90502, NL - 1006 BM Amsterdam, Niederlande,

Schlegel, Volker Techn. Universität Hamburg-Harburg, Arbeitsbereich Meerestechnik II, Eißendorfer Str. 42, 21071 Hamburg,

Schlüter, H.-J. Universität - GH Duisburg, FB 7 / Schiffstechnik, Lotharstraße, 47057 Duisburg,

Schmid, Werner Mathematisches Institut, Techn. Universität München, Postfach 20 24 20, 80290 München,

Schmidt, Alfred Inst. f. Angewandte Mathematik, Universität Freiburg, Hermann-Herder-Strasse 10, 79104 Freiburg,

Schneider, Reinhold FB Mathematik, Technische Hochschule Darmstadt, Schloßgartenstr. 7, 64289 Darmstadt,

Schüfer, Petra Universität Bremen, Fachbereich 3/Mathematik und Informatik, Postfach 330 440, 28334 Bremen,

Schulte, Stefan Siemens AG, Zentralabteilung, Forschung und Entwicklung, 91050 Erlangen,

Siebert, Kunibert G. Institut für Angewandte Mathematik, Hermann-Ludwigs-Universität, Hermann-Herder-Str. 10, 79104 Freiburg,

Specht, Ulf Lehr-und Forschungsgebiet der Mechanik, der RWTH-Aachen, TH-Postfach, 52056 Aachen,

Starke, Gerhard Institut f. Praktische Mathematik, Universität Karlsruhe, 76128 Karlsruhe,

Steinbach, Olaf Universität Stuttgart, Mathematisches Institut A, Pfaffenwaldring 57, 70569 Stuttgart,

Turek, Stefan Institut f. Angew. Mathematik, Universität Heidelberg, Im Neuenheimer Feld 294, 69120 Heidelberg,

Umlauf, Günter Institut fuer Algorithmen und , wissenschaftliches Rechnen (SCAI), Schloss Birlinghoven, 53754 Sankt Augustin,

Vuik, Kees Fac. of Technical Mathematics and Informatics, Delft University of Technology, Mekelweg 4, P.O. Box 5031, 2600 GA Delft, The Netherlands,

Wagner, Christian Institut für Computeranwendungen III, Universität Stuttgart, Pfaffenwaldring 27, 70569 Stuttgart,

Wagner, Frank Mathematisches Institut A, Techn. Universität München, Postf. 20 24 20, 80290 München,

Wappler, Jens Institut für Informatik und Praktische Mathematik, Universität Kiel, Olshausenstr. 40-60, 24098 Kiel,

Weiß, Dietmar Otto-von-Guericke-Universität Magdeburg, ISUT / Thermodynamik, Postfach 4120, 39016 Magdeburg,

Wesseling, Pieter Delft University of Technology, Faculty of Techn. Math. and Inform., Mekelweg 4, NL-2628 CD Delft, The Netherlands,

Wittum, Gabriel Institut für Computeranwendungen III, Universität Stuttgart, Pfaffenwaldring 27, 70550 Stuttgart,

Wrobel, Christian Institut für Computeranwendungen III, Universität Stuttgart, Pfaffenwaldring 27, 70550 Stuttgart,

Yserentant, Harry Mathematisches Institut, Universität Tübingen, Auf der Morgenstelle 10, 72076 Tübingen,

Zhou, Guoquiang Universität Heidelberg, Institut für angewandte Mathematik, Im Neuenheimer Feld 293/294, 69120 Heidelberg.

How to contact the authors by email:

Artlich, Stefan artlich@tu-harburg.d400.de

Bungartz, Hans-Joachim bungartz@informatik.tu-muenchen.de

Carstensen, Carsten carsten@ma.hw.ac.uk

Ferket, P.J.J. peterf@win.tue.nl

Feistauer, M. feist@karlin.mff.cuni.cz

Heise, Bodo heise@miraculix.numa.uni-linz.ac.at

Heinrich, Bernd heinrich@mathematik.tu-chemnitz.de

Hoppe, R.W.H. rohop@mathematik.tu-muenchen.de

Ihlenburg, Frank ihl@olgao.umd.edu

Katzer, Edgar edgar.katzer@mathematik.uni-magdeburg.d400.de

Kasper, Harald kasper@appel012.hydromech.uni-hannover.de

Klar, Axel klar@mathematik.uni-kl.de

Kolditz, O. kolditz@appel012.hydromech.uni-hannover.d400.de

Lube, Gert lube@namu13.gwdg.de

Makridakis, Ch. makr@athina.edu.uch.gr

Schippers, H. hschippers@nlr.nl

Schulte, Stefan stefan.schulte@zfe.siemens.de

Specht, Ulf usp@leibniz.lufmech.rwth-aachen.de

Vuik, C. c.vuik@math.tudelft.nl

Wagner, Frank wagnerf@mathematik.tu-muenchen.de

Zarrabi, Darius zarr@ifam.uni-hannover.d400.de

Notes on Numerical Fluid Mechanics (NNFM) Volume 51

Volume 30 Numerical Treatment of the Navier-Stokes Equations (W. Hackbusch / R. Rannacher, Eds.)

Volume 29 Proceedings of the Eighth GAMM-Conference on Numerical Methods in Fluid Mechanics (P. Wesseling, Ed.)

Volume 28 Vortical Solution of the Conical Euler Equations (K. G. Powell)

Volume 27 Numerical Simulation of Oscillatory Convection in Low-Pr Fluids (B. Roux, Ed.)

Volume 26 Numerical Solution of Compressible Euler Flows (A. Dervieux / B. van Leer / J. Periaux / A. Rizzi, Eds.)

Addresses of the Editors of the Series "Notes on Numerical Fluid Mechanics"

Prof. Dr. Ernst Heinrich Hirschel (General Editor)
Herzog-Heinrich-Weg 6
D-85604 Zorneding
Federal Republic of Germany

Prof. Dr. Kozo Fujii
High-Speed Aerodynamics Div.
The ISAS
Yoshinodai 3-1-1, Sagamihara
Kanagawa 229
Japan

Prof. Dr. Bram van Leer
Department of Aerospace Engineering
The University of Michigan
3025 FXB Building
1320 Beal Avenue
Ann Arbor, Michigan 48109-2118
USA

Prof. Dr. Keith William Morton
Oxford University Computing Laboratory
Numerical Analysis Group
8-11 Keble Road
Oxford OX1 3QD
Great Britain

Prof. Dr. Maurizio Pandolfi
Dipartimento di Ingegneria Aeronautica e Spaziale
Politecnico di Torino
Corso Duca Degli Abruzzi, 24
I-10129 Torino
Italy

Prof. Dr. Arthur Rizzi
Royal Institute of Technology
Aeronautical Engineering
Dept. of Vehicle Engineering
S-10044 Stockholm
Sweden

Dr. Bernard Roux
Institut de Mécanique des Fluides
Laboratoire Associé au C.R.N.S. LA 03
1, Rue Honnorat
F-13003 Marseille
France

Brief Instruction for Authors

Manuscripts should have well over 100 pages. As they will be reproduced photomechanically they should be produced with utmost care according to the guidelines, which will be supplied on request. In print, the size will be reduced linearly to approximately 75 per cent. Figures and diagrams should be lettered accordingly so as to produce letters not smaller than 2 mm in print. The same is valid for handwritten formulae. Manuscripts (in English) or proposals should be sent to the general editor, Prof. Dr. E. H. Hirschel, Herzog-Heinrich-Weg 6, D-85604 Zorneding.

MIX
Papier aus verantwortungsvollen Quellen
Paper from responsible sources
FSC® C105338

If you have any concerns about our products,
you can contact us on
ProductSafety@springernature.com

In case Publisher is established outside the EU,
the EU authorized representative is:
Springer Nature Customer Service Center GmbH
Europaplatz 3, 69115 Heidelberg, Germany

Printed by Libri Plureos GmbH
in Hamburg, Germany